Logic, Argumentation & Reasoning

Interdisciplinary Perspectives from the Humanities and Social Sciences

Volume 14

Series editor
Shahid Rahman

Logic, Argumentation & Reasoning explores the links between Humanities and the Social Sciences, with theories including, decision and action theory as well as cognitive sciences, economy, sociology, law, logic, and philosophy of sciences. It's two main ambitions are to develop a theoretical framework that will encourage and enable interaction between disciplines as well as to federate the Humanities and Social Sciences around their main contributions to public life: using informed debate, lucid decision-making and action based on reflection.

The series welcomes research from the analytic and continental traditions, putting emphasis on four main focus areas:

- Argumentation models and studies
- Communication, language and techniques of argumentation
- Reception of arguments, persuasion and the impact of power
- Diachronic transformations of argumentative practices

The Series is developed in partnership with the Maison Européenne des Sciences de l'Homme et de la Société (MESHS) at Nord - Pas de Calais and the UMR-STL: 8163 (CNRS).

Proposals should include:

- A short synopsis of the work or the introduction chapter
- The proposed Table of Contents
- The CV of the lead author(s)
- If available: one sample chapter

We aim to make a first decision within 1 month of submission. In case of a positive first decision the work will be provisionally contracted: the final decision about publication will depend upon the result of the anonymous peer review of the complete manuscript. We aim to have the complete work peer-reviewed within 3 months of submission.

The series discourages the submission of manuscripts that contain reprints of previous published material and/or manuscripts that are below 150 pages / 85,000 words.

For inquiries and submission of proposals authors can contact the editor-in-chief Shahid Rahman via: shahid.rahman@univ-lille3.fr or managing editor, Laurent Keiff at laurent.keiff@gmail.com.

More information about this series at http://www.springer.com/series/11547

Rafal Urbaniak · Gillman Payette
Editors

Applications of Formal Philosophy

The Road Less Travelled

Springer

Editors
Rafal Urbaniak
Institute of Philosophy,
Sociology and Journalism
University of Gdańsk
Gdańsk
Poland

and

Centre for Logic and Philosophy
of Science
Ghent University
Ghent
Belgium

Gillman Payette
Department of Philosophy
University of British Columbia
Vancouver, BC
Canada

ISSN 2214-9120 ISSN 2214-9139 (electronic)
Logic, Argumentation & Reasoning
Interdisciplinary Perspectives from the Humanities and Social Sciences
ISBN 978-3-319-58505-5 ISBN 978-3-319-58507-9 (eBook)
DOI 10.1007/978-3-319-58507-9

Library of Congress Control Number: 2017945681

Printed on acid-free paper

This Springer imprint is published by Springer Nature
The registered company is Springer International Publishing AG
The registered company address is: Gewerbestrasse 11, 6330 Cham, Switzerland

Contents

Chapter 1
Applied Formal Philosophy: Some Reflections on the Program

Gillman Payette and Rafal Urbaniak

What is mathematical philosophy? How is it different from *philosophy*? How is it *applied*? As to the first, in mathematical philosophy one uses mathematical tools to shed light on philosophical questions. Conceived in that way, all mathematical philosophy is applied: mathematical philosophy is just mathematics applied in philosophy. The pressing question is how mathematical philosophy is really different from *philosophy*?

To be sure mathematical philosophy is a *kind* of philosophy. However, some may think that it represents a special kind of approach to philosophy. And some think that the special-ness means that it is a better kind of philosophy, while others think that its special-ness means that it is inferior. We would like to offer a counter to both of these views.

Both of these views take their starting point from the same tendency in mathematical philosophy. Often the mathematical tools used end up being very interesting in themselves and take the attention of mathematical philosophers. This is one way to see what has happened to logic at the end of the XXth century. The tools for studying logic became what logicians actually studied. That is what one might describe as *pure* mathematical philosophy. Calling it 'pure' is not to deride or diminish such programs of study; it provides an important service, since it thoroughly investigates the limitations of tools that might be useful at some point. One example is that of Bayesian epistemology: while initially one might be tempted to apply probabilistic methods to (reformulations of) questions raised by classical epistemologists, a major stream of research in formal epistemology nowadays focuses on the issue of how

G. Payette
Department of Philosophy, University of British Columbia, Vancouver, BC, Canada
e-mail: gpayette@dal.ca

R. Urbaniak (✉)
Centre for Logic and Philosophy of Science, Ghent University, Ghent, Belgium
e-mail: rf.urbaniak@gmail.com

R. Urbaniak
Institute of Philosophy, Journalism and Sociology, University of Gdańsk, Gdańsk, Poland

R. Urbaniak and G. Payette (eds.), *Applications of Formal Philosophy*,
Logic, Argumentation & Reasoning 14, DOI 10.1007/978-3-319-58507-9_1

using such methods is to be justified and what the best probabilistic principles are, rather than on applying them to philosophical questions.

One should never be obtuse so as to dismiss pure, basic research. But that tendency towards pure research has given mathematical philosophy a bad, albeit prestigious, name. Some of the best analytic philosophers of our time are mathematical philosophers. So the field is looked on with some admiration in some circles. But for others that tendency has the opposite effect. Mathematical philosophy spins its wheels without ever doing any cutting—to mix a few metaphors. But that view is also mistaken. Mathematics has been used to cut with wild success in a number of areas of philosophy. Bertrand Russell's *An Introduction to Mathematical Philosophy* was probably one of the earliest attempts at application. Some might suggest that it should have been entitled "An Introduction to the Philosophy of Mathematics", but it was aptly named. What it did was lay out the way to deal with philosophical problems with mathematical precision. But the field that it then dealt with was mathematics, not ethics or epistemology—at least not directly. What early mathematical philosophy added was a way to approach philosophy by stating philosophical problems precisely; it first uses mathematically precise *languages* to precisely reformulate problems, and then deploys mathematical methods.

The obvious thought, then, is to figure out how to make language mathematically precise, and so we get formal logic and its application in the philosophy of language. That precision was also soon directed toward the foundations of science, but still under the guise of formal logic. After metaphysics became acceptable again, that too was an area where logical tools were applied to make the questions clearer. About the same time one could see ethics finally enter the crosshairs of logical analysis—although logics for imperatives had been around since the very beginnings of logical analysis. Thus, mathematical philosophy looked like it might be just that: apply logic to an area of philosophical study. The result was to look at mathematical philosophy as being coextensive with the "logical analysis" approach to philosophy. But that takes mathematical philosophy to be exhausted by logic, and we think that is a mistake.

The reason we think it is a mistake to see mathematical philosophy as exhausted by logic is because that perspective tempts one to reject mathematical philosophy on the basis that its program is to treat logical analysis as the right methodology for philosophy; it probably isn't. A tool should only be used when it will help. Sometimes logical analysis doesn't help; sometimes it makes things overly complicated. Or worse, sometimes it leads us to oversimplify things in a kind of procrustean manner and then draw the wrong conclusions.

What we want to do in philosophy is—paraphrasing Wilfred Sellars—understand how things, in the most general sense, hang together, in the most general sense. Logical analysis is very good in mathematics, but mathematics generally is better suited for philosophy, generally. How we want to model the questions we are asking is left wide open when we look to mathematics generally. But we are also left to consider whether applying mathematics really helps.

The kinds of answers we get when we apply mathematics to philosophical questions tend to be ones that shape the boundaries of investigation. When Arrow asked:

is welfare economics actually possible? He found an edge to that field.[1] And that is where philosophy should start too, at the edges. But that doesn't mean it is the only place for philosophy.

Indeed, when people say 'applied philosophy' the usual sense of 'applied' is captured by the application of philosophical theories to discussions well within the boundaries. Applied mathematical philosophy in this more usual sense of 'applied' is difficult. The problems are much more complex and conceptually difficult in the thick of things, and don't lend themselves to clear expression with mathematics. How do character and emotion affect the notion of correct moral action? Is race a real thing? We don't even know where to start with applying mathematics to those questions. But the working hypothesis of mathematical philosophy isn't that we ought to find the mathematical tools to apply to such questions. The approach to mathematical philosophy should be to stand ready with many tools at hand when those fields have found questions where mathematics might help.

That doesn't mean that mathematical philosophers must sit about waiting for the other fields to come to them. Certainly not; go out and find problems. But that kind of expedition must be guided by a kind of Hippocratic oath 'First, do not let mathematics do any philosophical harm.' So we can finally answer the pressing question: how does mathematical philosophy differ from philosophy? It isn't a *special* kind of philosophy. It is simply that *sometimes* in philosophy mathematics can help us answer the pressing philosophical questions we have. The hope is, dear reader, that you find the essays in this book to have followed that Hippocratic oath.

Acknowledgments Gillman Payette would like to thank the Social Sciences and Humanities Research Council of Canada for supporting this work through a Banting Postdoctoral Fellowship.

[1] This may be a poor example since Arrow was heavily influenced by one of his (undergraduate) mathematics teachers: Alfred Tarski. What influenced him were notions in the logic of relations.

Part I
Human Affairs

Chapter 2
The Logic of Guilt, Innocence and Legal Discourse

Andreas Kapsner

Abstract Even though the notions of guilt and innocence are two of the most important ones for any legal system, there are surprisingly deep and complicated confusions that entangle them. There is much clarifying work to be done here, part of which is to investigate the logical principles governing these concepts. In this paper, I build on Larry Laudan's analysis in his book *Truth, Error and Criminal Law*. Though I have great respect for the conceptual lines he draws, I will show that the logical relationships of these concepts is richer than Laudan makes out. In order to do so, I will use one of the oldest aids to logical reasoning, the square of oppositions. The square captures these logical relations well, albeit only with some idealizing assumptions in place. I will go on to argue that these idealizations are in harmony with a useful way to model legal deliberation in constructive logic.

Keywords Legal reasoning · Legal discourse · Guilt · Innocence · Constructive logic · Larry Laudan

2.1 Introduction

It is clear that the notions of guilt and innocence are two of the most important ones for any legal system. It might even seem that we could not claim to understand what the legal system is about at all if there should be any principled difficulty in understanding these concepts. But in fact it turns out that there are surprisingly deep and complicated confusions that entangle these two notions. There is much clarifying work to be done here, part of which is to investigate the logical principles governing these concepts.

In this essay, I will concentrate on the conceptualization of guilt and innocence in the jurisdiction of the so-called common law countries, especially the United States, where criminal trials are judged by lay juries. There are several reasons for this, the most important of which is the following: My paper is heavily influenced by, and in

A. Kapsner (✉)
Munich Center for Mathematical Philosophy, Munich, Germany
e-mail: Andreas.Kapsner@lrz.uni-muenchen.de

R. Urbaniak and G. Payette (eds.), *Applications of Formal Philosophy*,
Logic, Argumentation & Reasoning 14, DOI 10.1007/978-3-319-58507-9_2

part a commentary on, Larry Laudan's book *Truth, Error, and Criminal Law*, which focuses on the legal system of the U.S. [10].

As most readers of this volume will know, Laudan is one of the leading philosophers of science of our time, and the great joy in reading his work on legal matters is that he asks precisely the questions a philosopher would ask.[1] These are at times very different from the questions legal scholars are accustomed to ask and answer, and it is very illuminating to read an exposition of their thinking in a language that is much closer to home.[2] Though the book is altogether informal, by natural extension this language includes the notions of philosophical logic, and I hope to further push Laudan's project along by illustrating and extending his discussion of guilt and innocence with the aid of logical tools.

Laudan's main point about the usage of "guilt" and "innocence" by legal scholars and practitioners is that it is ambiguous: A factual reading (What happened?) is often conflated with an evidential reading (What does the evidence indicate happened?). His proposal is to disambiguate these readings by introducing new terminology that allows unequivocal reference to the different concepts at play.

I will show that the logical relationships of these concepts is richer than Laudan makes out. In order to do so, I will use one of the oldest aids to logical reasoning, the square of oppositions. The square captures these logical relations well, albeit only with some idealizing assumptions in place. I will go on to argue that these idealizations are in harmony with a useful way to model legal deliberation in constructive logic.

2.2 "Not Guilty" and "Innocent"

It would be unfair to say that it took a philosopher like Laudan to point out that there is widespread confusion about the legal meaning of the terms "guilt" and "innocence". Instead, this is something that legal scholars and practitioners themselves are quite aware of. Some of them, however, seem to believe that this kind of confusion is only prevalent *outside* of the legal profession.

---

[1]It is with pleasure that I follow the suggestion of an anonymous referee to mention Susan Haack's recent [7] as another example of an eminent philosopher taking up legal matters. In contrast to both Laudan's book and the present paper, she also addresses questions of proof outside of criminal trials, i.e. in private law. An interesting discussion of the difference between the notions of proof in a criminal trial and proof in a civil case is [17], which supplies a quite different analysis of the nature of proofs of guilt or innocence from the present paper.

[2]In turn, it is highly interesting to see the reaction of legal scholars to the book. Here is one example: "Enter Laudan, the distinguished epistemologist, training his sophisticated analytical apparatus on the legal system. [...] The result is an insightful, trenchant, exhilarating, bracing, disturbing exploration of the legal system that leaves no doubt that factual accuracy could be pursued considerably more ruthlessly, and that Laudan thinks it is a reproach to the legal system that it does not do so." [1], p. 199. It is intriguing, and maybe somewhat worrying as well, that the author of that review is not inclined to agree with Laudan on the last point.

Even if true, this would of course not mean that these confusions should be left unattended by legal professionals, especially in a legal system that relies on lay people (the jury members) to decide its most important questions about particular instances of guilt and innocence. In a very interesting piece, lawyer Hugh Duvall describes the point at which he usually addresses the problem in his closing argument, thereby relating the "most important piece of information I pass to the jury before it begins its deliberations."[3]

> It is the point at which I educate jury members to the distinction between their ability to return a verdict of "not guilty" and their inability to return a verdict of "innocent."
>
> [...]
>
> Even if you are very sure the defendant is guilty, but the state has not proven it "beyond a reasonable doubt," then it is your sworn duty to return a verdict of "not guilty." [...] I attempt to hammer home, relentlessly, that the jury's job has very little to do with the concept of "innocence." Its job is [...] one of analyzing what evidence the state has presented and determining whether it is enough to satisfy the jury that there is no reason to doubt the defendant's guilt. It is the jury's job to sniff out the reasonable possibility that the defendant may not be guilty.

The idea that Duvall nicely summarizes and that, according to Laudan, is "taught in every law school in the world",[4] is thus the following: Even though "guilty"/"guilt" and "innocent"/"innocence" are understood to be antonyms in vernacular English, one should not commit the error of thinking that a verdict of "not guilty" was the same as a proclamation of the innocence of the defendant. Indeed, a common law jury will never decide on the innocence of the defendant. As Duvall rightly notes, the only possible verdicts are "guilty" or "not guilty". His claim is that these notions are to be understood as evidentially constrained ones. "Guilty" is the appropriate verdict if the evidence produced by the prosecution suffices to meet a given standard of proof, normally "beyond reasonable doubt". "Not guilty" is the appropriate verdict if the evidence fails to meet this strict standard. Innocence, on the other hand, is not an evidentially constrained concept. A person is innocent of a crime if and only if she did not commit that crime, simple as that.[5]

Although the conceptual distinction is exactly on the right track, there are some problems with Duvall's claim that it is captured already in the legal use of the terms, or even that it can possibly be so captured without the introduction of further vocabulary.

Standing in the way of the latter possibility are seemingly mundane problems of easy expressibility: while the notion of "not guilty" has a natural contradictory, "guilty", there is a lack of a suitable or commonly used expression that is contradictory to "innocent". On the other hand, "innocent" is readily turned into a noun,

---

[3] See http://www.defendingoregon.com/innocent-v-guilty/, last accessed on July 5th 2017.

[4] Reference [10], p. 99.

[5] A note on pronouns: When they refer to unspecified persons (which, as it happens, is only the case when I talk about defendants), these will be male in odd numbered sections and female in even numbered sections.

"innocence", whereas "not guilty" does not allow such a transition.[6] These may seem like accidental and trivial problems, but when one tries to regiment the use of such terms, it is important to give people the linguistic resources to express all they want using the right concepts. Otherwise, the distinction will quickly be disregarded.

And indeed, one finds that common usage does not really respect the distinction between "not guilty" and "innocent" in the way Duvall wants to have it understood. Especially in press reports, one often reads "The defendant was pleading 'innocent'" or "The verdict was 'innocent'". One interesting explanation for this Duvall offers is that the journalists are well aware of their equivocation, but are willing to pay the price of inaccuracy for the gain of greater safety (ironically enough, greater *legal* safety): In writing "not guilty" one incurs the risk that somewhere along the way from draft to printing press, the word "not" is lost. Thus one risks great danger to the reputation of the defendant and, in form of legal charges, to the news company one works for. The meaning of "innocent", on the other hand, will survive many typos and virtually never turn into its opposite; thus, the above noted lack of a lexicalized contradictory comes as a virtue here.

This, so Duvall claims, is how a distinction that is clearly understood by the legal profession gets corrupted by the accidental needs of the publishing industry and ends up in widespread confusion on the side of the laymen. And indeed, the point about the press can be corroborated. The Associated Press Style Book, for example, advises: "Use *innocent*, rather than *not guilty*, in describing a defendant's plea or a jury's verdict, to guard against the word *not* being dropped inadvertently."[7]

2.3 New Vocabulary

Duvall correctly locates the most important source of confusion: To mistake assertions about the strength of evidence for a crime with assertions about the crime itself. However, Laudan's book brings in a lot of evidence that the distinction is not quite as clearly marked and adhered to in legal language as Duvall takes it to be.[8] Laudan takes this to warrant the introduction of some clarifying terminology. His solution is to use "guilt" and "innocence" as the natural antonyms that they are and

---

[6] There is "acquittal" but that carries a factual commitment to the actual outcome of a trial that I think neither is nor should be part of the meaning of "guilty", and should therefore not be part of the meaning of a candidate antonym either.

[7] Reference [6], p. 122.

[8] Apart from many uses of the terms by members of the legal profession that showcase the same ambiguity that common use does, the distinction is also used in other, quite different senses. Sometimes is used to mark a difference in culpability of the defendant. Here, the idea is that, even though someone may not be "innocent" because he in fact has done what the prosecution claims, he may indeed be "not guilty", because he did not act voluntarily (cases of insanity or hypnosis might fall under this category) or in self-defense. It seems clear that it would be better to have different terminology to mark this distinction, but I will for the most part ignore this usage and the cases that give rise to it in this paper.

to disambiguate the two ways of understanding them by giving them labels, namely "material" and "probatory".[9]

Material innocence is given if the defendant simply did not commit the crime. Likewise, *material guilt* is given if the defendant did commit the crime. *Probatory innocence*, on the other hand, is given if at the trial there is not enough evidence to meet the relevant standard of proof to lead to a conviction. *Probatory guilt*, finally, is given if there is enough inculpatory evidence to lead to a conviction. It is the probatory notions, not the material ones, that the verdict of a jury should be expressed in. The normal verdict "guilty" corresponds to "probatorily guilty", "not guilty" corresponds to "probatorily innocent". "Materially innocent" is what legal scholars, on Duvall's account, call simply "innocent", while "materially guilty" has no simple correlate in the legal vernacular.

As these concepts are at the core of this paper, I sum them up in a table for further reference:

Laudan	Traditional	Explication
Materially innocent	Innocent	Did not commit crime
Materially guilty		Did commit crime
Probatorily innocent	Not guilty (verdict)	ev. below standard of proof
Probatorily guilty	Guilty (verdict)	ev. above standard of proof

Maybe surprisingly, at least one legal scholar, Ronald Allen, applauds this call to linguistic reform by an outsider. He writes:

> [Laudan] makes a powerfully useful distinction between material guilt (did the guy actually do the crime) and probatory guilt (does the evidence satisfactorily prove that the guy actually did the crime) and shows, convincingly, how much present confusion rests upon the failure of legal scholars to have previously made and understood the significance of this distinction [1], p.199).

It seems clearly worthwhile, then, to further investigate these reframed concepts.

2.4 Logical Relationships

In particular, I am interested in the logical relationships holding between the material and the probatory notions. On this topic, Laudan notes:

> There is a salient asymmetry between the two pairs of distinctions. It consists in the fact that a) while a finding of [probatory guilt] sustains (fallibly) the assertion of [material guilt] (that is, the legal system justifiably assumes that someone proved to be guilty is genuinely

[9]Though Laudan does not discuss the above problems with expressibility, his solution meets them admirably.

guilty), b) a finding of [probatory innocence] (a "not-guilty" verdict) warrants no inference about [material innocence] ([10], p. 96).[10]

I take it to be admissible to interpret his talk of "sustaining an assertion" as a form of logical entailment (ignore the "fallibly" for a moment, I will come back to it).

Now, although the difference that Laudan points out might indeed be salient, the alleged asymmetry turns into quite a pleasing *symmetry* once we arrange and inspect the items adequately.

Indeed, the four concepts can be neatly mapped to the four corners of a *square of oppositions*, one of the oldest items in the logician's toolbox. Used traditionally to analyze modal and quantified statements,[11] it has recently regained scholarly attention and been applied in a variety of new areas.[12]

When we apply it to the topic at hand, Laudan's four new notions of guilt and innocence, we get the following:

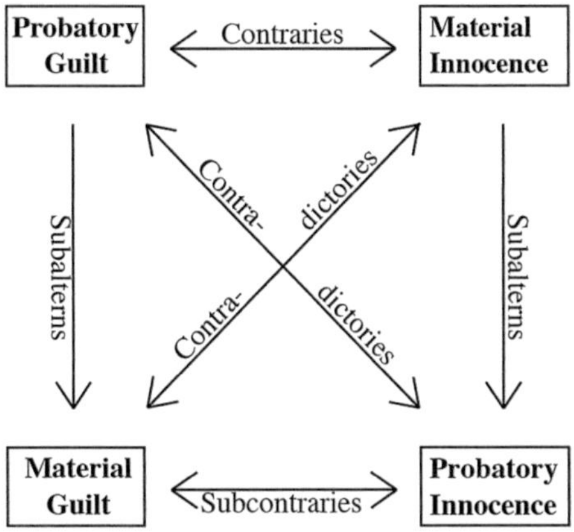

THE SQUARE OF GUILT AND INNOCENCE

The traditional explanation of the relations in this diagram were the following:

- Two propositions are *contradictory* iff (if and only if) they cannot both be true and they cannot both be false.
- Two propositions are *contraries* iff they cannot both be true but can both be false.
- Two propositions are *subcontraries* iff they cannot both be false but can both be true.
- A proposition is a *subaltern* of another iff it must be true if its superaltern is true, and the superaltern must be false if the subaltern is false.

---

[10]Notation altered. Laudan abbreviates his new notions with subscripted p's and g's, as in "guilt_p". This economy maybe pays off in a book length endeavor such as his, but I find the spelled out terms easier to read.

[11]See [13] for a history of the square as well as an account of certain problems involving non-denoting terms, which we can safely ignore, as they will not be relevant to the following application.

[12]See, for example, [3].

Given common (but not uncontroversial) assumptions about entailment, one can reformulate the last statement by saying that the superaltern (one of the upper corners) entails the subaltern (the corner straight below it).

Now, on to motivate the relations the square suggests: The diagonals deserve the least amount of explanation. We take it for granted that either the defendant committed the crime, or that she did not commit the crime.[13] Thus, material guilt and material innocence form a pair of contradictories. More debatable, we here assume that in all trials, the relevant standard of proof has either been reached or not. Thus, we take probatory guilt and probatory innocence to be contradictories as well, even though, admittedly, it is (at the very least) often difficult to *know* whether the standard of proof has been reached.[14]

That probatory guilt and material innocence are contraries means that we assume that against someone who genuinely did not commit a crime, there will not be enough evidence to meet the standard of proof. One could say that this standard is set as high as it is, in the US for example "beyond reasonable doubt", in an attempt to assure the contrariness of the two notions.

On the other hand, because of the high standard of proof it may well happen that someone who actually committed a crime will be acquitted, because her guilt could not be proven beyond all reasonable doubt. Thus, material guilt and probatory innocence might well hold at the same time, which is entailed by the subcontrary relation between them.

Lastly, looking at the vertical sides of the square, we see that probatory guilt entails material guilt, just as Laudan has pointed out. Also, as Laudan has noted, there is *no* implication from probatory innocence to material innocence. Probatory innocence just means that the evidence presented is not enough to put to rest all reasonable doubt. However, there is an implication in the other direction, for the same reason that justifies the contrariety arrow at the top of the square: If the defendant did not commit the crime, it will not be possible to present enough evidence to meet the standard of proof.

Now, it should be clear that there is some idealization involved in proclaiming these logical relations. Maybe a bit less clear is that these relations depend heavily on some core values that have shaped Western legal systems. I will first turn to these and then return to the question of idealization.

---

[13] As noted above, I put aside cases of self-defense, insanity etc. Putting them back into the picture would complicate things considerably (but not impossibly). The trouble is not just terminological, but is rooted in different standards of proof and distributions of the burden of proof attached to such cases. I will go on to show that these matters have a profound influence on the logical relationships between the concepts.

[14] I will briefly come back to the standard of proof below; for an in depth exploration, see Chap. 3 of Laudan's book.

2.5 Norms, Values and Legal Principles

The norms and values that are embedded in this square are indeed amongst the most central in our legal reasoning. They boil down to this: It is more harmful to society to convict innocent citizens than to let guilty criminals go free.

Laudan has collected a very nice succession of quotes that show how different thinkers have tried to quantify this difference in harm. The options range from *twice as harmful* to *a thousand times as harmful*,[15] hardly a range that indicates even a rough consensus. Indeed, there is even legal and philosophical scope for questioning this imbalance *in principle*[16]; nonetheless, few would dispute that it is deeply engrained in our traditional understanding of justice, no matter that we cannot flesh it out with exact numbers.

There are many ways in which the legal systems try to avoid punishing the innocent, even if that means that guilty criminals will go free. Among the most important ones are the following principles: The distribution of the *burden of proof*, the *presumption of innocence* and the notion of *proof beyond reasonable doubt*. If we squint at them a bit, it is possible to see the latter two principles rolled up into the first one, the distribution of the burden of proof. The point of taking this somewhat simplifying view is that we will able to do with one parameter in the logical model of legal litigation later on.

The *burden of proof* determines who has to prove what. To safeguard against false convictions, the burden of proof is on the prosecution in criminal cases.[17] From the outset the prosecution has to deliver proof of the defendant's guilt, not the defendant of his innocence. Both his testimony and his plea (we assume he pleads "Not guilty") will stand as long as what he says is not disproven.

---

[15]'Tis much more Prudence to acquit two Persons, tho' actually guilty, than to pass Sentence of Condemnation on one that is virtuous and innocent. Voltaire (1749).

It is better that five guilty persons should escape punishment than one person should die. Matthew Hale (1678).

It is better that ten guilty persons escape [punishment] than that one innocent suffer. William Blackstone (eighteenth century).

I should, indeed, prefer twenty guilty men to escape death through mercy, than one innocent to be condemned unjustly. John Fortesque (1471).

It is better a hundred guilty persons should escape than one innocent person should suffer. Benjamin Franklin (1785).

It is better . . . to acquit a thousand guilty persons than to put a single innocent man to death. Moses Maimonides (twelfth century) [10], p. 63.

[16]Laudan himself has some doubts, and legal scholar Daniel Epps has just published a thought provoking all out attack on the idea in the Harvard Law Review ([5]). He argues that it is a remnant of a time in which capital punishment was the only possible outcome of a trial that found defendants guilty of major crimes, and that it has outlived its social utility in a time when most verdicts are of a more reversible nature.

[17]This distribution of the burden of proof is as basic as a legal principle can get. Nonetheless, the British government recently gave out an information brochure that managed to get this basic point exactly backwards. See http://www.independent.co.uk/news/uk/home-news/ministry-of-justice-gets-law-terribly-wrong-in-its-guide-to-courts-10016086.html, last accessed on July 5th 2017.

However, the burden of proof does not necessarily have to stay on the prosecution. The course of a trial might shift the burden of proof away from them and onto the defense. For example, if the defendant pleads not guilty on grounds of insanity, then, depending on the country and state, he might have to bear the burden of proof for this claim.[18] It is the task of the judge to keep track of, decide on, and inform the jury about the burden of proof.

The *presumption of innocence* assures that at the beginning of the trial, the default assumption is that the defendant is innocent. Only if there is enough evidence to ensure his guilt should this presumption be given up. It is actually a matter of scholarly debate whether this is an independent principle or just a corollary of the burden of proof being on the prosecution. As Laudan writes, it seems to follow almost automatically, just as the presumption of innocence seems to entail the distribution of the burden of proof:

> If the state bears the full burden of proof, then, of course, one might say, the defendant is presumed innocent. Contrariwise, if the defendant is genuinely presumed innocent, then it naturally follows that the state must defeat that presumption by proving his guilt ([10], p.90).

Laudan's subsequent discussion shows that things are in fact not that simple,[19] but we will leave it at that.

The last item is the high *standard of proof* that is set at criminal trials. It can be seen as an elaboration on what exactly the burden of the burden of proof amounts to. It is not enough that the evidence makes it slightly more likely that the defendant is guilty than not. In the US, the evidence needs to prove this guilt "beyond all reasonable doubt". Though it is not completely clear what this means in practice, it is clear that it is meant to be an extremely high standard of proof.

There have been many attempts to somehow quantify this imbalance, for example in terms of probabilities, but the practical feasibility of this is evidently questionable. Not only is it hard in practice to know how exactly a given piece of evidence influences the probability of material guilt and innocence. It also seems that in order to set the right threshold, we should first have to determine how exactly the optimal ratio between false acquittal and false convictions should look; not a straightforward task, as hinted at above. How to determine the standard of proof, how to express it and how to interpret the expressions that have been given in the past is one of the major

---

[18]In many states of the U.S., this shift of the burden of proof was legislated after a famous assassination attempt on president Reagan in 1982. The attacker's plea of *not guilty on grounds of insanity* was successful, not because he could prove his insanity, but because the prosecution could not prove his sanity. After this unpopular outcome, many states changed their laws so that a man or a woman in similar circumstances would be convicted.

[19]Mark the name of the principle, for a start. If Duvall had been right and innocence had (obviously to legal professionals) nothing to do with the job of the jury, then why should the jurors presume something that is patently unrelated to their task? Of course, this may once again only be an allowance to accidental features, this time the above noted lack of a lexicalized noun form of "not guilty". The "presumption of not-guiltiness" might well have been too clumsy for the legislators, but to suppose that they willingly introduced an intense confusion into a clear conceptual outlay just because they could not find a proper way to express themselves seems a bit too drastic. Unlike journalists, legal professionals should have been unwilling to pay such a high price.

areas of puzzlement in legal epistemology. A good start on these matters is, again, Laudan's book; another logically informed discussion is [4]. For the rest of the paper, I will suppose that these problems are somehow surmountable and return to my claim that these values and principles are reflected in the square.

Without these principles, the square would indeed look quite different. Imagine a society, maybe one scared by terrorist attacks, that thinks it is more important not to let guilty criminals go free than to avoid convicting innocent people. In the legal system such a society might set up, a defendant would have to bear the burden of proof and have to prove, maybe beyond reasonable doubt, his innocence. If unsuccessful, he would be convicted on the basis of a presumption of guilt.

Not surprisingly, the notions in such a legal system would also form a logical square, only this time with the lower corners on top:

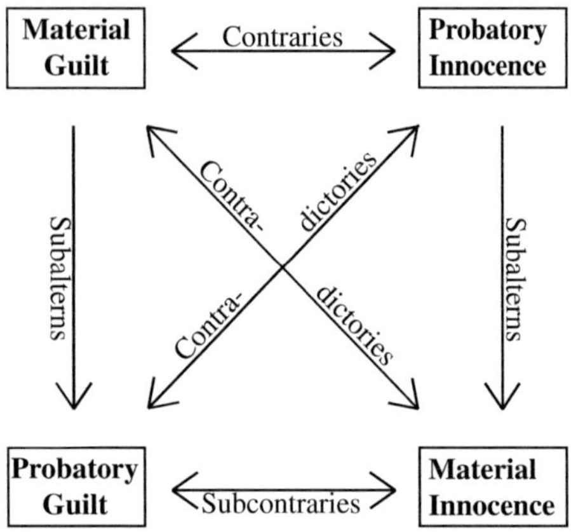

THE SQUARE OF Terror

Now, probatory guilt and material innocence could well exist together, when the innocent defendant is unable to reach the high standard of proof for his innocence. Likewise, it would be impossible to be materially guilty and probatorily innocent, that is, to be a criminal and escape punishment.

Of course, this square would likewise be subject to the idealizing assumptions I mentioned above, and which I will now discuss in more detail.

2.6 Idealizations

Let us then turn back to our original square and address the idealizations that are necessary to justify the logical relations in it. For example, the square invites us to draw an inference from the material innocence of a defendant to her probatory innocence. If this really would be such a sure conclusion, then why would we need skilled defense attorneys in the first place?

In truth, without a fair bit of idealization, there are no logical implications at all between the four corners, except for the diagonals. Indeed, if we were to drop the assumption that it will always be possible to tell whether the standard of proof has been met, even the contradictory arrows between probatory guilt and probatory innocence might become suspect. Note, though, that it isn't clear at all what we should say about the statement "The defendant is guilty" in a case in which we can't tell whether the standard of proof has been met. Is it neither true nor false? Might it be both?

Even if we leave that complication aside (as I will for the remainder of this paper), we will not be able to sustain the horizontal and vertical lines. It is clear that there can be situations where there is enough evidence against a materially innocent defendant to leave the jury no choice but to declare him probatorily guilty. Somewhat surprisingly, one can check that this case alone already gives a counterexample to all relations denoted by the horizontal and vertical arrows.

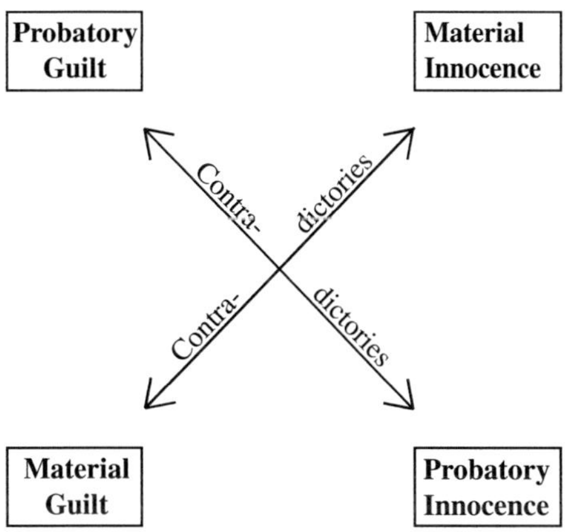

The Bare Square of Realism

However, the original arrows still hold up "fallibly", as Laudan says. The idealizing assumption, one that we might have to take back in practice, is for the most part[20] about the *quality* of the evidence.[21]

We can sustain the relations of the square if we suppose that the evidence we gather will be veridical. If this sounds too naive, maybe we can be a bit more sophisticated

[20] Among other things, we need to assume that the members of the jury understand the instructions (not as trivial an assumption as one might have thought, as the issues in this paper testify) and are able to draw the right conclusions from the evidence. Also, even purely veridical evidence can point in the wrong direction, if it is of the circumstantial sort and not counterbalanced by evidence for the innocence of the defendant.

[21] If we idealize away restrictions of quantity of evidence as well, then we only end up with *two* corners: If we assume not only that all evidence is veridical, but that we can collect all such veridical evidence pertinent to the case, then material guilt and probatory guilt will come to the same thing, as will material and probatory innocence. This "line of guilt and innocence" might be a blisteringly naive conception, but it is both an idealization and an ideal:

in our idealizing assumption and suppose that misleading pieces of evidence will often cancel each other out by contradicting each other, such that in most cases the veridical evidence will bear more weight at the end of the trial.

Even though this weaker assumption can sustain the relations in the square, the first version has the virtue that it opens up the possibility of an elegant logical model of the trial proceedings. The rest of the paper will aim to outline a framework that can be used to reason about the probative notions and the dialectical situation in trials, applying logical ideas I've developed earlier.[22]

2.7 Constructive Models for Legal Deliberation

The core of the following framework will belong to the family of constructive logics.[23] The best known constructive logic, of course, is intuitionistic logic, but it is not the only one that has been proposed; I have argued elsewhere at length that intuitionistic logic is not the most useful constructive logic once we leave the realm of mathematics behind and venture into more empirical areas, such as legal epistemology. What follows will instead be based on David Nelson's logic N_3.[24]

An immediate clarification seems appropriate: The square reports relations between evidential situations and actual facts (that is, between probatory and material notions). One might think that a constructive logic might be unsuitable in such a setting, due to the philosophical heritage of constructive logic. Constructive logicians have traditionally often been inspired by anti-realistic ideas, roughly that there is no realm of facts independent from our evidential situation. Though Laudan has found some legal scholars who seem to be endorsing such a view about criminal cases, it is clear that most would reject this. The facts of criminal deeds are not taken to be in some way ontologically dependent on the ability of a court to come to know of them.

(Footnote 21 continued)
It leaves out real-world complications, but it can serve as an epistemic ideal that we should set our legal institutions up to strive for. The square of guilt and innocence, in contrast, cannot, on its own, be said to embody such an ideal: It is quite easy to set up a legal institution that satisfies all the relations shown in the square: One that allows no evidence at all to be considered by the judge or jury and that leaves the burden of proof on the prosecution. It seems clear that this institution isn't even aiming at what it should. To make it a viable ideal, we should also stipulate that the state should make a serious attempt to find facts pertinent to the case and bring them to the attention of the court, and that this effort should not be (overly) thwarted.

[22] See Ref. [9].

[23] One referee remarked that the application of a constructive logic to the problem at hand is a rather obvious one. I was glad to read that, hoping that "obvious" implies "plausible". That said, I have not seen the application of a constructive logic to legal epistemology in print before, let alone the system I will propose below. Intuitionistic logic is discussed in [14]; however, the topic is not evidence but rather the nature of legal rules.

[24] See Ref. [11, 12, 15] for more on this logic.

Having such an anti-realist standpoint is, however, not a prerequisite for using a constructive logic. One might well be a logical pluralist who employs different logics at the material and at the probatory level (see [2] for one account of how this might play out; there are other possible ways as well, and I wish to make no commitment to one of those ways in particular).

One of the issues between the traditional realist / anti-realist debate we do not have to enter into when we take such a pluralist point of view is the question of truth. Truth need not be epistemically constrained, and those who like to take a naively realist view on truth are welcome to do so at the material level. I will, however, pay attention to the surrogate notion anti-realists like to give for truth, *correct assertibility*, without trying to use it to substitute truth. Material truth about a crime can co-exist peacefully with correct assertibility at trial, warranted by evidence. As we've seen, however, what determines the outcome of the trial is not so much material truth but rather whether the jury judges the assertions of the defense or those of the prosecution to be correct.

2.8 Nelson Logic

To get started on the logic, then, let me first introduce N_3 by way of Kripke models, which are rather similar to the Kripke models for intuitionistic logic. A model for N_3 will be a structure, $[W, \leq, v]$, where W is a non-empty set of information states,[25] and $\leq$ is a binary relation on those states which is reflexive, transitive and anti-symmetric, that is, a partial order. This is an accessibility relation on the information states that has a strong temporal flavor: One stage comes literally *after* the other. Intuitively, the stages track the progress of the trial and collect the evidence that has been presented so far.

The valuation function v assigns a truth value, 1 or 0, to each atomic statement p at each stage. We give both of the values 1 and 0 a substantive reading: 1 stands for "verified by the evidence", 0 for "falsified by the evidence".[26]

We allow v to be a *partial* function, so that statements might not receive either value at a given stage. This reflects the fact that at a stage of investigation, a statement might be neither verified nor falsified by the evidence presented so far.

The idealization I discussed above comes in when we impose the *heredity constraint* on our models. As all evidence is veridical, we will never have to go back and revise our account of what has been verified and what has been falsified by the evidence. Formally, this comes to:

For all p and all worlds w and w', if $w \leq w'$ and $w \Vdash_1 p$, then $w' \Vdash_1 p$, and
for all p and all worlds w and w', if $w \leq w'$ and $w \Vdash_0 p$, then $w' \Vdash_0 p$.

---

[25] Or "worlds", for those who are more accustomed to such talk.

[26] This is in contrast to the semantics of intuitionistic logic, in which one of the values marks the constructive notion, mathematical proof, and the other the mere *absence* of that notion.

Here is a diagram and a valuation that suits the requirements of N_3:

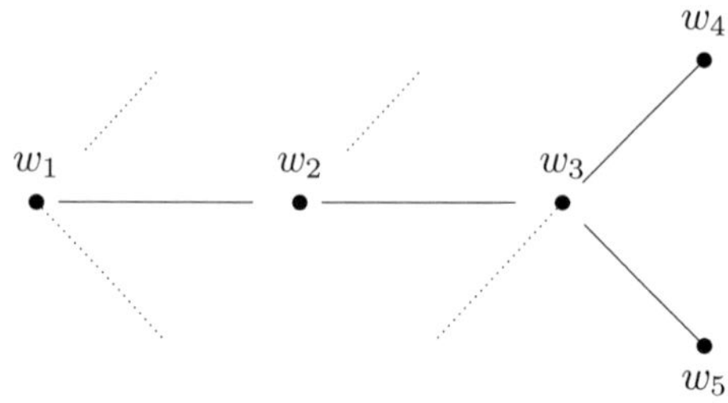

	w_1	w_2	w_3	w_4	w_5
p	-	-	1	1	1
q	-	0	0	0	0
r	1	1	1	1	1
s	-	-	-	1	0

Both 1 and 0 project forward, because both of them record a constructive achievement (verification and falsification respectively) that is taken to be permanent. There is a third option, here represented by "-": a gap, a mere absence of either verification or falsification. At the outset of the trial, none of the statements relating to the crime will receive a constructive value, i.e., all of them will be gaps.

Now, the logical connectives can be defined on this structure (we will stay at the propositional level). There are some possible choices here, but the connectives are not my main concern in this paper. Therefore, I simply list the clauses of N_3 and refer the interested reader to [9] for detailed discussion of the connectives:

$w \Vdash_1 A \wedge B$ iff $w \Vdash_1 A$ and $w \Vdash_1 B$
$w \Vdash_0 A \wedge B$ iff $w \Vdash_0 A$ or $w \Vdash_0 B$
$w \Vdash_1 A \vee B$ iff $w \Vdash_1 A$ or $w \Vdash_1 B$
$w \Vdash_0 A \vee B$ iff $w \Vdash_0 A$ and $w \Vdash_0 B$

$w \Vdash_1 -A$ iff $w \Vdash_0 A$
$w \Vdash_0 -A$ iff $w \Vdash_1 A$

$w \Vdash_1 A \supset B$ iff for all $x \geq w$, $x \nVdash_1 A$ or $x \Vdash_1 B$
$w \Vdash_0 A \supset B$ iff $w \Vdash_1 A$ and $w \Vdash_0 B$

Instead of investigating the plausibility of those clauses here, I want to focus on the definition of logical consequence that is appropriate for our application. In N_3, consequence is defined as follows:

$\Gamma \models A$ iff in every model and every $w \in W$, if $w \Vdash_1 B$ for every $B \in \Gamma$, then $w \Vdash_1 A$.

To wit, we want to go from verified statements to other verified statements. That makes sense in many situations; here, however, we have a slightly more complex situation. It is the consequence relation that will be reformulated to capture the burden of proof.

2.9 Modeling the Burden of Proof

The jury needs to decide which assertions and claims have been correctly made: Those by the defense or those by the prosecution. But the burden of proof dictates that these assertions have to be judged differently. The party bearing the burden of proof is indeed only making correct assertions when what it says is verified. The other party, on the other hand, is making correct claims *as long as what it says is not falsified*. That means that we are interested in two different sets of statements: The ones that are verified and receive a 1 by the valuation function, and those that are at least not falsified and thus do not receive value 0. Of course, the former is a subset of the latter.

To understand this difference is essential not only to decide which assertions have been made correctly. Also, to know what follows from a set of correct claims, one will have to take into account whether it was made under the burden of proof or not. Suppose, then, that we want to know which statements can correctly be asserted given the current distribution of the burden of proof and the correct assertions that have been made in the course of the trial. These earlier correct assertions will fall into two categories: Those made under the burden of proof, and those made not bearing the burden of proof.

I'll write Γ for the set of statements correctly asserted under the burden of proof, and Δ for the set of statements correctly asserted while not under the burden of proof. As the burden of proof can shift from the prosecution to the defense (and back) in the course of a trial, I will not identify the set of correct assertions by the prosecution with Γ and the set of correct assertions made by the defense with Δ. We will instead think of the burden of proof attached to single assertions: Either an assertion was made bearing the burden of proof, or not, and depending on that alone it will be sorted into Γ or Δ.

We will also have two classes of consequences we'll be interested in, depending on whether the next statement we wish to make will be made under the burden of proof or not. Thus, logical consequence will relate the ordered pair of sets of statements Γ and Δ to other pairs of sets of statements, Φ and Ψ. If the burden of proof will lie on the consequence, then Φ will contain a single statement and Ψ will be the empty set, and the other way around if the burden of proof will not lie on the consequence. By "the consequence" I simply mean the statement that is in the singleton Φ or Ψ. I will simply omit any of Γ, Δ, Φ and Ψ in case they are empty.

Now, I redefine logical consequence[27]:

> $\Gamma \mid \Delta \vDash_{hyb} \Phi \mid \Psi$ iff in every model and every $w \in W$, if $w \Vdash_1 B$ for every $B \in \Gamma$ and if $w \nVdash_0 C$ for every $C \in \Delta$, then $w \Vdash_1 D$ for every $D \in \Phi$ and $w \nVdash_0 E$ for every $E \in \Psi$.

The idea is that Γ collects all that we know is verified, because correctly uttered under the burden of proof, and Δ all that we know is at least not falsified, because uttered correctly by speakers not bearing the burden of proof. Whether or not the burden of proof is on the conclusion will decide whether we're interested in conclusions before or behind the bar. So, for example, the consequence A of an inference $\Gamma \mid \Delta \vDash_{hyb} A \mid$ will be such that we can utter it correctly if the burden of proof is on us, given that Γ and Δ collect verified and unfalsified statements, respectively.[28]

If, on the other hand, we know our assertion will not be made under the burden of proof, we'll be interested in the inferences of the form $\Gamma \mid \Delta \vDash_{hyb} \mid A$.

Again, for more information about the sorts of inferences this setup allows, I refer the interested reader to [9]. My main point in this section was to outline how Laudan's distinction between the material and the probatory level can motivate a constructive logic, and how it is possible to give a notion of logical consequence that allows space for the burden of proof to decide the correctness of assertion in adversarial situations like criminal trials.

2.10 Future Work

The logical setup above is, of course, only of interest as long as we reason about the probatory level.[29] As it is based on the same idealizing assumption as the square of oppositions, however, we can use the relations in the square as a set of bridge principles between these levels. Given the idealizing assumption, if the information

---

[27] I use, as in [9], $\vDash_{hyb}$, for "hybrid consequence", to denote this consequence relation.

[28] One may wonder whether the information recorded in Δ, which is essentially information about unfalsified statements, could be relevant at all if the conclusion has to be verified. Indeed, it does make a difference, as the following example will show. Let $A \vee B$ be the only element of Γ, and $-B$ the only element of Δ. Then we will find that $\Gamma \mid \Delta \vDash_{hyb} A \mid$, while $\Gamma \mid \nvDash_{hyb} A \mid$. Intuitively speaking, the burden of proof-bearing speaker who correctly asserted $A \vee B$ had to have either a verification of A or a verification of B. The latter case is excluded by the fact that someone else, not bearing the burden of proof, was able to utter $-B$.

Another matter worth mentioning is that the semantics, as it stands, allows for $A \wedge -A$ to be unfalsified. The rationale for that is just analogous to the reasoning that leads intuitionists to reject the validity of $A \vee -A$. To prove a disjunction, on the constructive view, you need to be able to prove one of the disjuncts. Likewise, on the view proposed here, to falsify a conjunction, you need to be able to falsify (at least) one of the conjuncts. Of course, to lead a party to contradict itself is a powerful move in legal discourse, no matter whether that party has to bear the burden of proof at the moment or not. That means that we should have a means of recognizing self-contradictory statements as defective in the logical system. Chap. 9 of [9] contains a mechanism for doing just that.

[29] The logic that is suitable for the material level might well be classical logic, or whichever logic one deems suitable to deal with facts about criminal events.

recorded at a stage of trial is enough to prove guilt beyond a reasonable doubt, then we can infer material guilt. Under the same assumption, the material innocence of the defendant will rule out all information stages that prove guilt beyond any reasonable doubt in the model of the possible ways the trial might unfold.

We saw that giving up the idealization meant the loss of the vertical and horizontal arrows of the square. In other words, the material and the probatory level will get dissociated, or at least, their connection will only be a fallible one, as Laudan had originally put it.

But even at the probatory level itself we will have to implement changes; evidence can not only be misleading, but also be discredited and outweighed by later counter evidence. What has been taken to be probatorily established at one stage of the trial might have become dubious at a later stage.

This will mean that the heredity constraint will have to be weakened in some way, or that some other way will have to be implemented to turn the logic above into a non-monotonic logic. Though there have been interesting attempts to base a non-monotonic logic on Nelson's N_3[30] that could be modified along the lines above, I think it is fair to say that a suitable system that can be adapted for the purpose at hand remains on the agenda for future work.

A very different set of questions concerns what interdisciplinary use could be made of the analysis. I believe that especially the material of the first part of this essay might be adapted for instructive uses. One could use the square of oppositions to teach people with no background in formal logic about the connections between guilt and innocence, e.g. at a law school. It would of course be highly beneficial if the jurors would understand the relations reported in the square, but in all its elegance I think it is too complicated to be of any use in pre-trial jury instructions.

In any case, I hope to have shown that Laudan's regimentation of the vocabulary allows new and fruitful routes of investigation that use both formal and informal methods.

Acknowledgments This research was financed by the research project ''New Logics for Verificationism'' PI 1082/1-102288852134 funded by the German Research Foundation. The paper has benefited form comments by two anonymous referees, Georgios Karageorgoudis and the audience at the fourth world congress on the square of oppositions, held in 2014 at the Vatican. I would like to express special thanks to João Marcos and Lothar Philipps. Prof. Philipps sadly passed away recently, and I would like to dedicate this paper to his memory.

References

1. Allen, R. (2010). Laudan, stein, and the limits of theorizing about juridical proof. *Law and Philosophy*, *29*(2), 195–230.
2. Beall, J., & Restall, G. (2006). *Logical pluralism*. Oxford: Oxford University Press.
3. Bèziau, J. Y., & Jacquette, D. (Eds.). (2012). *Around and beyond the square of opposition*. Berlin: Springer.

[30]See Ref. [8, 16].

4. Clermont, K. (2013). *Standards of decision in law*. Carolina Academic Press.
5. Epps, D. (2015). The consequences of error in criminal justice. *Harvard Law Review*, *128*(4), 1065–1151.
6. Goldstein, N. (Ed.). (2002). *Stylebook*. The Associated Press.
7. Haack, S. (2014). *Evidence matters. science, proof and truth in the law*. Cambridge: Cambridge University Press.
8. Jaspars, J. (1995). Partial up and down logic. *Notre Dame Journal of Formal Logic*, *36*(1), 134–157.
9. Kapsner, A. (2014). *Logics and falsifications*. Cham: Springer.
10. Laudan, L. (2006). *Truth, error and criminal law*. Cambridge: Cambridge University Press.
11. Nelson, D. (1949). *Constructible falsity. Journal of Symbolic logic*, *14*(1), 16–26.
12. Odintsov, S. (2008). *Constructive negations and paraconsistency*. Berlin: Springer.
13. Parsons, T. (2014). The traditional square of opposition. In by E. D. Zalta (Ed.), *The stanford encyclopedia of philosophy (Spring 2014 Edition)*. http://plato.stanford.edu/archives/spr2014/entries/square/.
14. Philipps, L. (1964). Rechtliche regelung und formale logik. *Archiv für Rechts- und Sozialphilosophie*, *1964*, 317–329.
15. Wansing, H. (1993). *The logic of information structures*. In Lecture notes in AI 681. Berlin: Springer.
16. Wansing, H. (1998). *Displaying modal logic*. Dordrecht: Kluwer Academic Publishers.
17. Wexler, S. (2010). Two forms of legal proof and the so-called blue bus problem. *International Zeitschrift*, *6*, 3.

Chapter 3
Counterfactuals, Logic Programming and Agent Morality

Luís Moniz Pereira and Ari Saptawijaya

Abstract This paper supplies a computational model, via Logic Programming (LP), of counterfactual reasoning of autonomous agents with application to morality. Counterfactuals are conjectures about what would have happened had an alternative event occurred. The first contribution of the paper is showing how counterfactual reasoning is modeled using LP, benefiting from LP abduction and updating. The approach is inspired by Pearl's structural causal model of counterfactuals, where causal direction and conditional reasoning are captured by inferential arrows of rules in LP. Herein, LP abduction hypothesizes background conditions from given evidence or observations, whereas LP updating frame these background conditions as a counterfactual's context, and then imposes causal interventions on the program through defeasible LP rules. The second contribution it to apply counterfactuals to agent morality using this LP-based approach. We demonstrate its potential for specifying and querying moral issues, by examining viewpoints on moral permissibility via classic moral principles and examples taken from the literature. Application results were validated on a prototype implementing the approach on top of an integrated LP abduction and updating system supporting tabling.

Some main parts of this chapter are to be found adapted and integrated into a subsequently elaborated book by the same authors [47], comprising many materials from the authors' previous works. This chapter defines and uses common notations and semantics as the book—which are standard. The chapter, however, is concentrated and self-contained; and especially geared to those concerned with just a logic formulation of counterfactuals in Logic Programming, and their use in computational morality, in a self-sufficient manner. The book is much more ample in scope, and combines a number of other computational logic aspects of morality, and of which counterfactual reasoning is just one, and where it is entwined with other facets present in the book. This chapter focusses only on the formulation, treatment and morality application of counterfactuals.

L.M. Pereira
NOVA Laboratory for Computer Science and Informatics (NOVA LINCS),
Universidade Nova de Lisboa, Lisbon, Portugal
e-mail: lmp@fct.unl.pt

A. Saptawijaya (✉)
Faculty of Computer Science, Universitas Indonesia, Depok, Jawa Barat, Indonesia
e-mail: saptawijaya@cs.ui.ac.id

R. Urbaniak and G. Payette (eds.), *Applications of Formal Philosophy*,
Logic, Argumentation & Reasoning 14, DOI 10.1007/978-3-319-58507-9_3

Keywords Abduction · Counterfactuals · Logic programming · Morality · Nonmonotonic reasoning

3.1 Introduction

Counterfactuals capture the process of reasoning about pasts events that did not occur, namely what would have happened had an event occurred; or, vice-versa, to reason about an event that did occur but supposing it had not. An example from [11]: *Lightning hits a forest and a devastating forest fire breaks out. The forest was dry after a long hot summer and many acres were destroyed.* One may think of a counterfactual about the forest fire, e.g., "if only there had not been lightning, then the forest fire would not have occurred". Counterfactuals have been widely studied, in philosophy [13, 24, 34] and psychology [11, 16, 35, 36, 39, 52]. They also have been studied from the computational viewpoint [8, 22, 41, 42, 63], where approaches in Logic Programming (LP), e.g., [8, 42, 63], are mainly based on probabilistic reasoning.

As our first contribution to this topic, we innovate upon LP abduction and updating in an implemented procedure for evaluating counterfactuals, taking the established approach of Pearl [41] as reference. Our approach concentrates on pure non-probabilistic counterfactual reasoning in LP—thus distinct from but complementing existing probabilistic approaches—by instead resorting to abduction and updating, in order to determine the logical validity of counterfactuals under so-called Well-Founded Semantics [62]. Nevertheless, the approach is adaptable to other semantics, e.g., Weak Completion Semantics [29] is employed in [45]. Our approach is therefore suited and applicable to instances when probabilities are not known or needed.

LP lends itself to Pearl's causal model of counterfactuals for two reasons: (1) The inferential arrow in an LP rule is suitable for expressing causal direction; and (2) LP is enriched with functionalities, such as abduction and defeasible reasoning with updates. They can be exploited to express Pearl's evaluation procedure for counterfactuals: LP abduction is employed to provide background conditions from observations made or evidence given, whereas defeasible logic rules allow us to represent adjustments to the current model—at selected points—via updates of (hypothetical) intervention.

Counterfactual thinking in moral reasoning has been investigated particularly via psychological experiments (see, e.g., [16, 36, 37, 39]), but it hasn't been explored very much in machine ethics. Machine ethics is an interdisciplinary field that emerges from the need of imbuing autonomous agents with the capacity for moral decision making to enable them to function in an ethically responsible manner via their own ethical decisions. As our second contribution, we apply counterfactual reasoning to machine ethics. The potential of LP for machine ethics has been reported in [25, 33, 46, 55], where the main characteristics of moral reasoning can appropriately be expressed by LP-based reasoning, such as abduction, integrity constraints, preferences, updating, and argumentation. The application of counterfactual reasoning to machine ethics—herein by resorting to our LP approach—is an extension of reasoning about counterfactuals in the wider context of the aforementioned well-developed LP-based non-monotonic reasoning methods, which are also important, appropriate, and promising for moral reasoning.

Counterfactual theories are very suggestive of a conceptual relationship to a form of debugging, namely in view of correcting moral blame, since people ascribe abnormal antecedents an increased causal power, and are also more likely to generate counterfactuals concerning abnormal antecedents. Two distinct processes can be identified when people engage in counterfactual thinking. For one, its frequent spontaneous triggers encompass bad outcomes and "close calls" (some harm that was close to happening). Second, such thinking comprises a process of finding antecedents which, if mutated, would prevent the bad outcome from arising. When people employ counterfactual thinking, they are especially prone to change abnormal antecedents, as opposed to normal ones. Following a bad outcome, people are likely to conceive of the counterfactual "if only [some abnormal thing] had not occurred, then the outcome would not have happened". See [52] for a review.

In this paper, counterfactuals are specifically engaged to distinguish whether an effect of an action is a cause for achieving a morally dilemmatic goal or merely a side-effect of that action. The distinction is essential for establishing moral permissibility from the viewpoints of the Doctrines of Double Effect and of Triple Effect, as scrutinized herein through several off-the-shelf classic moral examples from the literature. Note that, the application of counterfactuals in these examples neither aims at defending the two doctrines nor resolving the dilemmas appearing in the examples, as even philosophers' opinions are split on them. Instead, its purpose is to show that counterfactuals, supported by our LP approach, are capable and appropriate for expressing different viewpoints on permissibility according to both doctrines based on views argued in the literature. By materializing these doctrines in concrete moral dilemmas, the results of counterfactual evaluation are readily comparable to those from the literature. Abstaining from probability permits focusing on the naturalized logic of human counterfactual moral reasoning. Moreover, people naturally do not compute formal probabilities, nor are probabilities always available, when making moral decisions via counterfactuals. Thus one can benefit from counterfactuals for inferring intentions through a probabilistic model to explain moral permissibility [32]. Note that, even though the LP technique introduced in this paper is relevant for modeling counterfactual moral reasoning, its use is general and not specific to morality.

The paper is organized as follows. Section 3.2 reviews basic notation in LP, abduction in LP, and Pearl's structure-based counterfactuals. We discuss how causation and intervention in Pearl's approach can be expressed in LP, and subsequently detail a LP approach to evaluate counterfactuals in Sect. 3.3. The application of counterfactuals to machine ethics is elaborated in Sect. 3.4. Section 3.5 frames our contributions in the context of related work. We conclude in Sect. 3.6, by touching upon prospects of counterfactuals in expressing moral issues, thereby opening up further opportunities of application in machine ethics, within a combination of abduction and updating.

3.2 Preliminaries

By an alphabet $\mathcal{A}$ of a language $\mathcal{L}$ we mean a union of countable, pairwise disjoint sets of constants, function symbols, and predicate symbols. Moreover, an alphabet is assumed to contain a countable set of variable symbols. A term over $\mathcal{A}$ is defined recursively as either a variable, a constant or an expression of the form $f(t_1, \ldots, t_n)$, where f is a function symbol of $\mathcal{A}$, and t_is are terms. An *atom* over $\mathcal{A}$ is an expression of the form $p(t_1, \ldots, t_n)$, where p is a predicate symbol of $\mathcal{A}$, and t_is are terms. A *literal* is either an atom a or its negation $not\ a$. Literals of the latter form are called default literals. The *negation complement of a literal* L is denoted by $compl(L)$, where the negation complement of a positive literal a and its negation $not\ a$ are defined as $compl(a) = not\ a$ and $compl(not\ a) = a$, respectively.

A term (respectively, atom and literal) is said to be a *ground* if it does not contain variables. The set of all ground terms (respectively, ground atoms) of $\mathcal{A}$ is called the Herbrand universe (respectively, Herbrand base) of $\mathcal{A}$.

Definition 1 (*Logic Program*) A *(normal) logic program* is a countable set of rules of the form:

$$H \leftarrow L_1, \ldots, L_m$$

where H is an atom, $m \geq 0$, and L_is $(1 \leq i \leq m)$ are literals.[1]

The comma operator in rules is read as conjunction. A normal logic program is called *definite* if none of its rules contains default literals. Following the standard convention, rules of the form $H \leftarrow$ are alternatively written as H. A rule of this form is called a *fact*.

The alphabet $\mathcal{A}$ used to write program P is assumed to precisely comprise all the constants, the function and predicate symbols that explicitly appear in P. By Herbrand universe (respectively, base) of P we mean the Herbrand universe (respectively, base) of $\mathcal{A}$. We denote the Herbrand base of P by $\mathcal{H}_P$. By a ground logic program we mean the set of ground rules obtained from P by substituting in all possible ways each of the variables in P by elements of its Herbrand universe.

We define next three-valued Herbrand interpretations and models of logic programs.[2] It permits representing incomplete knowledge, where some atoms are neither true nor false, but rather undefined. Let F be a set of atoms, $F = \{a_1, \ldots, a_n\}$, and $not\ F = \{not\ a_1, \ldots, not\ a_n\}$.

---

[1]In the sequel, unless otherwise specified, we generally write *logic programs*, or simply *programs*, to refer to *normal logic programs*.

[2]In the sequel, we simply write *interpretations* and *models* to refer to *Herbrand interpretations* and *Herbrand models*, respectively.

Definition 2 (*Three-valued Interpretation*) A *three-valued interpretation* I of a logic program P is a set of literals

$$I = T \cup not\ F$$

such that $T \subseteq \mathcal{H}_P$, $F \subseteq \mathcal{H}_P$ and $T \cap F = \emptyset$.

In a three-valued interpretation, the set T (respectively, F) is the set of atoms that are true (respectively, false) in I, and the truth value of the remaining atoms is undefined.

We may view an interpretation I of a program P as a function $I : \mathcal{H}_P \to \mathcal{V}$, where $\mathcal{V} = \{0, 0.5, 1\}$, defined by:

$$I(A) = \begin{cases} 0 & \text{if } not\ A \in I \\ 1 & \text{if } A \in I \\ 0.5 & \text{otherwise} \end{cases}$$

Models are defined as usual, and based on a truth valuation function.

Definition 3 (*Truth Valuation*) If I is an interpretation, the *truth valuation* $\hat{I}$ corresponding to I is a function $\hat{I} : \mathcal{F} \to \mathcal{V}$, where $\mathcal{F}$ is the set of ground literals, conjunctions of literals, and rules formed over the language. It is defined as follows:

If L is a ground atom, then $\hat{I}(L) = I(L)$.
If L is a default literal, i.e., $L = not\ A$, then $\hat{I}(L) = 1 - \hat{I}(A)$.
If S and T are conjunctions of literals, then $\hat{I}((S, T)) = min(\hat{I}(S), \hat{I}(T))$.
If $H \leftarrow B$ is a rule, where B is a conjunction of literals, then:

$$\hat{I}(H \leftarrow B) = \begin{cases} 1 & \text{if } \hat{I}(B) \leq \hat{I}(H) \\ 0 & \text{otherwise} \end{cases}$$

For any $F \in \mathcal{F}$, the values 0, 0.5 and 1 of $\hat{I}(F)$ correspond to the truth values false, undefined and true, respectively. We write $I \models F$, for $F \in \mathcal{F}$, iff $\hat{I}(F) = 1$.

Definition 4 (*Model*) An interpretation I is called a *model* of a program P iff for every ground instance $H \leftarrow B$ of a rule in program P we have $\hat{I}(H \leftarrow B) = 1$.

We define some orderings among interpretations and models as follows.

Definition 5 (*Classical Ordering* [50]) If I and J are two interpretations then we say that $I \preceq J$ if $I(A) \leq J(A)$ for any ground atom A. If $\mathcal{I}$ is a collection of interpretations, then an interpretation $I \in \mathcal{I}$ is called *minimal* in $\mathcal{I}$ if there is no interpretation $J \in \mathcal{I}$ such that $J \preceq I$ and $J \neq I$. An interpretation I is called *least* in $\mathcal{I}$ if $I \preceq J$, for any other interpretation $J \in \mathcal{I}$. A model M is called *minimal* (respectively, *least*) if it is minimal (respectively, least) among all models of P.

Definition 6 (*Fitting Ordering* [17]) If I and J are two interpretations then we say that $I \preceq_F J$ iff $I \subseteq J$. If $\mathcal{I}$ is a collection of interpretations, then an interpretation

$I \in \mathcal{I}$ is called *F-minimal* in $\mathcal{I}$ if there is no interpretation $J \in \mathcal{I}$ such that $J \preceq_F I$ and $J \neq I$. An interpretation I is called *F-least* in $\mathcal{I}$ if $I \preceq_F J$, for any other interpretation $J \in \mathcal{I}$. A model M is called *F-minimal* (respectively, *F-least*) if it is F-minimal (respectively, F-least) among all models of P.

Note that the classical ordering is related with the *degree of truth* of their atoms, whereas the Fitting ordering is related with the *degree of information*. Under the latter ordering, the undefined value is less than both values true and false, providing that true and false being incompatible.

In [61], it is shown that every definite program has a unique least model, which determines the so-called *least model semantics* of a definite program. Other semantics for more general programs, allowing default literals in the body of a rule, have been proposed. In [21], *Stable Model Semantics* is introduced. Informally, when one assumes that some set of (hypothetical) default literals are all true, and all the others are false, some consequences follow according to the semantics of definite programs. If the consequences completely corroborate the hypotheses made, then they form a stable model.

Despite its advantages, that it provides semantics for more general programs than its predecessors and is closely related to autoepistemic logic and default theory (see [20] and [9]), Stable Model Semantics has some drawbacks. Some programs may have no stable models, e.g., the program $p \leftarrow not\ p$. Even for programs with stable models, their semantics do not always lead to the expected intended semantics (see [3] for a discussion).

The *Well-Founded Semantics* [62], which we refer to in this paper, addresses the difficulties encountered with the Stable Model Semantics. It has been shown in [51] that the Well-Founded Semantics is also equivalent to major formalizations of non-monotonic reasoning.

The Well-Founded Semantics can be viewed as *three-valued Stable Model Semantics* [49]. In order to formalize the notion of three-valued stable models, the language of programs is expanded with the additional propositional constant u with the property of being undefined in every interpretation. It is therefore assumed that every interpretation I satisfies:

$$\hat{I}(\mathsf{u}) = \hat{I}(not\ \mathsf{u}) = 0.5$$

A *non-negative* program is a program whose rules' bodies are either atoms or u. It is proven in [49] that every non-negative logic program has a unique least three-valued model.

The next definition extends the Gelfond-Lifschitz operator [21] to a three-valued operator Γ^*.

Definition 7 (*Γ^*-operator*) Let P be a logic program and I be its three-valued interpretation. The *extended GL-transformation of P modulo I* is the program $\frac{P}{I}$ obtained from P by performing the following operations:

Remove from P all rules which contain a default literal $L = not\ A$ such that $\hat{I}(L) = 0$;

Replace in the remaining rules of P those default literals $L = not\ A$ which satisfy $\hat{I}(L) = 0.5$ by u;
Remove from all remaining rules those default literals $L = not\ A$ which satisfy $\hat{I}(L) = 1$.

Since the resulting program $\frac{P}{I}$ is non-negative, it has a unique three-valued least model J. We define $\Gamma^*(I) = J$.

Definition 8 (*Well-Founded Semantics*) A three-valued interpretation I of a logic program P is a three-valued stable model of P if $\Gamma^*(I) = I$. The *Well-Founded Semantics* of P is determined by the unique F-least three-valued stable model of P, and can be obtained by the bottom-up iteration of Γ^* starting from the empty interpretation.

Example 1 Consider program P:

$$\begin{aligned} a &\leftarrow not\ b. \\ b &\leftarrow not\ a. \\ c &\leftarrow not\ d. \\ d &\leftarrow not\ e. \\ p &\leftarrow a. \\ p &\leftarrow b. \end{aligned}$$

Let $I_0 = \emptyset$ be the empty interpretation.

The least three-valued model of $\frac{P}{I_0}$:

$$\begin{aligned} a &\leftarrow \mathsf{u}. \\ b &\leftarrow \mathsf{u}. \\ c &\leftarrow \mathsf{u}. \\ d &\leftarrow \mathsf{u}. \\ p &\leftarrow a. \\ p &\leftarrow b. \end{aligned}$$

is $\Gamma^*(I_0) = \{not\ e\}$.
Let $I_1 = \Gamma^*(I_0)$. The least three-valued model of $\frac{P}{I_1}$:

$$\begin{aligned} a &\leftarrow \mathsf{u}. \\ b &\leftarrow \mathsf{u}. \\ c &\leftarrow \mathsf{u}. \\ d &\leftarrow . \\ p &\leftarrow a. \\ p &\leftarrow b. \end{aligned}$$

is $\Gamma^*(I_1) = \{d, not\ e\}$.
Let $I_2 = \Gamma^*(I_1)$. The least three-valued model of $\frac{P}{I_2}$:

$$\begin{aligned} a &\leftarrow \mathsf{u}. \\ b &\leftarrow \mathsf{u}. \\ d &\leftarrow . \\ p &\leftarrow a. \\ p &\leftarrow b. \end{aligned}$$

is $\Gamma^*(I_2) = \{d, not\ c, not\ e\}$.
Let $I_3 = \Gamma^*(I_2)$. The least three-valued model of $\frac{P}{I_3}$:

$$\begin{aligned} a &\leftarrow \mathsf{u}. \\ b &\leftarrow \mathsf{u}. \\ d &\leftarrow . \\ p &\leftarrow a. \\ p &\leftarrow b. \end{aligned}$$

is $\Gamma^*(I_3) = \{d, not\ c, not\ e\}$.

Therefore, the well-founded model of P is $I_3 = \{d, not\ c, not\ e\}$, where d is true, c and e are both false, and a, b and p are undefined.

In the sequel, we write the well-founded model of program P as $WFM(P)$.

3.2.1 Abductive Logic Programs

Abduction is a reasoning method where one chooses from available hypotheses those that best explain the observed evidence, in a preferred sense. That is, the best explanation can generally be determined through some integrity constraints or preference rules [14, 44], rather than simply taking a minimal explanation. In LP, an abductive hypothesis (*abducible*) is a positive literal Ab or its negation complement Ab^* (syntactically an atom, but denoting literal $not\ Ab$), whose truth value is not initially assumed. An *abductive logic program* (ALP) is one allowing abducibles in the body of rules.

We next define an abductive framework in LP [30], which includes integrity constraints for restricting abduction, under the Well-Founded Semantics [62]. The definitions in this section are adapted from those of [4].

Definition 9 (*Integrity Constraint*) An *integrity constraint* is a rule in the form of a denial:

$$\bot \leftarrow L_1, \ldots, L_m.$$

where $\bot/0$ is a reserved predicate symbol in $\mathcal{L}$, $m \geq 1$, and L_is $(1 \leq i \leq m)$ are literals.

Definition 10 (*Abductive Framework*) An *abductive framework* is a triple $\langle P, \mathcal{A}, \mathcal{I} \rangle$, where $\mathcal{A}$ is the set of abducibles, P is a logic program over $\mathcal{L} \setminus \{\bot\}$ such that there is no rule in P whose head is in $\mathcal{A}$, and $\mathcal{I}$ is a set of integrity constraints.

Definition 11 (*Abductive Scenario*) Let F be an abductive framework $\langle P, \mathcal{A}, \mathcal{I}\rangle$. An *abductive scenario* of F is a tuple $\langle P, \mathcal{A}, \mathcal{S}, \mathcal{I}\rangle$, where $\mathcal{S} \subseteq \mathcal{A}$ and there is no $A \in \mathcal{S}$ such that $compl(A) \in \mathcal{S}$, i.e., $\mathcal{S}$ is consistent.

The consistency of an abductive scenario can be imposed by an integrity constraint $\bot \leftarrow Ab, Ab^*$.

Let observation O be a set of literals, analogous to a query in LP. Abducing an explanation for O amounts to finding consistent abductive solutions to a goal, whilst satisfying the integrity constraints, where abductive solutions consist in the semantics obtained by replacing in P the abducibles in $\mathcal{S}$ by their truth value. We define formally abductive solutions under the Well-Founded Semantics below.

Given an abductive scenario $\langle P, \mathcal{A}, \mathcal{S}, \mathcal{I}\rangle$ of an abductive framework $\langle P, \mathcal{A}, \mathcal{I}\rangle$, we first define $P_{\mathcal{S}}$ as the smallest set of rules that contains the fact A if $A \in \mathcal{S}$; and $A' \leftarrow \mathsf{u}$ for each $A' \in \mathcal{A}$, $A' \notin \mathcal{S}$ and $A' \neq compl(A)$. Alternatively, and obviously equivalent, instead of adding to $P_{\mathcal{S}}$ the rule $A' \leftarrow \mathsf{u}$, one may simply replace the corresponding A' with u both in P and $\mathcal{I}$.

Definition 12 (*Abductive Solution*) Let F=$\langle P, \mathcal{A}, \mathcal{I}\rangle$ and $\langle P, \mathcal{A}, \mathcal{S}, \mathcal{I}\rangle$ be an abductive scenario of F. The consistent set of abducibles $\mathcal{S}$ is an *abductive solution* to F if $\bot$ is false in $M_s = WFM(P \cup P_{\mathcal{S}} \cup \mathcal{I})$.

We further say that $\mathcal{S}$ is an *abductive solution for query Q* if Q is true in M_s, written $M_s \models Q$.

Abduction in LP can be accomplished by a top-down query-oriented procedure for finding a query solution by need. The solution's abducibles are leaves in its procedural query-rooted call-graph, i.e., the graph is recursively generated by the procedure calls from literals in bodies of rules to heads of rules, and thence to the literals in a rule's body. The correctness of this top-down computation requires the underlying semantics to be relevant, as it avoids computing a whole model (to warrant its existence) in finding an answer to a query. Instead, it suffices to use only the rules relevant to the query—those in its procedural call-graph—to find its truth value. The Well-Founded Semantics (WFS) enjoys this relevancy property, i.e., it permits finding only relevant abducibles and their truth value via the aforementioned top-down query-oriented procedure. Those abducibles not mentioned in the solution are indifferent to the query.

Example 2 A library in a city is closed with two possible explanations: it is closed on the weekend, or when it is not the weekend, there are no librarians working. These days, librarians are often absent from their work because they are participating in a strike. On the other hand, a museum in that city is only closed when there is a special visit by important guests. This example can be expressed by an abductive framework $\langle P, \mathcal{A}, \mathcal{I}\rangle$, where $\mathcal{A} = \{weekend, weekend^*, strike, strike^*, specialVisit, specialVisit^*\}$, P is the program below, and $\mathcal{I} = \emptyset$:

$$closed\_library \leftarrow weekend \qquad closed\_library \leftarrow weekend^*, absent$$
$$absent \leftarrow strike \qquad closed\_museum \leftarrow specialVisit$$

Consider the query $Q = \{closed\_library\}$. This query induces the call-graph with $closed\_library$ as its root and, through procedure calls, it ends with two leaves: one leaf containing abducible $\{weekend\}$, and the other containing $\{weekend^*, strike\}$. That is, there are two abductive solutions for query Q, viz., $\mathcal{S}_1 = \{weekend\}$ and $\mathcal{S}_2 = \{weekend^*, strike\}$. Note that the abducible $specialVisit$ is not mentioned in either solution: its truth value is irrelevant for query Q because the WFS enjoys the relevancy property, namely it finds only relevant abducibles (and their truth values) through the aforementioned procedure calls, driven by the considered query Q.

Example 3 Recall Example 2 plus new information: librarians may also be free from working in case the library is being renovated. The rule below is added to program P, where $renov$ and $renov^*$ are new abducibles in $\mathcal{A}$:

$$absent \leftarrow renov$$

The same query $Q = \{closed\_library\}$ now returns an additional solution, viz. $\mathcal{S}_3 = \{weekend^*, renov\}$. Let us further suppose that the municipal authority permits any renovations to take place only on weekends, expressed as an integrity constraint in $\mathcal{I}$:

$$\mathcal{I} = \{\bot \leftarrow weekend^*, renov\}$$

That is, $\mathcal{S}_3 = \{weekend^*, renov\}$ is now ruled out from the abductive solutions.

3.2.2 *Pearl's Structure-Based Counterfactuals*

Pearl [41] proposes a structural theory of counterfactuals based on a probabilistic causal model and a calculus of intervention (viz., do-calculus). A causal model M consists of two sets of variables U (*background variables*) and V (*endogenous variables*), and a set F of functions that decides how values are assigned to each variable $V_i \in V$. The variables in U are background knowledge that have no explanatory mechanism encoded in model M. The values of all variables in V are uniquely determined by every instantiation $U = u$ of the background knowledge.

Every causal model M can be associated with a directed graph, called the *causal diagram* of M. This diagram identifies the endogenous and background variables (U and V, resp.) that have direct influence on each V_i.

Procedure 1 Given evidence e, the probability of the counterfactual sentence "Y would be y had X been x" can be evaluated in a three-step process:

1. ***Abduction***: Update the probability $P(u)$ by the evidence e to obtain $P(u \mid e)$. This step explains the past circumstance $U = u$ in the presence of evidence e.
2. ***Action***: Modify M by the action $do(X = x)$. This step minimally adjusts model M by a hypothetical intervention via the external action $do(X = x)$ to comply with the antecedent condition of the counterfactual.

3. ***Prediction***: Compute the probability of $Y = y$ in the modified model. In this step the consequence of the counterfactual is predicted based on the evidential understanding of the past (Step 1), and the hypothetical modification performed in Step 2.

In summary, the approach determines the probability of the counterfactual's consequence $Y = y$ by performing an intervention to impose the counterfactual's antecedent $X = x$ (other things being equal), given evidence e about $U = u$.

3.3 LP-Based Counterfactuals

Our LP approach is based on an existing procedure of counterfactuals evaluation, viz., the aforementioned Pearl's three-step procedure. Since the idea of each step in our LP approach mirrors each one in Pearl's, our approach immediately compares to Pearl's. Our approach also benefits from the epistemic adequacy of Pearl's approach. We apply the idea of Pearl's three-step procedure to logic programs, but leave out probability assignments, and instead employ LP abduction and updating to determine the logical validity of counterfactuals.

Two important ingredients in Pearl's approach of counterfactuals are his causal model and intervention. *Causation* denotes a specific relation of cause and effect. Causation can be captured by LP rules, where the inferential arrow in a logic rule represents causal direction. LP abduction is thus appropriate for inferring causation, providing an explanation to a given observation. That said, LP abduction is not immediately sufficient for counterfactuals. Consider a simple logic program $P = \{b \leftarrow a\}$. Whereas abduction permits obtaining explanation a to observation b, the evaluation of counterfactual "if a had not been true, then b would not have been true" cannot immediately be evaluated from the conditional rule $b \leftarrow a$, for if its antecedent is false the counterfactual would be trivially true. That justifies the need for an *intervention*. That is, it requires explicitly imposing the desired truth value of a, and subsequently checking whether the predicted truth value of b consistently follows from this intervention. As described in Pearl's approach, such an intervention establishes a required adjustment so as to ensure that the counterfactual's antecedent be met. The additional notion of an intervention permits the value of the antecedent to differ from its actual one, whilst maintaining the consistency of the modified model. We resort to LP abduction and updating to express causal source and intervention, respectively.

3.3.1 *Causal Model and LP Abduction*

With respect to an abductive framework $\langle P, \mathcal{A}, \mathcal{I} \rangle$, an observation O is analogous to Pearl's evidence e. That is, program P has rules that have O as their conclusion, and hence O does not belong to the set of abducibles A. Recall that in Pearl's approach, a model M consists of set U of background variables, whose values are conditional on case-considered observed evidence. These background variables are not causally explained in M, as they have no parent nodes in the causal diagram of M. In terms of LP abduction, they correspond to a set of abducibles $E \subseteq A$ that provide abductive explanations for observation O. Indeed, these abducibles have no preceding causal explanatory mechanism, as they have no rules in the program with the abducibles as their conclusions. In a nutshell, an abductive framework $\langle P, \mathcal{A}, \mathcal{I} \rangle$ that provides an abduced explanation $E \subseteq A$ to the available observation O mirrors the structure of Pearl's model M with its specific U supporting an explanation to the current observed evidence e.

3.3.2 Intervention and LP Updating

Besides abduction, our approach also benefits from LP updating, which is supported by well-established theory and properties, cf. [1, 2]. It allows a program to be updated by asserting or retracting rules, thus changing the state of the program. LP updating is therefore appropriate for representing changes and dealing with incomplete information.

The specific role of LP updating in our approach is twofold: (1) updating the program with the preferred explanation to the current observation, thus fixing in the program the initial abduced background context of the counterfactual being evaluated; (2) facilitating an apposite adjustment to the causal model by hypothetical updates of causal intervention on the program, affecting defeasible rules. Both roles are sufficiently accomplished by *fluent* (i.e., state-dependent literal) updates, rather than rule updates. In the first role, explanations are treated as fluents. In the second, reserved predicates are introduced as fluents for the purpose of intervention upon defeasible rules. For the latter role, fluent updates are particularly more appropriate than rule updates (e.g., intervention by retracting rules), because intervention is hypothetical only. Removing rules from the program would be overkill, as the rules might be needed to elaborate justifications and introspective debugging.

3.3.3 Evaluating Counterfactuals in LP

The procedure to evaluate counterfactuals in LP essentially takes the three-step process of Pearl's approach as its reference. That is, each step in the LP approach captures the same idea of its corresponding step in Pearl's.

The key idea of evaluating counterfactuals with respect to an abductive framework, at some current state (discrete time) T, is as follows. In step 1, abduction is performed to explain the factual observation.[3] The observation corresponds to the evidence that both the antecedent and the consequence literals of the present counterfactual were factually false.[4] There can be multiple explanations available for an observation; choosing a suitable one among them is a pragmatic issue, which can be dealt with by integrity constraints or preferences [14, 44]. The explanation fixes the abduced context in which the counterfactual is evaluated by updating the program with the explanation.

In step 2, defeasible rules are introduced for atoms forming the antecedent of the counterfactual. Given the past event E, that renders its corresponding antecedent literal false, held at factual state $T_E < T$, its causal intervention is realized by a hypothetical update H at state $T_H = T_E + \Delta_H$, such that $T_E < T_H < T_E + 1 \leq T$. That is, a hypothetical update strictly takes place between two factual states, thus $0 < \Delta_H < 1$. In the presence of defeasible rules this update permits hypothetical modification of the program to consistently comply with the antecedent of the counterfactual.

In step 3, the WFM of the hypothetical modified program is examined to verify whether the consequence of the counterfactual holds true at state T. One can easily reinstate to the current factual situation by canceling the hypothetical update, e.g., via a new update of H's complement at state $T_F = T_H + \Delta_F$, such that $T_H < T_F < T_E + 1$.

Based on the aforementioned ideas and analogously to the three-step process of Pearl's, our approach is defined below, abstracting from the above state transition detail (cf. Section 3.3.5 for a concrete discussion of this state transition). The following definitions are needed by the procedure.

Definition 13 A set of integrity constraints is *satisfied* in $WFM(P)$ iff none is false in $WFM(P)$. That is, the body of an integrity constraint is either false or undefined [43].

We next rephrase Definition 12 about abductive solutions and relate them to explanations of observations. As our counterfactual procedure is based on the Well-Founded Semantics, the standard logical consequence relation $P \models F$ used in the definition below presupposes the Well-Founded Model of P in verifying the truth of formula F, i.e., whether F is true in $WFM(P)$.

Definition 14 Given an abductive framework $\langle P, \mathcal{A}, \mathcal{I} \rangle$ and an observation O, a consistent abductive solution $E \subseteq \mathcal{A}$ is an *explanation* for observation O iff $P \cup E \models O$ and $\mathcal{I}$ is satisfied in $WFM(P \cup E)$, where all abducibles not appearing in E have been replaced by u, both in P and $\mathcal{I}$.[5]

---

[3]We assume that people are using counterfactuals to convey truly relevant information rather than to fabricate arbitrary subjunctive conditionals (e.g., "If I had been watching, then I would have seen the cheese on the moon melt during the eclipse"). Otherwise, implicit observations must simply be made explicit observations, to avoid natural language conundrums or ambiguities [23].

[4]This interpretation is in line with the corresponding English construct, cf. [27], commonly known as *third conditionals*.

[5]This replacement of abducible $A \notin E$ with u in P and $\mathcal{I}$ is an alternative but equivalent to adding $A \leftarrow \mathsf{u}$ into $P \cup E$, as foreseen by Definition 12.

Procedure 2 Let $\langle P, \mathcal{A}, \mathcal{I} \rangle$ be an abductive framework, where program P encodes the modeled situation on which counterfactuals are evaluated. Consider a counterfactual "if Pre had been true, then $Conc$ would have been true", where Pre and $Conc$ are finite conjunctions of literals.

1. ***Abduction***: Compute an explanation $E \subseteq \mathcal{A}$ for the observation $O = O_{Pre} \cup O_{Conc} \cup O_{Oth}$, where:
   - $O_{Pre} = \{compl(L_i) \mid L_i \text{ is in } Pre\}$,
   - $O_{Conc} = \{compl(L_i) \mid L_i \text{ is in } Conc\}$, and
   - O_{Oth} is other (possibly empty) observations: $O_{Oth} \cap (O_{Pre} \cup O_{Conc}) = \emptyset$.

   Update program P with E, obtaining program $P \cup E$.
2. ***Action***: For each literal L in conjunction Pre, introduce a pair of reserved meta-predicates $make(B)$ and $make\_not(B)$, where B is the atom in L. These two meta-predicates are introduced for the purpose of establishing causal intervention: they are used to express hypothetical alternative events to be imposed. This step comprises two stages: (a) *Transformation*:
   - Add rule $B \leftarrow make(B)$ to program $P \cup E$.
   - Add $not\ make\_not(B)$ to the body of each rule in P whose head is B. If there is no such rule, add rule $B \leftarrow not\ make\_not(B)$ to program $P \cup E$.

   Let $(P \cup E)_\tau$ be the resulting transform. (b) *Intervention*:
   Update program $(P \cup E)_\tau$ with literal $make(B)$ or $make\_not(B)$, for $L = B$ or $L = not\ B$, resp. Assuming that Pre is consistent, $make(B)$ and $make\_not(B)$ cannot be imposed at the same time. Let $(P \cup E)_{\tau,\iota}$ be the program obtained after these hypothetical updates of intervention.
3. ***Prediction***: Verify whether $(P \cup E)_{\tau,\iota} \models Conc$ and $\mathcal{I}$ is satisfied in $WFM((P \cup E)_{\tau,\iota})$.

This three-step procedure defines *valid* counterfactuals.

Definition 15 Let $\langle P, \mathcal{A}, \mathcal{I} \rangle$ be an abductive framework, where program P encodes the modeled situation on which counterfactuals are evaluated. The counterfactual

"If Pre had been true, then $Conc$ would have been true"

is *valid* given observation $O = O_{Pre} \cup O_{Conc} \cup O_{Oth}$ iff O is explained by $E \subseteq \mathcal{A}$, $(P \cup E)_{\tau,\iota} \models Conc$, and $\mathcal{I}$ is satisfied in $WFM((P \cup E)_{\tau,\iota})$.

Since the Well-Founded Semantics supports top-down query-oriented procedures for finding solutions, checking validity of counterfactuals, i.e., whether their conclusion $Conc$ follows (step 3), given the intervened program transform (step 2) with respect to the abduced background context (step 1), in fact amounts to checking in a derivation tree whether query $Conc$ holds true while also satisfying $\mathcal{I}$.

Example 4 Recall the example in the introduction. Let us slightly complicate it by having two alternative abductive causes for the forest fire, viz., storm (which implies lightning hitting the ground) or barbecue. Storm is accompanied by strong wind that causes the dry leaves falling onto the ground. Note that dry leaves are important for forest fire in both cases. This example is expressed by abductive framework $\langle P, \mathcal{A}, \mathcal{I} \rangle$, using abbreviations b, d, f, g, l, s for *barbecue, dry leaves, forest fire, leaves on the ground, lightning*, and *storm*, resp., where $\mathcal{A} = \{s, b, s^*, b^*\}$, $\mathcal{I} = \emptyset$, and P as follows:

$$f \leftarrow b, d. \quad f \leftarrow b^*, l, d, g. \quad l \leftarrow s. \quad g \leftarrow s. \quad d.$$

The use of b^* in the second rule of f is intended so as to have mutual exclusive explanations.

Consider the counterfactual "if only there had not been lightning, then the forest fire would not have occurred", where $Pre = not\ l$ and $Conc = not\ f$.

1. ***Abduction***: Besides $O_{Pre} = \{l\}$ and $O_{Conc} = \{f\}$, say that g is observed too: $O_{Oth} = \{g\}$. Given $O = O_{Pre} \cup O_{Conc} \cup O_{Oth}$, there are two possible explanations: $E_1 = \{s, b^*\}$ and $E_2 = \{s, b\}$. Consider a scenario where the minimal explanation E_1 (in the sense of minimal positive literals) is preferred as the update for P, so we obtain $P \cup E_1$. Note, program $P \cup E_1$ corresponds to a state with $WFM(P \cup E_1) = \{d, s, g, l, f, not\ b\}$. This updated program reflects the evaluation context of the counterfactual, where all literals of Pre and $Conc$ were false in the initial factual situation.
2. ***Action***: The transformation results in program $(P \cup E_1)_{\tau}$:

$$f \leftarrow b, d. \quad f \leftarrow b^*, l, d, g. \quad g \leftarrow s. \quad d.$$
$$l \leftarrow make(l) \quad l \leftarrow s, not\ make\_not(l)$$

   Program $(P \cup E_1)_{\tau}$ is updated with $make\_not(l)$ as the required intervention. It engenders program $(P \cup E_1)_{\tau,\iota}$ corresponding to a new state with $WFM((P \cup E_1)_{\tau,\iota}) = \{d, s, g, make\_not(l), not\ make(l), not\ b, not\ l, not\ f\}$.
3. ***Prediction***: We verify that $(P \cup E_1)_{\tau,\iota} \models not\ f$, and $\mathcal{I} = \emptyset$ is trivially satisfied in $WFM((P \cup E_1)_{\tau,\iota})$.

We thus conclude that, for this E_1 scenario, the given counterfactual is valid.

Example 5 In the other explanatory scenario of Example 4, where E_2 (instead of E_1) is preferred to update P, the counterfactual is no longer valid, because $WFM((P \cup E_2)_{\tau,\iota}) = \{d, s, g, b, make\_not(l), not\ make(l), not\ l, f\}$, and thus $(P \cup E_2)_{\tau,\iota} \not\models not\ f$. Indeed, the forest fire would still have occurred but due to an alternative cause: barbecue. Skeptical and credulous counterfactual evaluations could be defined, i.e., by evaluating the presented counterfactual for each abduced background context. Given that step 2 can be accomplished by a one-time transformation, such skeptical and credulous counterfactual evaluations require only executing step 3 for each background context fixed in step 1.

Another form related to counterfactuals is *semifactuals*, i.e., one that combines a counterfactual antecedent and an enduring factual consequence [11], with a typical form of statement "Even if ...". Other comparable linguistic constructs also exist, e.g., "No matter if ...", "Though ...", etc. The LP procedure for counterfactuals (Procedure 2) can easily be adapted for evaluating semifactuals. Like in counterfactuals, the antecedent of a semifactual is supposed false in the factual situation. But different from counterfactuals, the consequence of a semifactual should instead be factually ensured *true* (rather than false).

Consider the semifactual "even if Pre had been true, $Conc$ would still have been true". Its LP evaluation follows Procedure 2 with the only modification on the definition of O_{Conc} in Step 1, i.e., for semifactuals, O_{Conc} is defined as $O_{Conc} = \{L_i \mid L_i \text{ is in } Conc\}$, to warrant its consequence factually true. The validity condition for semifactuals is the same as for counterfactuals, cf. Definition 15.

Example 6 Recall Example 5, where $E_2 = \{s, b\}$ is preferred. Consider semifactual "even if there had not been lightning, the forest fire would still have occurred", where $Pre = not\ l$ and $Conc = f$. This semifactual is valid, because given the same $WFM((P \cup E_2)_{\tau,\iota})$ as in Example 5, we now have $(P \cup E_2)_{\tau,\iota} \models Conc$, i.e., $(P \cup E_2)_{\tau,\iota} \models f$.

3.3.4 Some Properties

Since the idea of each step in the LP approach mirrors the corresponding one in Pearl's, the LP approach therefore immediately compares to Pearl's, its epistemic adequacy and properties are analogous to those of Pearl's.

In [34], salient properties of counterfactuals in logic are argued for (and in [45] for the Weak Completion Semantics), including three counterfactual fallacies that distinguish the counterfactual conditional from the material one. Our approach also satisfies these three properties, as shown below as an illustration of our theory. Other counterfactual properties, such as reflexivity, modus tollens, disjunction in the antecedent, combination of sentences, etc., are postponed for future work; ascertaining their satisfaction is not in the purview of the present paper.

Let $Pre \triangleright Conc$ represent counterfactual statement "if Pre had been true, then $Conc$ would have been true".

Property 1 *Fallacy of strengthening the antecedent:*
$A \triangleright B$ does not imply $A \wedge C \triangleright B$.

Example 7 Recall Example 4, where E_1 is adopted. We have shown that counterfactual $not\ l \triangleright not\ f$ is valid. Let us strengthen its antecedent with "there had been a barbecue", obtaining counterfactual $not\ l \wedge b \triangleright not\ f$. For this new counterfactual, $O_{Pre} = \{l, not\ b\}$, whereas the other observations (O_{Conc} and O_{Oth}) are the same as in Example 4. The only explanation of $O = O_{Pre} \cup O_{Conc} \cup O_{Oth}$ is $E_{pr_1} = \{s, b^*\}$. The transform $(P \cup E_{pr_1})_\tau$ is as follows:

$$f \leftarrow b, d. \quad f \leftarrow b^*, l, d, g. \quad g \leftarrow s. \quad d.$$
$$l \leftarrow make(l) \quad l \leftarrow s, not\ make\_not(l)$$
$$b \leftarrow make(b) \quad b \leftarrow not\ make\_not(b)$$

The required interventions $make\_not(l)$ and $make(b)$ update this program, obtaining $(P \cup E_{pr_1})_{\tau,\iota}$ with $WFM((P \cup E_{pr_1})_{\tau,\iota})$ computed to be

$$\{d, s, g, make\_not(l), not\ make(l), not\ l, make(b), not\ make\_not(b), b, f\}.$$

Observe that, the intervention $make(b)$ hypothetically updates the truth value of b from false (in $P \cup E_{pr_1}$) to true (in $(P \cup E_{pr_1})_{\tau,\iota}$). Since $(P \cup E_{pr_1})_{\tau,\iota} \not\models not\ f$, counterfactual $not\ l \wedge b \triangleright not\ f$ is not valid.

Property 2 *Fallacy of contraposition:*
$A \triangleright B$ does not imply not $B \triangleright$ not A.

Example 8 Recall the abductive framework of Example 4, but with $O_{Oth} = \emptyset$, rendering $O = \{l, f\}$, and thus the two explanations E_1 and E_2 do not change. Therefore, E_1 being preferred, the counterfactual $not\ l \triangleright not\ f$ is valid, as shown in that example. Consider its contraposition: $f \triangleright l$. Its corresponding observation $O_{pr_2} = \{not\ f, not\ l\}$ admits a single explanation $E_{pr_2} = \{s^*, b^*\}$. Through intervention $make(f)$ is imposed on the transform $(P \cup E_{pr_2})_\tau$, we obtain $not\ l \in WFM((P \cup E_{pr_2})_{\tau,\iota})$. Thus, $(P \cup E_{pr_2})_{\tau,\iota} \not\models l$, and $f \triangleright l$ is not a valid counterfactual.

Property 3 *Fallacy of transitivity:*
$A \triangleright B$ and $B \triangleright C$ do not imply $A \triangleright C$.

Example 9 Let P_t be program:

$$marry \leftarrow pregnant \quad criticized \leftarrow not\ marry, pregnant$$

in the abductive framework $\langle P_t, \mathcal{A}_t, \mathcal{I}_t \rangle$, using abbreviation c, m, p for $criticized$, $marry$, and $pregnant$, resp., where $A_t = \{p, p^*\}$ and $I_t = \emptyset$. Consider counterfactuals C_1, C_2, C_3:
$(C_1)\ not\ p \triangleright not\ m$; $(C_2)\ not\ m \triangleright c$; and $(C_3)\ not\ p \triangleright c$.

We can verify that C_1 is valid, given the only explanation $E_{C_1} = \{p\}$ to the observation $O_{C_1} = \{p, m\}$ and the intervention $make\_not(p)$, since $not\ m \in WFM((P \cup E_{C_1})_{\tau,\iota})$, and thus $(P \cup E_{C_1})_{\tau,\iota} \models not\ m$. We can similarly verify that C_2 is valid, since $(P \cup E_{C_2})_{\tau,\iota} \models c$, where $E_{C_2} = \{p\}$ explains $O_{C_2} = \{m, not\ c\}$, and the imposed intervention is $make\_not(m)$. But C_3 is not valid, since $(P \cup E_{C_3})_{\tau,\iota} \not\models c$, given that $not\ c \in WFM((P \cup E_{C_3})_{\tau,\iota})$, where $E_{C_3} = \{p\}$ explains $O_{C_3} = \{p, not\ c\}$ with intervention $make\_not(p)$.

3.3.5 Implementation

We have developed a prototype, QUALM, that implements the procedure on top of an existing integrated system of LP abduction and updating supporting tabling, in XSB Prolog [58], cf. [45] and its subsequent refinement [47]. QUALM allows specifying predicates that are subject to intervention, e.g., predicate l in Example 4. This information is useful for the transformation stage, in step 2 of the procedure.

In QUALM, the state transition of the program, as a consequence of program updating (asserting or retracting fluents for our case), is facilitated by timestamps that are internally managed. By convention the program is initially inserted at state $T = 1$. The state subsequently progresses to $T = 2$.

Observations are explained by posing a top-level query, e.g., ?- $query((l, f, g), E_1)$ provides explanation E_1 to the observation $O = \{l, f, g\}$ of Example 4. Thanks to WFS that underlies XSB, QUALM enjoys the relevancy property (cf. Section 3.2.1) in computing explanations to observations. In order to fix $E_1 = \{s, b^*\}$ as the abduced context in evaluating counterfactual at the present state $T = 2$, both fluents s and b^*, that held at the factual state $T_{E_1} = 1$, are asserted (via QUALM's reserved predicate $updates(L)$ to assert fluents in list L). These updates render them as facts in the updated program $P \cup E_1$.

A causal intervention "there had not been lightning" is enacted by the hypothetical update of fluent $make\_not(l)$ via query ?- $updates([make\_not(l)])$. As described in Sect. 3.3.3, this update strictly takes place between two consecutive factual states; in this case between $T_{E_1} = 1$ and the current state $T = 2$. QUALM internally assigns a fraction of timestamp, say 0.01, just after T_{E_1}, viz., the hypothetical update $make\_not(l)$ is imposed at state $T_H = 1.01$. It thus simulates an intervention via an update in the past, while keeping the present state at $T = 2$. After this update, the validity of the present counterfactual (at $T = 2$) can be checked by testing its conclusion, e.g., ?- $query(f, E)$ to query whether forest fire would have occurred after the hypothetical update. QUALM answers 'no', verifying the counterfactual's validity that the forest fire would not have occurred.

To reinstate the current factual situation from a counterfactual mode, a hypothetical update can be canceled by updating the program with its fluent complement, e.g., ?- $updates([not\ make\_not(l)])$, occurring at a fraction of time after T_H (also internally assigned by QUALM), e.g., at $T_F = T_H + 0.01 = 1.02$. The program overrides the hypothetical update $make\_not(l)$ that was enacted at $T_H = 1.01$, which is equivalent to retracting it. Consequently, the intervention is no longer imposed on the program.

3.4 Counterfactuals in Morality

People typically reason about what they should or should not have done when they examine decisions in moral situations. It is therefore natural for them to engage counterfactual thoughts in such settings. Counterfactual thinking has been investigated in the context of moral reasoning, notably by experimental studies in psychology. For example, psychologists wish to understand the kind of counterfactual alternatives people tend to imagine in contemplating moral behaviors [36] and the influence of counterfactual thoughts in moral judgment [39]. As argued in [16], the function of counterfactual thinking about morality is used in two processes: the evaluative and the reflective. Through evaluation, counterfactuals help correct wrong behavior in the past, thus guide future moral decisions. Reflection, on the other hand, permits momentary experiential simulation of possible alternatives, thereby allowing careful consideration before a moral decision is made, and to subsequently justify it.

Morality and normality judgments typically correlate. Normality mediates morality with causation and blame judgments. The controllability in counterfactuals mediates between normality, blame and cause judgments. The importance of control, namely the possibility of counterfactual intervention, is highlighted in theories of blame that presume someone to be responsible only if they had some control of the outcome [64].

The potential of LP to machine ethics has been reported in [25, 33, 46] and with emphasis on LP abduction and updating in [55]. Here we investigate how moral issues can be expressed with counterfactual reasoning by resorting to an LP approach. Particularly, we look into its application for examining viewpoints on moral permissibility, exemplified by classic moral dilemmas from the literature on the Doctrines of Double Effect (DDE) [38] and of Triple Effect (DTE) [31].

DDE was first introduced by Thomas Aquinas in his discussion of the permissibility of self-defense [7]. The current version of this principle emphasizes the permissibility of an action that causes a harm by distinguishing whether this harm is a mere *side-effect* of bringing about a good result, or rather an *intended means* to bringing about the same good end [38]. According to the Doctrine of Double Effect, the former action is permissible, whereas the latter is impermissible. In [26], DDE has been utilized to explain the consistency of judgments, shared by subjects from demographically diverse populations, on a number of variants of the classic trolley problem [18]: *A trolley is headed toward five people walking on the track, who are unable to get off the track in time. The trolley can nevertheless be diverted onto a side track, thereby preventing it from killing the five people. However, there is a man standing on the side track. The dilemma is therefore whether it is morally permissible to divert the trolley, killing the man but saving the five.* DDE permits diverting the trolley since that action does not intend to harm the man on the side track in order to save the five.

Counterfactuals may provide a general way to examine DDE in moral dilemmas, by distinguishing between a *cause* and a *side-effect* as a result of performing an action to achieve a goal. This distinction between causes and side-effects may explain the permissibility of an action in accordance with DDE. That is, *if some morally wrong effect E happens to be a cause for a goal G that one wants to achieve by performing an action A, and not a mere side-effect of A, then performing A is impermissible.* This is expressed by the counterfactual form below, in a setting where action A is performed to achieve goal G:

$$If \text{ not } E \; had\; been\; true, \; then \text{ not } G \; would\; have\; been\; true.$$

The evaluation of this counterfactual form identifies permissibility of action A from its effect E, by identifying whether the latter is a necessary cause for goal G or a mere side-effect of action A. That is, if the counterfactual proves valid, then E is instrumental as a cause of G, and not a mere side-effect of action A. Since E is morally wrong, achieving G that way, by means of A, is impermissible; otherwise, not. Note that the evaluation of counterfactuals in this application is considered from the perspective of agents who perform the action, rather than from others' (e.g., observers).

There have been a number of studies, both in philosophy and psychology, on the relation between causation and counterfactuals. The *counterfactual process view* of causal reasoning [37], for example, advocates counterfactual thinking as an essential part of the process involved in making causal judgments. This relation between causation and counterfactuals can be important for providing explanations in cases involving harm, which underlie people's moral cognition [59] and trigger other related questions, such as "Who is responsible?", "Who is to blame?", "Which punishment would be fair?", etc. Herein, we explore the connection between causation and counterfactuals, focusing on agents' deliberate action, rather than on causation and counterfactuals in general. More specifically, our exploration of this topic links it to the Doctrines of Double Effect and Triple Effect and dilemmas involving harm, such as the trolley problem cases. Such cases have also been considered in psychological experiments concerning the role of gender and perspectives (first vs. third person) in counterfactual thinking related to moral reasoning, see [39]. The reader is referred to [13] and [28] for a more general and broad discussion on causation and counterfactuals.

We exemplify an application of this counterfactual form in two off-the-shelf military cases from [57]—abbreviations in parentheses: terror bombing (teb) vs. tactical bombing (tab). The former refers to bombing a civilian target (civ) during a war, thus killing civilians (kic), in order to terrorize the enemy (ror), and thereby get them to end the war (ew). The latter case is attributed to bombing a military target (mil), which will effectively end the war (ew), but with the foreseen consequence of killing the same number of civilians (kic) nearby. According to DDE, terror bombing fails permissibility due to a deliberate element of killing civilians to achieve the goal of ending the war, whereas tactical bombing is accepted as permissible.

Example 10 We first model terror bombing with ew as the goal, by considering the abductive framework $\langle P_e, \mathcal{A}_e, \mathcal{I}_e \rangle$, where $\mathcal{A}_e = \{teb, teb^*\}$, $\mathcal{I}_e = \emptyset$ and P_e:

$$ew \leftarrow ror \quad ror \leftarrow kic \quad kic \leftarrow civ \quad civ \leftarrow teb$$

$$ew \leftarrow ror \qquad ror \leftarrow kic \qquad kic \leftarrow civ \qquad civ \leftarrow teb$$

We consider the counterfactual "if the civilians had not been killed, then the war would not have ended", or equivalently using the above counterfactual form "if $not\ kic$ were the case, then it would *not* have been the case that ew".

According to our analysis $Pre = not\ kic$ and $Conc = not\ ew$. The observation $O = \{kic, ew\}$, with O_{Oth} being empty, has a single explanation $E_e = \{teb\}$. The rule $kic \leftarrow civ$ transforms into $kic \leftarrow civ, not\ make\_not(kic)$. Given intervention $make\_not(kic)$, the counterfactual is valid, because $not\ ew \in WFM((P_e \cup E_e)_{\tau,\iota})$, and thus $(P_e \cup E_e)_{\tau,\iota} \models not\ ew$. Given intervention $make\_not(kic)$, the counterfactual is valid, because $WFM((P_e \cup E_e)_{\tau,\iota}) = \{teb, civ, make\_not(kic), not\ kic, not\ ror, not\ ew\}$, and thus $(P_e \cup E_e)_{\tau,\iota} \models not\ ew$. Given the only explanation $E_e = \{teb\}$ to the observation $O = \{kic, ew\}$ (with O_{Oth} being empty), the new rule $kic \leftarrow civ, not\ make\_not(kic)$ obtained from $kic \leftarrow civ$, and the intervention $make\_not(kic)$, one can verify that the counterfactual is valid, because $not\ ew$ is true in $WFM((P_e \cup E_e)_{\tau,\iota}) = \{teb, civ, make\_not(kic), not\ kic, not\ ror, not\ ew\}$. That means the morally wrong kic is instrumental in achieving the goal ew: it is a cause for ew by performing teb and not a mere side-effect of teb. Hence teb is DDE morally impermissible.

Example 11 Tactical bombing with the same goal ew can be modeled by the abductive framework $\langle P_a, \mathcal{A}_a, \mathcal{I}_a \rangle$, where $\mathcal{A}_a = \{tab, tab^*\}$, $\mathcal{I}_a = \emptyset$ and P_a:

$$ew \leftarrow mil \qquad mil \leftarrow tab \qquad kic \leftarrow tab$$

Given the same counterfactual, we now have $E_a = \{tab\}$ as the only explanation to the same observation $O = \{kic, ew\}$. Note that the transform contains rule $kic \leftarrow tab, not\ make\_not(kic)$, which is obtained from $kic \leftarrow tab$. By imposing the intervention $make\_not(kic)$, one can verify that the counterfactual is not valid, because $ew \in WFM((P_a \cup E_a)_{\tau,\iota})$, and thus $(P_a \cup E_a)_{\tau,\iota} \not\models not\ ew$. Therefore, the morally wrong kic is just a side-effect in achieving the goal ew. Hence tab is DDE morally permissible.

In this tactical bombing example, we could alternatively employ a semifactual: "even if civilians had not been killed, the war would still have ended". It will be interesting to explore in the future the applicability of semifactuals to machine ethics, to identify indifferent actions.

Example 12 Consider two countries a and its ally, b, that concert a terror bombing, modeled by the abductive framework $\langle P_{ab}, \mathcal{A}_{ab}, \mathcal{I}_{ab} \rangle$, where $\mathcal{A}_{ab} = \{teb, teb^*\}$, $\mathcal{I}_{ab} = \emptyset$ and P_{ab} below. The abbreviations $kic(X)$ and $civ(X)$ refer to 'killing civilians by country X' and 'bombing a civilian target by country X'. As usual in LP, underscore (_) represents an anonymous variable.

$$ew \leftarrow ror \qquad ror \leftarrow kic(\_)$$
$$kic(X) \leftarrow civ(X) \qquad civ(\_) \leftarrow teb$$

Being represented as a single program (rather than a separate knowledge base for each agent), this scenario should appropriately be viewed as if a joint action performed by a single agent. Therefore, the counterfactual of interest is "if civilians had not been killed by a and b, then the war would not have ended". That is, the antecedent of the counterfactual is a conjunction: $Pre = not\ kic(a) \wedge not\ kic(b)$. One can easily verify that $not\ ew \in WFM((P_{ab} \cup E_{ab})_{\tau,\iota})$, where $E_{ab} = \{teb\}$. Thus, $(P_{ab} \cup E_{ab})_{\tau,\iota} \models not\ ew$ and the counterfactual is valid: the concerted teb is DDE impermissible.

This application of counterfactuals can be challenged by a more complex scenario, to distinguish moral permissibility according to DDE vs. DTE. DTE [31] refines DDE particularly on the notion about harming someone as an intended means. That is, DTE distinguishes further between doing an action *in order* that an effect occurs and doing it *because* that effect will occur. The latter is a new category of action, which is not accounted for in DDE. Though DTE also classifies the former as impermissible, it is more tolerant to the latter (the third effect), i.e., it treats as permissible those actions performed just *because* instrumental harm will occur.

Kamm [31] proposed DTE to accommodate a variant of the trolley problem, viz., the *Loop Case* [60]: *A trolley is headed toward five people walking on the track, and they will not be able to get off the track in time. The trolley can be redirected onto a side track, which loops back towards the five. A fat man sits on this looping side track, whose body will by itself stop the trolley. Is it morally permissible to divert the trolley to the looping side track, thereby hitting the man and killing him, but saving the five?* Diverting the trolley in this case strikes most moral philosophers as permissible [40]. Referring to a psychology study [26], 56% of its respondents judged that diverting the trolley in this case is also permissible. To this end, DTE may provide the justification, that it is permissible because it will hit the man, and not in order to intentionally hit him [31]. Nonetheless, DDE views diverting the trolley in the Loop case as impermissible.

We use counterfactuals to capture the distinct views of DDE and DTE in the Loop case.

Example 13 We model the Loop case with the abductive framework $\langle P_o, \mathcal{A}_o, \mathcal{I}_o \rangle$, where sav, div, hit, tst, mst stand for *save the five, divert the trolley, man hit by the trolley, train on the side track* and *man on the side track*, resp., with sav as the goal, $\mathcal{A}_o = \{div, div^*\}$, $\mathcal{I}_o = \emptyset$, and P_o:

$$sav \leftarrow hit \quad hit \leftarrow tst, mst \quad tst \leftarrow div \quad mst.$$

DDE views diverting the trolley as impermissible, because this action redirects the trolley onto the side track, thereby hitting the man. Consequently, it prevents the trolley from hitting the five. To come up with the impermissibility of this action, it is required to show the validity of the counterfactual "if the man had *not* been hit by the trolley, the five people would *not* have been saved". Given observation $O = O_{Pre} \cup O_{Conc} = \{hit, sav\}$, its only explanation is $E_o = \{div\}$. Note that rule $hit \leftarrow tst, mst$ transforms into $hit \leftarrow tst, mst, not\ make\_not(hit)$, and the required intervention is $make\_not(hit)$. The counterfactual is therefore valid, because $not\ sav \in WFM((P_o \cup E_o)_{\tau,\iota})$, hence $(P_o \cup E_o)_{\tau,\iota} \models not\ sav$. This means

hit, as a consequence of action div, is instrumental as a cause of goal sav. Therefore, div is DDE morally impermissible.

DTE considers diverting the trolley as permissible, since the man is already on the side track, without any deliberate action performed in order to place him there. In P_o, we have the fact mst ready, without abducing any ancillary action. The validity of the counterfactual "if the man had not been on the side track, then he would not have been hit by the trolley", which can easily be verified, ensures that the unfortunate event of the man being hit by the trolley is indeed the consequence of the man being on the side track. The lack of deliberate action (exemplified here by pushing the man—psh for short) in order to place him on the side track, and whether the absence of this action still causes the unfortunate event (the third effect) is captured by the counterfactual "if the man had not been pushed, then he would not have been hit by the trolley". This counterfactual is not valid, because the observation $O = O_{Pre} \cup O_{Conc} = \{psh, hit\}$ has no explanation $E \subseteq \mathcal{A}_o$, i.e., $psh \notin \mathcal{A}_o$, and no fact psh exists either. This means that even without this hypothetical but unexplained deliberate action of pushing, the man would still have been hit by the trolley (just because he is already on the side track). Though hit is a consequence of div and instrumental in achieving sav, no deliberate action is required to cause mst, in order for hit to occur. Hence div is DTE morally permissible.

Next, we consider a more involved trolley example.

Example 14 Consider a variant of the Loop case, viz., the *Loop-Push Case* (see also Extra Push Case in [31]). Differently from the Loop case, now the looping side track is initially empty, and besides the diverting action, an ancillary action of pushing a fat man in order to place him on the side track is additionally performed. This case is modeled by the abductive framework $\langle P_p, \mathcal{A}_p, \mathcal{I}_p \rangle$, where $\mathcal{A}_p = \{div, psh, div^*, psh^*\}, \mathcal{I}_p = \emptyset$, and P_p:

$$sav \leftarrow hit \quad hit \leftarrow tst, mst \quad tst \leftarrow div \quad mst \leftarrow psh$$

Recall the counterfactuals considered in the discussion of DDE and DTE of the Loop case:

- "If the man had not been hit by the trolley, the five people would not have been saved." The same observation $O = \{hit, sav\}$ provides an extended explanation $E_{p_1} = \{div, psh\}$. That is, the pushing action needs to be abduced for having the man on the side track, so the trolley can be stopped by hitting him. The same intervention $make\_not(hit)$ is applied to the same transform, resulting in a valid counterfactual: $(P_p \cup E_{p_1})_{\tau,\iota} \models not\ sav$, because $not\ sav \in WFM((P_p \cup E_{p_1})_{\tau,\iota})$.
- "If the man had not been pushed, then he would not have been hit by the trolley." The relevant observation is $O = \{psh, hit\}$, explained by $E_{p_2} = \{div, psh\}$. Whereas this counterfactual is not valid in DTE of the Loop case, it is valid in the Loop-Push case. Given rule $psh \leftarrow not\ make\_not(psh)$ in the transform and intervention $make\_not(psh)$, we verify that $(P_p \cup E_{p_2})_{\tau,\iota} \models not\ hit$, as $not\ hit \in WFM((P_p \cup E_{p_2})_{\tau,\iota})$.

From the validity of these two counterfactuals it can be inferred that, given the diverting action, the ancillary action of pushing the man onto the side track causes him to be hit by the trolley, which in turn causes the five to be saved. In the Loop-Push, DTE agrees with DDE that such a deliberate action (pushing) performed in order to bring about harm (the man hit by the trolley), even for the purpose of a good or greater end (to save the five), is likewise impermissible.

3.5 Related Work

In Pearl's approach, intervention is realized by superficial revision, by imposing the desired value to the intervened node and cutting it from its parent nodes. This is also the case in our approach, by means of hypothetical updates affecting defeasible rules. Other subtle ways of intervention may involve deep revision, which can be realized in LP. It is beyond the scope of the paper, but amply discussed in [45]. Unlike Pearl's, our approach is non-probabilistic, which corresponds to assigning probability to abductive explanations (or variables in U of Pearl's causal model) of 0 or 1.

A formalization of our procedure is reported in [45]—albeit based on different semantics (WCS vs. WFS)—along with some properties specific to our LP-based approach. The present paper complements [45] in the sense that we provide an implemented procedure employing our WFS-based LP abduction and updating, realized in our prototype QUALM. That is, it lays emphasis more on the LP engineering aspect for relating the role of LP abduction and updating to Pearl's causal model and hypothetical intervention, and in realizing the procedure in QUALM. Moreover, this paper also shows how counterfactuals apply to examine moral issues, which is not touched on at all in [45]. Due to the similarity of common features to WFS and WCS, the Propositions and Proofs in [45] can be transposed to the WFS setting, which we do not repeat here, given the distinct emphasis just made salient about each of these two otherwise conceptually similar complementary approaches.[6]

LP abduction and revision are employed in [15] to evaluate indicative conditionals, but not counterfactual conditionals. LP abduction is employed through a rewrite system to find solutions for an abductive framework; the rewrite system intuitively captures the natural semantics of indicative conditionals. Rule revisions are additionally used to satisfy conditions whose truth-value is unknown and which cannot be explained by abduction.

In [42], counterfactuals are evaluated using contradiction removal semantics of LP. The work is based on Lewis's counterfactuals [34], where a model of a logic

[6]Both WFS and WCS are 3-valued semantics that differ in dealing with close world assumption (CWA) and rules with positive loops (e.g., $p \leftarrow p$). WFS enforces CWA, i.e., atom a that has no rule is interpreted as false, whereas in WCS undefined. Nevertheless, they can be transformed one to another: adding rules $a \leftarrow u$ and $u \leftarrow not\ u$ for a reserved atom u renders a unknown in WFS; alternatively, adding $a \leftarrow false$ enforces CWA in WCS. In this paper, positive loops are not needed and do not appear throughout examples we consider.

program represents a world in Lewis's concept. The semantics defines the most similar worlds by removing contradictions from the associated program, obtaining the so-called maximal non-contradictory submodels of the program. It does not concern itself with LP abduction and updating; both being relevant for our work, which is based on Pearl's concept rather than Lewis's, without the need of a world distance measure.

Probabilistic LP (PLP) language P-log with the stable model semantics is employed, in [8], to encode Pearl's Probabilistic Causal Model (PCM), without involving abduction. It does not directly encode Pearl's three-step process, but focuses on P-log probabilistic approach to compute the probability of a counterfactual query. Our work does not deal with probability, only logic, though it mirrors Pearl's three-step process via LP abduction and updating. Our approach is also not based on stable model semantics, but instead on WFS with its relevancy property, which is more appropriate for LP abduction as argued earlier. In [63], Pearl's PCM is encoded using PLP CP-logic, without involving abduction. Whereas P-log has its own *do*-operator to achieve intervention in its probabilistic reasoning, CP-logic achieves it by eliminating rules. Similar to P-log, our approach introduces meta-predicates *make* and *make_not* to accomplish intervention via defeasible rules and fluent updates, without eliminating rules, as CP-logic does.

Several logic-based approaches have been employed in machine ethics research, e.g., [5, 6, 10, 19, 48]. While some approaches provide implementations in LP, such as in [5, 6, 19], they have not exploited LP-based reasoning features and recent techniques in LP systems that appear essential and promising for moral reasoning and decision making. The approach in [19] mainly just emphasizes the use of default negation in defeasible rules to capture non-monotonic reasoning, whereas the use of LP in [5, 6] is constrained to its purpose of learning rules from cases. Clearly, the potential of Logic Programming goes beyond that.

LP abduction is used in [46] to model moral reasoning in various scenarios of the trolley problem, both from DDE and DTE viewpoints, sans counterfactuals. Abducibles are used to represent decisions, e.g., diverting the trolley, pushing the man, etc. Impermissible actions are ruled out using an integrity constraint, and a posteriori preferences are eventually enacted to come up with a moral decision from the remaining alternatives of action. The subsequent work [25] refines it with uncertainty of actions and consequences in several scenarios of the trolley problem by resorting to P-log.

The use of causation, based on structural approach, to define and model issues related to morality, such as blame and responsibility, is discussed in [12, 24]. The interest of the present work is to bring counterfactuals (rather than causation), inspired by the structural approach, into a wider context of LP-based non-monotonic reasoning, given the lack of pure non-probabilistic counterfactual reasoning in LP, and to foster the interplay of various LP-based reasoning for the application of machine ethics (cf. [25, 46]), particularly in addressing moral permissibility by referring to the Doctrines of Double Effect and Triple Effect. It is nevertheless interesting to explore the applications of LP-based probabilistic reasoning to study degrees of blame and moral responsibility.

One of the difficulties in using an integrity constraint to express impermissibility is that it requires the representation to be crafted sufficiently in detail in order for the integrity constraint to be applicable. The examples in the present paper have not exploited the full potential of integrity constraints yet. While we use counterfactuals to examine permissibility (so we are not bound to have a subtle problem representation), integrity constraints can be used for other purposes, e.g., if LP programs for teb and tab examples are combined, integrity constraint: $\bot \leftarrow teb, tab$ can be introduced to choose among mutually exclusive abducibles, teb or tab. The decision to have separate models for them in this paper is solely for clearer presentation. Nevertheless, integrity constraints should be treated carefully by counterfactuals, because an intervention may render integrity constraints unsatisfiable, and hence their body's support may need to be abductively revised in order to re-impose satisfaction.

Side-effects in abduction have been investigated in [44] through the concept of inspection points; the latter are construed in a procedure by 'meta-abducing' a specific abducible $abduced(A)$ whose function is only checking that its corresponding abducible A is indeed already adopted elsewhere. Therefore, the consequence of the action that triggers this 'meta-abducing' is merely a side-effect. Indeed, inspection points may be employed to distinguish a cause from a mere side-effect, and thus may provide an alternative or supplement to counterfactuals employed for the same purpose.

3.6 Conclusion and Future Work

This paper presents a formulation of counterfactual evaluation by means of LP abduction and updating. The approach corresponds to the three-step process in Pearl's structural theory, but omits probability to concentrate on a naturalized logic. We addressed how to examine (non-probabilistic) moral reasoning about permissibility, employing this LP approach to distinguish between causes and side-effects as a result of agents' actions to achieve a goal.

Counterfactuals may also be suitable to address moral justification, via 'compound counterfactuals': *Had I known what I know today, then if I were to have done otherwise, something preferred would have followed.* Such counterfactuals, typically imagining alternatives with worse effect—so-called *downward counterfactuals* [35], may provide moral justification for what was done due to deficiencies in current knowledge. This is accomplished by evaluating what would have followed if the intent had been otherwise, other things (including present knowledge) being equal. It may justify that what would have followed is no morally better than the actual ensued consequence. QUALM can evaluate such compound counterfactuals, thanks to its implemented incremental tabling of fluents [53]. Because fluents are tabled and time-stamped, events in the past subjected to hypothetical updates of intervention can readily be accessed. Indeed, these hypothetical updates take place without requiring any undoing of other fluent updates in the states of those past events or up to the current one. More recent updates are kept in tables and readily provide the current knowledge. We are investigating the application of compound counterfactuals, e.g., to justify an exception for an action to be permissible, that may lead to agents' argumentation following Scanlon's contractualism [56].

Acknowledgements We thank the anonymous reviewers for their constructive comments and suggestions. Both authors acknowledge the support from Fundação para a Ciência e a Tecnologia (FCT/MEC) NOVA LINCS PEst UID/CEC/04516/2013. Ari Saptawijaya acknowledges the support from FCT/MEC with the doctoral grant SFRH/BD/72795/2010. We thank Emmanuelle-Anna Dietz for the fruitful discussions.

References

1. Alferes, J. J., Brogi, A., Leite, J. A. & Pereira, L. M. (2002). Evolving logic programs. In *Proceedings of European Conference on Artificial Intelligence (JELIA 2002)*(Vol. 2424, pp. 50–61). LNCS, Springer.
2. Alferes, J. J., Leite, J. A., Pereira, L. M., Przymusinska, H., & Przymusinski, T. (2000). Dynamic updates of non-monotonic knowledge bases. *Journal of Logic Programming*, *45*(1–3), 43–70.
3. Alferes, J. J., & Pereira, L. M. (1996). *Reasoning with logic programming* (Vol. 1111). LNAI, Berlin: Springer.
4. Alferes, J. J., Pereira, L. M., & Swift, T. (2004). Abduction in well-founded semantics and generalized stable models via tabled dual programs. *Theory and Practice of Logic Programming*, *4*(4), 383–428.
5. Anderson, M. & Anderson, S. L. (2008). EthEl: Toward a principled ethical eldercare robot. In *Proceedings AAAI 2008 Fall Symposium on AI in Eldercare.*
6. Anderson, M., Anderson, S. L. & Armen, C. (2006). MedEthEx: a prototype medical ethics advisor. In *Proceedings of 18th Innovative Applications of Artificial Intelligence Conference (IAAI 2006).*
7. Aquinas, T. (1998). Summa Theologica II-II, Q.64, art. 7, Of Killing. In W. P. Baumgarth & R. J. Regan, Eds, *On Law, Morality, and Politics*. Hackett.
8. Baral, C. & Hunsaker, M. (2007). Using the probabilistic logic programming language P log for causal and counterfactual reasoning and non-naive conditioning. In *Proceedings of 20th International Joint Conference on Artificial Intelligence (IJCAI).*
9. Bidoit, N., & Froidevaux, C. (1991). General logic databases and programs: default logic semantics and stratification. *Journal of Information and Computation*, *91*(1), 15–54.
10. Bringsjord, S., Arkoudas, K., & Bello, P. (2006). Toward a general logicist methodology for engineering ethically correct robots. *IEEE Intelligent Systems*, *21*(4), 38–44.
11. Byrne, R. M. J. (2007). *The rational imagination: How people create alternatives to reality.* Cambridge, MA: MIT Press.
12. Chockler, H., & Halpern, J. Y. (2004). Responsibility and blame: A structural-model approach. *Journal of Artificial Intelligence Research*, *22*, 93–115.
13. Collins, J., Hall, N., & Paul, L. A. (Eds.). (2004). *Causation and counterfactuals*. Cambridge, MA: MIT Press.
14. Dell'Acqua, P., & Pereira, L. M. (2007). Preferential theory revision. *Journal of Applied Logic*, *5*(4), 586–601.
15. Dietz, E.-A., Hölldobler, S. & Pereira, L. M. (2015). On indicative conditionals. In *Procs. 1st International Workshop on Semantic Technologies (IWOST)* (Vol. 1339), CEUR Workshop Proceedings.
16. Epstude, K., & Roese, N. J. (2008). The functional theory of counterfactual thinking. *Personality and Social Psychology Review*, *12*(2), 168–192.
17. Fitting, M. (1985). A Kripke-Kleene semantics for logic programs. *Journal of Logic Programming*, *2*(4), 295–312.

18. Foot, P. (1967). The problem of abortion and the doctrine of double effect. *Oxford Review*, *5*, 5–15.
19. Ganascia, J.-G. (2007). Modelling ethical rules of lying with answer set programming. *Ethics and Information Technology*, *9*(1), 39–47.
20. Gelfond, M. (1987). On stratified autoepistemic theories. In *Proceedings of 6th National Conference on Artificial Intelligence (AAAI)*.
21. Gelfond, M. & Lifschitz, V. (1988). The stable model semantics for logic programming. In *Proceedings of 5th International Logic Programming Conference*. MIT Press.
22. Ginsberg, M. L. (1986). Counterfactuals. *Artificial Intelligence*, *30*(1), 35–79.
23. Grice, Paul. (1991). *Studies in the way of words*. Cambridge, MA: Harvard University Press.
24. Halpern, J. Y., & Hitchcock, C. (2015). Graded causation and defaults. *British Journal for the Philosophy of Science*, *66*, 413–457.
25. Han, T. A., Saptawijaya, A. & Pereira, L. M. (2012). Moral reasoning under uncertainty. In *Proceedings of 18th International Conference on Logic for Programming, Artificial Intelligence and Reasoning (LPAR)* (Vol. 7180, pp. 212–227). LNCS, Springer.
26. Hauser, M., Cushman, F., Young, L., Jin, R. K., & Mikhail, J. (2007). A dissociation between moral judgments and justifications. *Mind and Language*, *22*(1), 1–21.
27. Hewings, M. (2013). *Advanced grammar in use with answers: A self-study reference and practice book for advanced learners of english*. New York, NY: Cambridge University Press.
28. Hoerl, C., McCormack, T., & Beck, S. R. (Eds.). (2011). *Understanding counterfactuals, understanding causation: Issues in philosophy and psychology*. Oxford, UK: Oxford University Press.
29. Hölldobler, S. & Kencana Ramli, C. D. P. (2009). Logic programs under three-valued Łukasiewicz semantics. In *Proceedings of 25th International Conference on Logic Programming (ICLP)* (Vol. 5649, pp. 464–478). LNCS, Springer.
30. Kakas, A., Kowalski, R., & Toni, F. (1992). Abductive logic programming. *Journal of Logic and Computation*, *2*(6), 719–770.
31. Kamm, F. M. (2006). *Intricate ethics: Rights, responsibilities, and permissible harm*. Oxford, UK: Oxford University Press.
32. Kleiman-Weiner, M., Gerstenberg, T., Levine, S. & Tenenbaum, J. B. (2015). Inference of intention and permissibility in moral decision making. In *Proceedings of 37th Annual Conference of the Cognitive Science Society*.
33. Kowalski, R. (2011). *Computational logic and human thinking: How to be artificially intelligent*. New York, NY: Cambridge University Press.
34. Lewis, D. (1973). *Counterfactuals*. Cambridge, MA: Harvard University Press.
35. Markman, K. D., Gavanski, I., Sherman, S. J., & McMullen, M. N. (1993). The mental simulation of better and worse possible worlds. *Journal of Experimental Social Psychology*, *29*, 87–109.
36. McCloy, R., & Byrne, R. M. J. (2000). Counterfactual thinking about controllable events. *Memory and Cognition*, *28*, 1071–1078.
37. McCormack, T., Frosch, C., & Burns, P. (2011). The relationship between children's causal and counterfactual judgements. In C. Hoerl, T. McCormack, & S. R. Beck (Eds.), *Understanding Counterfactuals*., Understanding causation Oxford, UK: Oxford University Press.
38. McIntyre, A. (2004). Doctrine of double effect. In E. N. Zalta (Eds.), *The Stanford Encyclopedia of Philosophy*. Center for the Study of Language and Information, Stanford University (Fall 2011 edition) http://plato.stanford.edu/archives/fall2011/entries/double-effect/.
39. Migliore, S., Curcio, G., Mancini, F., & Cappa, S. F. (2014). Counterfactual thinking in moral judgment: an experimental study. *Frontiers in Psychology*, *5*, 451.
40. Otsuka, M. (2008). Double effect, triple effect and the trolley problem: Squaring the circle in looping cases. *Utilitas*, *20*(1), 92–110.
41. Pearl, J. (2009). *Causality: Models reasoning and inference*. Cambridge, MA: Cambridge University Press.
42. Pereira, L. M., Aparício, J. N. & Alferes, J. J. (1991). Counterfactual reasoning based on revising assumptions. In *Proceedings of International Symposium on Logic Programming (ILPS 1991)* (pp. 566–577). MIT Press.

43. Pereira, L. M., Aparício, J. N. & Alferes, J. J.(1991). Hypothetical reasoning with well founded semantics. In *Proceedings of 3rd Scandinavian Conference on Artificial Intelligence*. IOS Press.
44. Pereira, L. M., Dell'Acqua, P., Pinto, A. M. & Lopes, G. (2013). Inspecting and preferring abductive models. In K. Nakamatsu & L. C. Jain (Eds.), *The Handbook on Reasoning-Based Intelligent Systems* (pp. 243–274). World Scientific Publishers.
45. Pereira, L. M., Dietz, E.-A. & Hölldobler, S. (2016). Abductive framework for counterfactual reasoning in logic programming. Available from http://userweb.fct.unl.pt/~lmp/publications/online-papers/counterfactuals.pdf.
46. Pereira, L. M. & Saptawijaya, A. (2011). Modelling morality with prospective logic. In M. Anderson & S. L. Anderson (Eds.), *Machine Ethics* (pp. 398–421) Cambridge, U. K.
47. Pereira, L. M. & Saptawijaya, A. (2011). *Programming Machine Ethics*, volume 26 of *Studies in Applied Philosophy, Epistemology and Rational Ethics*. Springer, Switzerland, 2016.
48. Powers, T. M. (2006). Prospects for a Kantian machine. *IEEE Intelligent Systems*, *21*(4), 46–51.
49. Przymusinska, H. & Przymusinski, T. C. (1990). Semantic issues in deductive databases and logic programs. In *Formal Techniques in Artificial Intelligence: A Sourcebook* (pp 321–367). North-Holand.
50. Przymusinski, T. C. (1989a). Every logic program has a natural stratification and an iterated least fixed point model. In *Proceedings of 8th ACM Symposium on Principles Of Database Systems (PODS)* (pp. 11–21).
51. Przymusinski, T. C. (1989b). Three-valued non-monotonics formalisms and logic programming. In *Proceedings of 1st International Conference on Principles of Knowledge Representation and Reasoning (KR)*.
52. Roese, N. J. (1997). Counterfactual thinking. *Psychological Bulletin*, *121*(1), 133–148.
53. Saptawijaya, A. & Pereira, L. M. (2013). Incremental tabling for query-driven propagation of logic program updates. In *Proceedings of 19th International Conference on Logic Programming, Artificial Intelligence and Reasoning (LPAR)* (Vol. 8312, pp. 694–709). LNCS, Springer.
54. Saptawijaya, A. & Pereira, L. M. (2014a). Joint tabling of logic program abductions and updates (Technical Communication of ICLP 2014). *Theory and Practice of Logic Programming*, *14*(4–5). Online Supplement. Available from http://arxiv.org/abs/1405.2058.
55. Saptawijaya, A. & Pereira, L. M. (2014b). Towards modeling morality computationally with logic programming. In *PADL 2014* (Vol. 8324, pp. 104–119). Springer.
56. Scanlon, T. M. (1998). *What we owe to each other*. Cambridge, MA: Harvard University Press.
57. Scanlon, T. M. (2008). *Moral dimensions: Permissibility, meaning blame*. Cambridge, MA: Harvard University Press.
58. Swift, T., & Warren, D. S. (2012). XSB: Extending Prolog with tabled logic programming. *Theory and Practice of Logic Programming*, *12*(1–2), 157–187.
59. Tetlock, P. E., Visser, P. S., Singh, R., Polifroni, M., Scott, A., Elson, S. B., et al. (2007). People as intuitive prosecutors: The impact of social-control goals on attributions of responsibility. *Journal of Experimental Social Psychology*, *43*, 195–209.
60. Thomson, J. J. (1985). The trolley problem. *The Yale Law Journal*, *279*, 1395–1415.
61. van Emden, M. H., & Kowalski, R. (1976). The semantics of predicate logic as a programming language. *Journal of the ACM*, *4*(23), 733–742.
62. van Gelder, A., Ross, K. A., & Schlipf, J. S. (1991). The well-founded semantics for general logic programs. *Journal of the ACM*, *38*(3), 620–650.
63. Vennekens, J., Bruynooghe, M. & Denecker, M. (2010). Embracing events in causal modeling: Interventions and counterfactuals in CP-logic. In *JELIA 2010* (Vol. 6341, pp. 313–325). LNCS. Springer.
64. Weiner, B. (1995). *Judgments of responsibility: A foundation for a theory of social conduct*. New York, NY: The Guilford Press.

Chapter 4
The Wisdom of the Multitude: Diversity Versus Size

Peter C. Stone and Koji Kagotani

Abstract The Condorcet Jury Theorem (CJT), together with a large and growing literature of ancillary results, suggests two conclusions regarding collective wisdom. First, large committees outperform small committees, other things equal. Second, heterogeneous committees can, under the right circumstances, outperform homogeneous ones, again other things equal. But this literature has done little to bring these two conclusions together. This paper employs simulations to compare the respective contributions of size and difference to optimal committee performance. It demonstrates that the contributions depend dramatically upon *bias*. In the presence of low bias, committee composition matters little. In the presence of high bias, it can matter a great deal; optimal committee performance, however, does not vary dramatically between low- and high-bias committees.

Keywords Bias · Collective wisdom · Condorcet Jury Theorem · Diversity · Monte Carlo simulations

4.1 Introduction

In the *Politics*, Aristotle offers the following famously cryptic defense of the "wisdom of the multitude:"

> The principle that the multitude ought to be in power rather than the few best might seem to be solved and to contain some difficulty and perhaps even truth. For the many, of whom each individual is not a good man, when they meet together may be better than the few good, just as a feast to which many contribute is better than a dinner provided out of a single purse. For each individual among the many has a share of excellence and practical wisdom, and when they meet together, just as they become in a manner one man, who has many feet, and hands, and senses, so too with regard to their character and thought. Hence the many are better judges than a single man of music and poetry; for some understand one part, and some another, and among them they understand the whole (*Politics* 1281a40–1281b9) [14].

P.C. Stone (✉)
Department of Political Science, Trinity College Dublin, Dublin, Ireland
e-mail: pstone@tcd.ie

K. Kagotani
Faculty of Economics, Osaka University of Economics, Osaka, Japan
e-mail: kagotani@osaka-ue.ac.jp

R. Urbaniak and G. Payette (eds.), *Applications of Formal Philosophy*,
Logic, Argumentation & Reasoning 14, DOI 10.1007/978-3-319-58507-9_4

Here Aristotle appears to appeal to at least two different ways in which the multitude could display wisdom.[1] On the one hand, the many could display wisdom simply by virtue of its numbers, just as a large group of poor people might together possess more money than Donald Trump. On the other hand, the many could display wisdom by virtue of the various perspectives it contains; person A might be better placed to judge the words to a song, while person B might be better placed to judge its music, and so A and B might together form a more accurate judgment about the song than either could separately. In other words, the wisdom of the multitude might depend upon its *size*, or the *differences* it contains, or both. These two factors need not travel together; a small and diverse group is one thing, an army of clones quite another. But this fact raises the following challenge for anyone sharing Aristotle's faith in collective wisdom: does the multitude display wisdom due to size, difference, or both? And if both, when will one factor matter more than the other?

Democratic theorists have made great strides in answering these questions using the Condorcet Jury Theorem (CJT). The CJT established in the late eighteenth century that when it comes to decision-making, size matters. Other things equal, a large committee of sufficiently competent voters will outperform a small one. (Throughout this paper, we will use the term "committee" to refer to any collective decision-making body.) But a growing literature suggests that difference matters as well as size. Other things equal, a heterogeneous committee, with people representing different backgrounds and perspectives, will outperform a committee of clones. This heterogeneity may reflect differences in race, gender, socioeconomic class, or a variety of other factors. Recent generalizations of the CJT model demonstrate the truth of this claim, although they also note that heterogeneity only produces beneficial effects under moderately restrictive circumstances. This paper uses these generalizations to compare the respective merits of size and difference. It then makes policy recommendations based upon this comparison, and draws conclusions from the comparison regarding collective wisdom.

Section 4.2 reviews existing results on the contributions size and diversity make to decision-making. These results rely upon a generalized form of the CJT model. Section 4.3 derives expressions for measuring committee success in a collective decision-making environment characterized by diversity. Section 4.4 employs simulations to generate comparative statics for the respective contributions of size and diversity within this model. This enables comparisons of these contributions under a variety of assumptions. Section 8 concludes by exploring the policy and philosophical implications of the results.

[1]For more detailed analyses of this passage, see [7, 22, 30, 31].

4.2 Size and Diversity

In his classic work *Essai sur l'Application de l'Analyse à la Probabilité des Décisions Rendues à la Pluralité des Voix* (1785), Condorcet imagined a committee charged with making some decision using simple majority rule. The decision is dichotomous, with the committee facing a choice between options a and b. One option is unambiguously better than the other, although the identity of the better option is obviously not known in advance. All committee members share a common utility function which is maximized if the correct option is chosen. The committee is comprised of $n = 2k + 1$ members (i.e., n is odd), each of whom decides correctly with fixed common probability p, with $\frac{1}{2} < p < 1$. The probability that one committee member votes correctly is independent of the probability that any other committee member votes correctly. Under these conditions, Condorcet demonstrated that the following two results hold: 1) the probability that the committee majority will decide correctly is higher than p; and 2) the probability that the committee majority will decide correctly approaches 1 as the size of the committee increases to infinity [9]. Committee size matters, other things equal. Indeed, for a fixed value of p, a larger committee will always outperform a smaller one [5].

The assumptions underlying the CJT are admittedly rigid. The theorem assumes, for example, that voting takes place without deliberation (which would presumably lead to violations of the independence assumption). These assumptions can be relaxed without undermining the central results, so long as this is done with caution. Grofman et al., [16], for example, showed that majorities still outperform the average individual in a group even if p varies by individual, so long as the distribution of values of p is symmetric. Boland [6] demonstrated that the distribution of values of p does not matter, so long as the average value of p is sufficiently high. Paroush [25] has demonstrated the importance of Boland's result, by showing that having the average value of p exceed $\frac{1}{2}$ is not a sufficient condition for the CJT's result under certain distributions of voter competence. Owen et al., [23], however, demonstrated that an average value of p exceeding $\frac{1}{2}$ is sufficient for the asymptotic result (i.e., for the probability of majority success to approach 1 as committee size increases). And [17] establishes a lower bound for committee competence levels with minimal assumptions regarding the distribution of p values.

Committee performance might also be influenced by a number of additional factors, factors not explicitly addressed by Condorcet. The committee may, for example, know in advance that one option is *ex ante* more likely than the other to be correct. Assume, without loss of generality, that option a is correct with probability $\pi \geq \frac{1}{2}$. Then the option maximally likely to be correct when all information is taken into account may not be the same as the option maximally likely to be correct when only the individual judgments are considered. In particular, the optimal voting rule will have a presumption in favor of a, with b winning only if a sufficiently large supermajority favors it. Alternatively, the committee's utility function could treat differently a wrongful rejection of a and a wrongful rejection of b. A jury might be more concerned about convicting the innocent than acquitting the guilty. Suppose the group loses utility e_a by wrongfully rejecting a and e_b by wrongfully rejecting b. (The group receives utility 0 if it chooses correctly.) Then the rule that maximizes the probability of successful choice, in terms of committee member judgments, may not be the rule that maximizes expected utility. The committee might prefer to sacrifice some probability of correctly choosing b in exchange for a smaller increase in the probability of correctly choosing a, because the cost of failing to choose a correctly is sufficiently high. Once again, this translates into a voting rule with a presumption in favor of a.[2]

In practice, then, there is a close link between the *ex ante* probability that each option is correct and the costs associate with each type of mistake. A committee with $\pi \geq \frac{1}{2}$ and $e_a = e_b$ will maximize expected utility with a certain supermajority voting rule that favors a. For every such value of π, there exists a committee with unequal values of e_a and e_b and $\pi = \frac{1}{2}$ that optimizes performance using the same voting rule. As a result, the literature has tended to treat these two cases together, e.g., [2, 20, 21].[3]

All of these results suggest that Condorcet's original conclusion—size matters in decision making—is highly robust. Other results suggest that difference also matters to decision making—that it pays to have decisions made by people who are not clones of one another. There has, however, been less work done to sustain this conclusion in the CJT environment.[4] This may be because it has proven difficult to translate the contribution difference makes into terms that can be captured within the CJT framework. One possible way of capturing this contribution is to assume that homogeneity leads to violations of the independence assumption. Given this assumption, heterogeneity improves decision-making by decreasing the average correlation level between individual judgments. This in turn results in better committee decision-making [19].[5] Another approach is to hold the average value of p constant, and then compare homogeneous and heterogeneous committees. This is the approach taken

---

[2]Stone [28] considers only the absolute magnitude of the loss, and so treats e_a and e_b as positive. Here we treat these expressions as losses, and so assume $e_a, e_b < 0$.

[3]See also [8], which demonstrated that a committee whose members have different prior beliefs regarding the correct option is equivalent to a committee whose members have the same prior beliefs but different attitudes towards risk.

[4]Surowiecki [29] explores the contribution of difference in non-voting environments. [24] deals with difference in voting systems, but without reference to the CJT.

[5]Cf. [11, 12], who consider various forms of the independence assumption and investigate which forms make possible CJT-like results. Dietrich and Spiekermann assume voter homogeneity throughout their discussion, however.

by [15, 18]. But this approach leaves no room for the marginal contribution of diversity. Other things equal, if a committee adds one more member, it should add one with the highest possible competence level. There is thus no positive advantage to including people of different backgrounds on the Kanazawa/Fey approach, assuming independence is assured.

A third approach is to model the differential contributions made by different groups explicitly. This can be accomplished by introducing the idea of *bias* into the decision. Assume again that option a is correct *ex ante* with probability π. Assume further that the committee can be composed of members from two different groups, X and Y. Members of X select a when a is correct with probability q, and b when b is correct with probability p, with $q > p > \frac{1}{2}$. Members of Y select a when a is correct with probability p, and b when b is correct with probability q. Thus, X-members are biased in favor of a, whereas Y-members have an equal and opposite bias in favor of b. The term "bias" is not ideal. It has connotations of irrationality that we do not intend. We have considered using other terms for this variation in competence levels with the state of the world. This variation is easily confused with variation in competence levels between individuals. We believe that "bias" is the best term available for avoiding this confusion. Finally, allow the error costs e_a and e_b to vary. All committee members, regardless of group, receive utility 0 for a correct choice, pay cost e_a if a is wrongfully rejected, and pay cost e_b if b is wrongfully rejected.

Stone [28] demonstrates that given these assumptions (and assuming the committee votes via majority rule) the optimal committee composition will be a function of p, q, π, e_a, and e_b. The expected utility-maximizing committee will have $k + s^*$ X-members and $k - s^* + 1$ Y-members on it, where s^* is the smallest integral value of s between $-k$ and k for which the inequality

$$\frac{e_a}{e_b}\frac{\pi}{1-\pi} > \left[\frac{p(1-p)}{q(1-q)}\right]^s$$

fails to hold. (There may be no such value, in which case an all-X member committee will be optimal.) Should the inequality hold for $s = k$, then an all X-member committee would be optimal. And should it fail to hold for $s = -k$, then an all Y-member committee would maximize expected utility. Moreover, this condition requires that $q > 1 - p$, a condition ensured whenever $p > \frac{1}{2}$.[6] (If $q \leq 1 - p$, then heterogeneity never increases, and sometimes decreases, expected committee utility.) Heterogeneity thus may, but not necessarily will, make a difference for the better. In the simplest case, where $\pi = \frac{1}{2}$ and $e_a = e_b$, the optimal committee will be maximally heterogeneous, with a bare majority of Y-members on it. (A bare majority of X-members will perform equally well.)

Stone's model suggests that committee decision-making can be improved either by taking steps to ensure optimal committee composition or by increasing committee size. Stone's original result, however, does not consider the respective magnitudes

[6]This condition is minimal but not negligible. If it were not satisfied, then an X-member would be more likely to select a when b is correct than when a is correct!.

of these two approaches. When does size matter a lot, and when does difference matter a lot? Under what conditions will taking steps to ensure diversity prove more cost-effective than simply involving more people? These are the questions explored in this paper.

4.3 Comparing Size and Diversity

Committee performance can be assessed either in terms of the probability that the committee decides correctly via majority rule, or in terms of the expected utility of committee decision-making. (The one is equal to the other whenever the costs associated with the two types of error are the same.) We will begin, therefore, by deriving expressions for both of these measures.

Assume a committee with $k + s$ X-members and $k - s + 1$ Y-members (with $s > 0$). Then conditional upon the state of the world (say, a), committee decision making functions as a pair of sets of Bernoulli trials. It is equivalent to making multiple tosses of two coins, one of which has a different bias towards heads than the other. Coin 1 gets tossed $k + s$ times, and lands heads up with probability q, while coin 2 gets tossed $k - s + 1$ times, and comes up heads with probability p.

We assume throughout that committee members vote sincerely. This is not an innocuous assumption, even under the assumption (standard in the CJT literature) of a common utility function [1]. Ban-Yashar and Milchtaich [3], however, demonstrates that sincere voting will always be rational in the CJT environment whenever the optimal voting rule is used. This is easy to find in many cases—as noted before, simple majority rule will be optimal whenever $\pi = \frac{1}{2}$ and $e_a = e_b$—but more difficult to identify at other times.[7]

Let x be the number of correct votes by the X-members, y be the number of correct votes by the Y-members, and $z = x + y$ be the total number of correct votes in a committee. Then the expected counts of correct votes by different types of members are calculated given the state of the world: $E(x \mid a) = (k + s)q$, $E(x \mid b) = (k + s)p$, $E(y \mid a) = (k - s + 1)p$, and so forth. One can use these to calculate the (unconditional) expected counts of correct votes in the committee: $E(z) = (k + s)[\pi q + (1 - \pi)p] + (k - s + 1)[\pi p + (1 - \pi)q]$.

From here, one can proceed to calculate C_a (or C_b), the probability that the committee decides correctly via majority rule, conditional on a (or b) being correct. Then one can also calculate C, the unconditional probability of a correct committee vote:

[7]Ben-Yashar and Nitzan [4] provides a general formula for deriving the optimal voting rule in the CJT environment. Stone [28] (pp. 407–408) applies this formula to the case of committees whose members have equal and opposite biases, but the resulting expression defies simple analysis.

$$\begin{aligned} C &= \pi C_a + (1-\pi)C_b \\ &= \pi \sum_{i=s}^{k} \left[\binom{k+s}{i} q^i (1-q)^{k+s-i} \left[\sum_{j=k+1-i}^{k-s+1} \binom{k-s+1}{j} p^j (1-p)^{k-s+1-j} \right] \right] \\ &\quad + \pi \sum_{i=k+1}^{k+s} \binom{k+s}{i} q^i (1-q)^{k+s-i} \\ &\quad + (1-\pi) \sum_{i=s}^{k} \left[\binom{k+s}{i} p^i (1-p)^{k+s-i} \left[\sum_{j=k+1-i}^{k-s+1} \binom{k-s+1}{j} q^j (1-q)^{k-s+1-j} \right] \right] \\ &\quad + (1-\pi) \sum_{i=k+1}^{k+s} \binom{k+s}{i} p^i (1-p)^{k+s-i}. \end{aligned} \tag{4.1}$$

There are more X-members than Y-members. If fewer than s X-members vote correctly, then it is impossible for the committee to decide right—there are not enough Y-members to form a correct majority. If the number of correct X-members is between s and k (inclusive), then the committee votes correctly so long as a sufficient number of Y-members vote correctly. And if $k+1$ or more X-members vote correctly, it simply doesn't matter how the Y-members vote—a correct majority is guaranteed.

Finally, let u be the utility derived from a committee vote. This vote will yield utility 0 with probability $\pi C_a + (1-\pi)C_b$; utility e_a with probability $\pi(1-C_a)$; and utility e_b with probability $(1-\pi)(1-C_b)$. The expected utility $E(u)$ of a committee vote is thus:

$$e_a \pi(1-C_a) + e_b(1-\pi)(1-C_b). \tag{4.2}$$

With these preliminaries out of the way, it becomes possible to compare the respective contributions of size and diversity both to C and to $E(u)$. As mentioned before, whenever $e_a = e_b$ the expected utility of the committee is maximized if and only if the probability of correct committee choice is maximized.[8] For this reason, it makes sense to focus upon investigating this special case.

As discussed in the previous section, the heterogeneous committee achieves the highest probability of a correct committee vote under certain conditions [28]. This implies that as the parameter s increases beyond a certain point, the probability of a correct committee vote decreases. However, the probability of a correct committee decision C is determined by the multiple parameters n, p, q, s and π, with the effect of each parameter being a function of the other four. Even though we can explore the sign of the impact analytically, it is very difficult to examine the relative impact of diversity, bias, and size (as well as prior beliefs) on performance. If we can repeat the committee's decision-making a certain number of times (for example, 10,000 times),

[8] If $e_a = e_b$, then $E(u) = e_a[\pi(1-C_a) + (1-\pi)(1-C_b)] = e_a(1-C)$. Because $e_a < 0$, this means that we can find a maximum of $E(u)$ by maximizing C.

we can calculate the probability of the correct committee vote given a set of the parameters, which is close to the objective probability of correct collective choice. Hence, we use the Monte Carlo simulation approach to explore the relationship between the relative impact of each parameter and optimal committee performance.

4.4 Monte Carlo Simulation

Our simulation procedures are as follows. First, the state of the world $w \in \{a, b\}$ is drawn from the bernoulli distribution with $\Pr(w = a) = \pi$. Second, the count of correct votes in the committee is generated as the sum of a pair of binomial distributions $\mathrm{B}(\mathrm{X}, q) + \mathrm{B}(\mathrm{Y}, p)$ if the state of the world a is chosen, and $\mathrm{B}(\mathrm{X}, p) + \mathrm{B}(\mathrm{Y}, q)$ otherwise. Third, if the committee majority (at least $k + 1$ members) votes correctly, this is called committee success. Fourth, we iterate this procedures 10,000 times for a given set of parameters n, p, q, s, and π, and then calculate the ratio of committee success, which is called here the probability of committee success. As a further simplifying assumption, we begin with $\pi = \frac{1}{2}$ and relax it later. We address the three theoretical issues below.

> **1. Homogeneity versus heterogeneity**. If $\pi \geq \frac{1}{2}$, then the optimal committee will always have an X-member majority. Without loss of generality, then, we will assume the committee will have such a majority. We will then compare the maximally homogeneous committee (i.e., 100% X-member) with the maximally heterogeneous committee (i.e., a bare X-member majority). We will also consider all committee compositions within this range. When $\pi = \frac{1}{2}$ and $e_a = e_b$, the optimal committee will be maximally heterogeneous. This may not be the case, however, if either condition fails to hold.
>
> **2. Bias (the relative sizes of q and p)**. Committee composition could potentially matter much more when the biases of committee members are large than when they are small. The size of a bias is larger the greater the distance between q and p. We shall consider a high-bias and a low-bias committee.
>
> **3. Committee size**. Obviously, the higher the value of n, the better the committee will perform. We will take this into account by comparing several values of n. Specifically, we will consider $n = 11, 21, 51, 101, 201$ and 401. All of these are realistic committee sizes for at least certain classes of decisions.

Figure 4.1 shows the simulation results for the low-bias committee ($p = 0.51$ and $q = 0.55$) and the high-bias committee ($p = 0.51$ and $q = 0.85$). The committee performance is measured by the probability of correct committee vote. To provide the intuitive understanding of the patterns, we use the degree of homogeneity $\frac{k+s}{n} \times 100$ to examine the committee composition. When we increase s from 1 to $k + 1$, the committee becomes progressively more homogeneous. It changes from approximately 50 percent X-members to 100 percent X-members. The parameter $\pi = 0.5$ means that two states of the world a and b are equally likely to occur.

We observe two results. First, more homogeneity worsens high-bias committee performance, which is consistent with the theoretical arguments by [28]. The difference is particularly striking for high values of n. For these values, heterogeneous

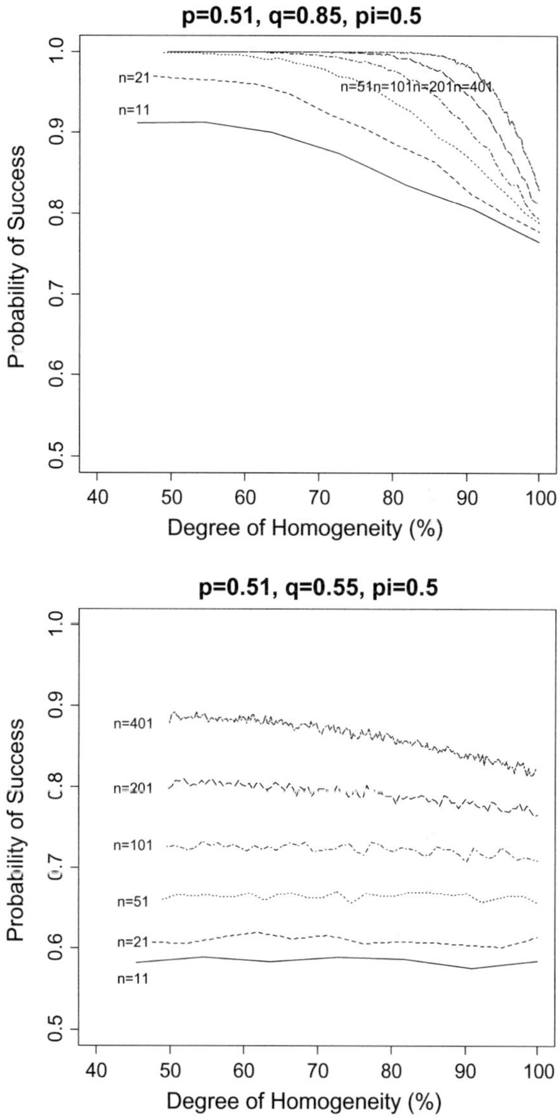

Fig. 4.1 The high-bias committee and the low-bias committee

high-bias committees perform nearly perfectly, while homogeneous high-bias committees still have a long way to go before C approaches 1. In contrast, heterogeneity matters little for the performance of the low-bias committee of relatively small size ($n \leq 101$) because the probability of committee success remains the same over different values of the parameters. For the low bias committee of relatively large size ($n > 101$), more homogeneity does somewhat constrain the committee performance.

Second, increasing committee size enhances performance regardless of homogeneity level. With $\pi = 0.5$, the maximally heterogenous committee clearly performs the best in all cases except for low-bias committees with small numbers of members. In these cases, heterogeneity makes a negligible difference.

It is not too surprising that the high-bias committees routinely outperform the low-bias committees here, for all values of n. *Ex ante* (i.e., before state of the world is selected), both X-members and Y-members have an expected competence level of 0.68 in the high-bias model, but only 0.53 in the low-bias model. It is therefore difficult to distinguish the effects generated by having a more-or-less competent committee from the effects generated by bias size. To address this problem, we ran simulations for two further models (Fig. 4.2). In both of these models, X- and Y-members have the same *ex ante* competence level, equal to 0.625. In the modified high-bias model, $q = 0.74$ and $p = 0.51$, while in the modified low-bias model, $q = 0.65$ and $p = 0.6$.

The results generated here are similar but not identical. First, once again the heterogeneous committee is clearly best in the presence of high bias while committee composition hardly matters for the low-bias committee. Second, when the *ex ante* probability of success is the same for both committees, with the heterogeneous composition, the probability of success in the low-bias committee is almost the same as that in the high-bias committee. In contrast to the previous results, as the committee composition becomes homogeneous, the low-bias committee clearly begins to outperform the high-bias committee. Third, increasing committee size once again increases the probability of committee success. In the high-bias committee, increasing the size raises the probability of committee success and reduces the relative effect of homogeneity. The heterogeneous committee success rate rapidly approaches 1, and so its comparative advantage lessens as the homogeneous committee gets large as well. (It has nowhere to go.) On the other hand, the size of the committee matters for the relatively small committee because the probability of committee success is almost one for the committee with more than 100 members.

Next we relax our assumption regarding the states of the world and assume $\pi = 0.75$, so that state of the world a is significantly more likely to occur than state of the world b (Fig. 4.3). We employ the same values of p and q as in the first set of simulations and consider the results. First, both the high-bias and the low-bias committees demonstrate interesting results regarding the composition of the committee. The high-bias committee with more than 50 members generates results similar to those previously seen. That is, the best committee is the most heterogeneous and more homogeneity dampens the committee performance. Things are somewhat different for the high-bias committees with 11 and 21 members. The reason for this is that when $\pi = 0.75$, $s^* = 2$. The optimal committee will thus have 3 more X-members than Y-Members on it. This makes a significant difference for small high-bias committees. The optimal 11- and 21-member committees will have 7 and 12 X-members, respectively; put another way, the percentage of these committees comprised of X-members will be 63 and 57%. Maximal heterogeneity is thus no longer optimal heterogeneity, even though homogeneity still does worse.

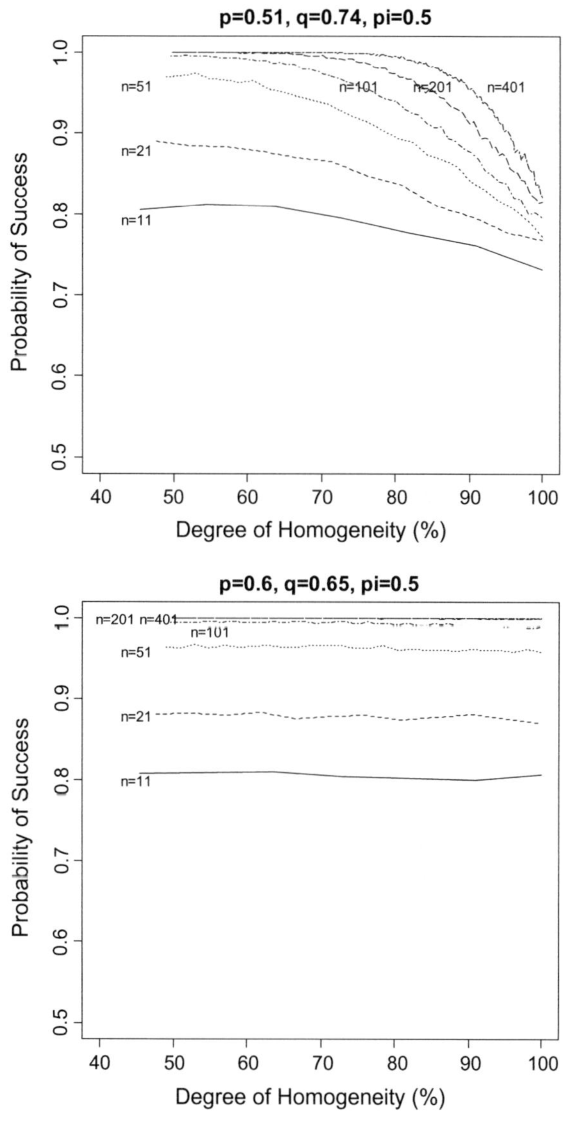

Fig. 4.2 The high-bias committee and the low-bias committee

A similar, but much stronger, result obtains for the low-bias committee. For this case, $s^* = 114$. This means that, for every low-bias committee of less than maximal size, the optimal composition will involve no Y-members at all. Even for the 401-member committee, the optimal committee will have a 314-87 majority in favor of X-members. It will be 78% X-member. The simulation results clearly reflect this, as the homogeneous low-bias committee outperforms any heterogeneous one (albeit marginally), except when $n = 401$. These results are caused by the assumption on the states of the world $\pi = 0.75$. Since state of the world a is more likely to occur,

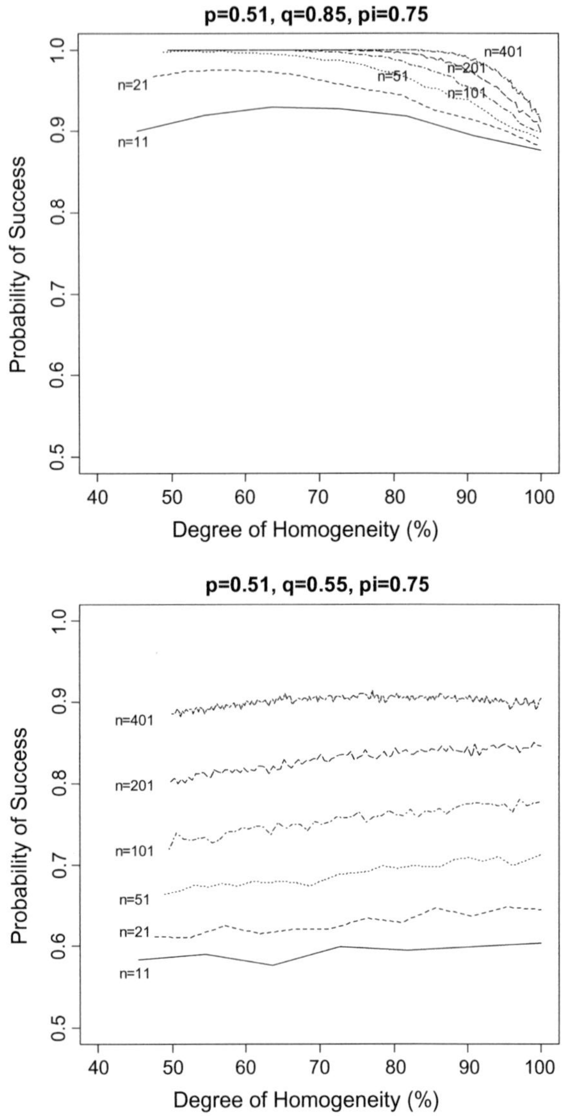

Fig. 4.3 The high-bias committee and the low-bias committee

the committee can improve its performance by including more X-members who are more likely to detect state of world a.

As in the first set of simulations, however, it is difficult to distinguish the effect of the bias size from the effect of the overall competence level. For this reason, we ran one final pair of simulations with $\pi = 0.75$ but with the same competence levels as in the second set of simulations (Fig. 4.4). Once again, optimal committee composition will not be maximally heterogeneous. For the high-bias committee, $s^* = 4$, and for

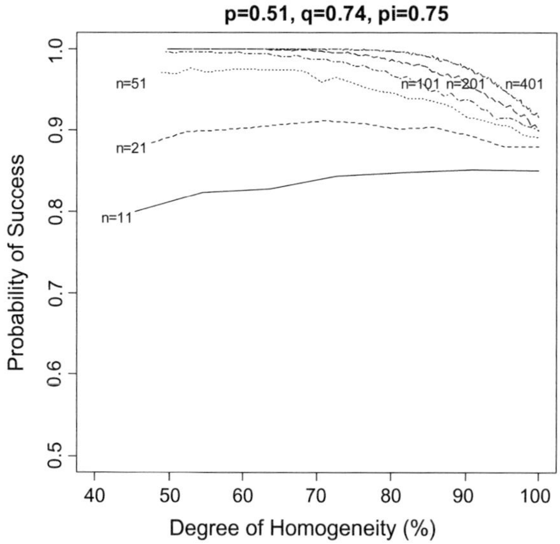

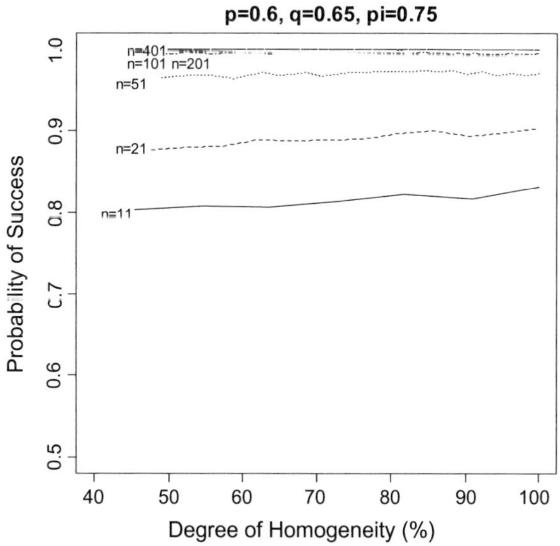

Fig. 4.4 The high-bias committee and the low-bias committee

the low-bias committee, $s^* = 21$. This means that, even for the high-bias committee, the optimal committee will consist almost entirely of X-members for low values of n. And the optimal low-bias committee will have no Y-members at all until $n = 51$.

Otherwise, the results are essentially the same as in the second set of simulations. For high values of n, the optimal committee performs extremely well for both high- and low-bias committees. Indeed, when n is at least 101, optimal committee performance is near-perfect. But the difference between the committee with the best composition and the committee with the worst composition is far more pronounced in the presence of high bias than in the presence of low bias. With low bias, composition makes a relatively small contribution to performance. But with high bias, even a 401-member homogeneous committee will have a significant chance of failure, while a 401-member heterogeneous committee will almost never fail.

4.5 Conclusion

The simulation results impart several important lessons regarding committee performance in the presence of bias. Most importantly, heterogeneity is more likely to make a difference, for better or for worse, when the committee members have high bias. By "high bias," we mean a large difference between the values of p and q. In the presence of high bias, committee composition will matter much more. The optimal committee will significantly outperform committees with less-than-optimal composition. This will be true regardless of the value of s^*. If the optimal committee is homogeneous, then a homogeneous high-bias committee will significantly outperform a heterogeneous one. If the optimal committee is heterogeneous, then the opposite will occur. This difference will tend to shrink as n gets large for one obvious reason. As n gets large, the optimal committee will decide correctly with a probability near one. And so for large values of n, the difference between the best committee's performance (which hovers near 1) and that of the worst will of necessity grow smaller. Note, however, that even for the largest high-bias committees, poor committee composition will significantly worsen committee performance. A 401-member high-bias committee that is composed well will perform with near-perfection, but a 401-member high-bias committee that is composed poorly will lag significantly behind.

All of this has important implications for policymaking. In particular, it suggests that, if a policy maker has control over the values of p and q, she should prefer, other things equal, for the values to be close together rather than far apart. The precise manner in which this works remains a subject for further study. Such study could reveal, for example, how much the difference between p and q matters relative to the absolute sizes of the two variables. The present simulation results suggest that both factors will considerably influence committee success.

Normally, however, policy makers have no control over the existing biases within the population. All they control is 1) how many people from that population get selected and 2) which people they select. (Policy makers could, of course, try to find people with low levels of bias, an intuitively obvious idea that our study supports.) If the policy maker has no control over p and q, then the relative attention she gives to committee size versus committee composition should vary according to the size of the bias. In the presence of high bias, committee composition matters a great deal. In the presence of low bias, however, committee composition matters little, and the only way to improve committee performance significantly is to increase n. Given that selection criteria can prove costly, especially if employed on a large scale, this conclusion could matter significantly in policymaking contexts. Policy makers facing decisions in which the population is strongly divided may wish to select smaller committee within which diversity of opinion is ensured. With low bias, policy makers may wish to forego consideration of diversity and simply assemble a large committee.

On a more philosophical level, the results bear out both parts of Aristotle's claim. Size matters, and so does difference. Size matters much more reliably than difference. Even a clone army will decide very well if the army grows large enough. Still, as difference increases, so does the possibility of serious error if the difference goes unacknowledged. Acknowledging it properly, however, may prove more difficult than traditional democratic theory has envisaged. In the face of deeply-divided societies, democratic theorists have typically pointed to one of two possible forms of acknowledgment. Sometimes, they call for *proportional* representation of different social groups on committees. If a group comprises $x\%$ of a population, then the decision-making bodies representing that population should draw $x\%$ of its members that group. This is the ideal of *descriptive representation* [26]. At other times, however, differences between groups prove serious enough that democrats recognize the need for *equal* representation of different social groups, regardless of their size. In effect, this solution treats the two groups as two separate societies, not two parts of a single society. This proposal has historically been realized in the *jury de medietate linguae*, a practice established in medieval England whereby a Jew (and later, any foreigner) appearing in court could insist upon having six of his own compatriots sit on the twelve-member jury judging him [10].

The results in this paper do not provide strong support for either of these traditional democratic ideas. If a population is composed of two groups with strongly different biases, then optimal decision-making requires the presence of both. The level of presence required, however, need not be equal, nor will it have any relationship to the respective sizes of the two groups in the general population.[9] Instead, it will depend upon a variety of factors, including the size of the bias, the *ex ante* probability that one option is right, and the error costs associated with each option on the table. Within the current democratic theory literature, this approach is probably closest to that of [13], who defend the representation of different "discourses" within society in deliberative democratic procedures. This representation does not, for Dryzek and Niemeyer, depend upon the number of people within the larger society practicing each discourse. This approach may present insuperable obstacles to institutionalization, but it may well be the most sound from an epistemic perspective.

Acknowledgements Earlier versions of this paper were presented at the 6th Munich-Sydney-Tilburg Conference on Models and Decisions, Munich, April 10–12, 2013, and at the 3rd Annual Conference of the European Political Science Association, Barcelona, June 20–22, 2013. We would like to thank participants at both events for helpful comments and suggestions.

[9]For a more detailed defence of this claim, see [27].

References

1. Austen-Smith, D., & Banks, J. S. (1996). Information aggregation, rationality, and the Condorcet jury theorem. *American Political Science Review*, *90*(1), 34–45.
2. Ben-Yashar, R., Koh, W.T., & Nitzan, S. (2009). Is specialization desirable in commitee decision making? (Department Working Paper, 16 June 2009, Economics Department, Bar Ilan University, Ramat Gan, Israel).
3. Ben-Yashar, R., & Milchtaich, I. (2007). First and second best voting rules in committees. *Social Choice and Welfare*, *29*(3), 453–486.
4. Ben-Yashar, R. C., & Nitzan, S. I. (1997). The optimal decision rule for fixed-size committees in dichotomous choice situations: The general result. *International Economic Review*, *38*(1), 175–186.
5. Berend, D., & Sapir, L. (2005). Monotonicity in Condorcet's jury theorem. *Social Choice and Welfare*, *24*(1), 83–92.
6. Boland, P. J. (1989). Majority systems and the Condorcet jury theorem. *The Statistician*, *38*(3), 181–189.
7. Bookman, J. (1992). The wisdom of the many: an analysis of the arguments of books iii and iv of Aristotle's Politics. *History of Political Thought*, *13*(1), 1–12.
8. Chwe, M. S.-Y. (1999). Minority voting rights can maximize majority welfare. *American Political Science Review*, *93*(1), 85–97.
9. Condorcet, M. (1995). An essay on the application of analysis to the probability of decisions rendered by a plurality of votes, 1785. In I. McLean & A.B. Urken (Eds.), *Classics of social choice* (pp. 91–112). Ann Arbor, MI: University of Michigan Press.
10. Constable, M. (1994). *The law of the other: The mixed jury and changing conceptions of citizenship, law, and knowledge*. Chicago, IL: University of Chicago Press.
11. Dietrich, F., & Spiekermann, K. (2013a). Epistemic democracy with defensible premises. *Economics and Philosophy*, *29*(1), 87–120.
12. Dietrich, F., & Spiekermann, K. (2013b). Independent opinions? On the causal foundations of belief formation and jury theorems. *Mind*, *122*(487), 655–685.
13. Dryzek, J. S., & Niemeyer, S. (2008). Discursive representation. *American Political Science Review*, *102*(4), 481–493.
14. Everson, S. (Ed.). (1996). *Aristotle: The politics and the constitution of Athens*. New York: Cambridge University Press.
15. Fey, M. (2003). A note on the Condorcet jury theorem with supermajority voting rules. *Social Choice and Welfare*, *20*, 27–32.
16. Grofman, B., Owen, G., & Feld, S. (1983). Thirteen theorems in search of the truth. *Theory and Decision*, *15*, 261–278.
17. Hawthorne, J. (2001). Voting in search of the public good: The probabilistic logic of majority judgments. Department of Philosophy, University of Oklahoma. (unpublished manuscript).
18. Kanazawa, S. (1998). A brief note on a further refinement of the Condorcet jury theorem for heterogeneous groups. *Mathematical Social Sciences*, *35*(1), 69–73.
19. Ladha, K. K. (1992). The Condorcet jury theorem, free speech, and correlated votes. *American Journal of Political Science*, *36*(3), 617–634.
20. Nitzan, S., & Paroush, J. (1984). Are qualified majority rules special? *Public Choice*, *42*(3), 257–272.
21. Nitzan, S., & Paroush, J. (1994). A general theorem and eight corollaries in search of correct decision. *Theory and Decision*, *17*, 211–220.
22. Ober, J. (2013). Democracy's wisdom: An Aristotelian middle way for collective judgment. *American Political Science Review*, *107*(1), 104–122.
23. Owen, G., Grofman, B., & Feld, S. L. (1989). Proving a distribution-free generalization of the Condorcet jury theorem. *Mathematical Social Sciences*, *17*(1), 1–16.
24. Page, S. E. (2008). *The difference: How the power of diversity creates better groups, firms, schools, and societies*. Princeton, N.J.: Princeton University Press.

25. Paroush, J. (1997). Stay away from fair coins: A Condorcet jury theorem. *Social Choice and Welfare*, *15*, 15–20.
26. Pitkin, H. F. (1967). *The concept of representation*. Berkeley, CA: University of California Press.
27. Stone, P. (2012). Many voices, different voices. In G. Delannoi (Ed.), *Direct democracy and sortition, number 56 in les cahiers du cevipof*. CEVIPOF: Paris, France.
28. Stone, P. (2015). Introducing difference into the Condorcet jury theorem. *Theory and Decision*, *78*(3), 399–409.
29. Surowiecki, J. (2005). *The wisdom of crowds*. New York: Anchor Books.
30. Waldron, J. (1995). The wisdom of the multitude: Some reflections on book 3, chapter 11 of Aristotle's Politics. *Political Theory*, *23*(4), 563–584.
31. Wilson, J. L. (2011). Deliberation, democracy, and the rule of reason in Aristotle's Politics. *American Political Science Review*, *105*(2), 259–274.

Chapter 5
A Logic for Human Actions

Clayton Peterson

Abstract The present paper introduces an action logic able to model human actions. We begin by providing an analysis of the proof-theory of action logics from the perspective of category theory. Action logics are classified as different types of monoidal deductive systems with respect to their categorical structure. This enables us to correlate the properties of the logical connectives with the type of deductive system that is used. We then provide a philosophical analysis of action connectives and, in light of our analysis, show which type of deductive system is required to model human actions. According to the usual distinction between *actions* and *propositions* in dynamic logic, we distinguish between an action logic, representing the formal structure of actions, and a propositional action logic, expressing the formal structure of the language we use to talk about actions.

Keywords Action negation · Action disjunction · Deductive systems · Monoidal logics

5.1 Introduction

The concept of *action* is of obvious philosophical significance given its place in many disciplines, including law, ethics, psychology, computer science, physics and metaphysics. As propositional and predicate logics are meant to represent the formal structure of sentences, philosophers wondered whether it was possible to develop some kind of *logic of actions* that could represent the formal structure of actions. *Action logics* began with the work of Kanger [36] and Pörn [64], but the notion of action in deontic contexts was also studied at length by von Wright [75]. Following Segerberg et al. [70], *action logics* in philosophy can roughly be divided in two broad classes. On the one hand, the work of Kanger and Pörn lead, with the help of Belnap and Perloff [11, 79], to what is nowadays known as STIT logics (the acronym stands for *seeing to it that*). Dynamic logic on the other hand was developed by Pratt [61, 62] and was introduced in deontic contexts by Meyer [54, 55]. Common to both

C. Peterson (✉)
Munich Center for Mathematical Philosophy, Munich, Germany
e-mail: clayton.peterson@lrz.uni-muenchen.de

R. Urbaniak and G. Payette (eds.), *Applications of Formal Philosophy*,
Logic, Argumentation & Reasoning 14, DOI 10.1007/978-3-319-58507-9_5

dynamic and STIT logics is the fact that these frameworks have often been presented as tools that can help to analyze and clarify some notions that are relevant to legal and ethical discourse, either from a philosophical or a computational point of view.

The aim of the present paper is to provide an analysis of action logics used in deontic contexts. This analysis will be made from a philosophical standpoint rather than a computational one. Our main objective is to develop an action logic that is relevant to *human* actions, and this is not, as Segerberg [69, p. 377] pointed out, the aim of the logics that are used for programming. Category theory will be used as a foundational framework to analyze the proof-theory of action logics. Building on previous work (cf. [58, 59]), we will show how action logics can be classified according to their categorical structure. One benefit of this mode of analysis is that the properties of the logical connectives can be correlated to the categorical structure of the logic that is used. In this respect, if we know which properties should be satisfied by the logical connectives, then we also know which type of monoidal deductive system should be used to model human actions.

In the first section, we begin by presenting the categorical framework, exploiting the intimate relationship between monodial categories and deductive systems. This is followed by a review of the main action logics one can find within the literature, including dynamic logic, Kleene algebras, Pratt's action logic, linear logic, Boolean algebras of actions, STIT logics and Lucas's formalization of von Wright's conception of actions. We will bring to light the relations between these logics by using the properties of their action connectives and show how they are different variations of monoidal deductive systems. Then, we provide a philosophical analysis of action connectives and propose to distinguish between an action logic $\mathcal{AL}$, which represents the structure of actions, and a propositional action logic $\mathcal{PAL}$, which expresses the propositional structure of the language we use to talk about actions.

5.2 Logic Categorically Conceived

5.2.1 Preliminaries

Categorical logic began with Lawvere's [46] work, who showed how logical systems can be defined on the grounds of category theory using pairs of adjoint functors. While a functor is a mapping between two categories that respects identities and compositions, an adjunction is a pair of functors respecting some naturality conditions.[1] From this perspective, logical connectives can be seen as functors having specific properties and, in this respect, logical systems can be defined on the grounds of specific functors by focusing on the properties of the connectives.

The starting point of our analysis is the proof-theoretical perspective of category theory, where deductive systems are defined in such a way that they can be proven

[1] See Mac Lane [52, pp. 13, 16, 80-2] for the definitions of functors, natural transformations and adjoint functors.

to be instances of monoidal categories. This analysis is inspired by Lambek's [42–44] work but differs from his initial approach. While Lawvere wanted to expose the categorical structure of logical systems, Lambek did the converse and tried to show the logical structure of category theory. Indeed, Lambek's idea was to define categorical notions in terms of *deductive systems* (for an overview, see [45]), hence providing a logical analysis of category theory and showing how categorical notions can be defined using specific rules and axiom schemata.

Lambek saw a conceptual equivalence between categorical and logical notions and thus defined the former in terms of the latter. That being said, it can be argued that category theory is epistemologically superior to logic.[2] Indeed, category theory can be used as a foundational framework in many disciplines, including physics [6], computer science [10], linguistics [17] and logic [35]. Hence, category theory offers an abstract framework that enables one to model different disciplines. We adopt Lawvere's standpoint and assume category theory as a foundational framework for logic, which is accomplished by adapting Lambek's notion of deductive systems. The interest of doing so is that it enables us to provide a proper classification of logical systems, relating the properties of the logical connectives to the categorical structure of the logic. In light of the philosophical analysis of actions connectives provided in Sect. 5.4, we will thus be able to determine which type of deductive system should be used to model human actions given the properties we want to be satisfied.

5.2.2 Deductive Systems and Categories

The basic idea behind this framework is to define logical systems in such a way that their categorical structure is explicitly exposed via the rules and axiom schemata that are used.[3] A *deductive system* is understood as a collection of formulas and a collection of equivalence classes of proofs (or deductions) satisfying the identity axiom (1) and the rule (cut).[4]

$$\frac{}{\varphi \longrightarrow \varphi}\,(1) \qquad\qquad \frac{\varphi \longrightarrow \psi \qquad \psi \longrightarrow \rho}{\varphi \longrightarrow \rho}\,(\text{cut})$$

Thus defined, a deductive system satisfies the definition of a category. Following Mac Lane [52, pp. 7–8], a category $\mathcal{D}$ is composed of

1. $\mathcal{D}$-*objects*;
2. $\mathcal{D}$-*arrows*;
3. an operation assigning to each arrow a domain and a codomain within the $\mathcal{D}$-objects;

[2] See also Marquis [53] for arguments in favor of this point.

[3] This section is built on previous work. See also [59].

[4] It should be noted that there is a distinction between a *proof* of ψ from φ, represented by the consequence relation $\varphi \longrightarrow \psi$ (where φ is a formula), and the *derivation* of a proof, corresponding to its demonstration.

4. composition of arrow gf for each pair $f : x \longrightarrow y$ and $g : y \longrightarrow z$ that respects associativity, i.e. $h(gf) = (hg)f$;
5. an identity arrow 1_y for each $\mathcal{D}$-object y such that $1_y f = f$ and $g 1_y = g$ for each pair $f : x \longrightarrow y$ and $g : y \longrightarrow z$ (the identity law).

A *deductive system* can be shown to be a category $\mathcal{D}$, where objects are propositions and arrows are equivalence classes of proofs. Given that we are working with equivalence classes of proofs, we will omit the names of the arrows in the presentation.[5] An arrow $\varphi \longrightarrow \psi$ represents an equivalence class of proofs (or deductions) of ψ from φ. If there is such an arrow, then ψ is a logical consequence of (or is deducible from) φ within the deductive system. The reflexivity of the consequence relation is captured via the identity arrow (1) for each formula φ and we can show that it respects the identity law. Similarly, it is easy to show that the consequence relation is transitive and that composition of proofs respects associativity.

So far, only the consequence relation is defined and there are no logical connectives. But one can easily add some via the definition of a monoidal deductive system. The first connective to consider is conjunction. A *monoidal deductive system* **M** is defined as a deductive system satisfying (t) for increasing monotony, (a) for the associativity of $\otimes$ and (r) and (l), which make 1 into the unit of the conjunction $\otimes$. The double line means that the rule can be applied both ways.

$$\frac{\varphi \longrightarrow \psi \qquad \rho \longrightarrow \tau}{\varphi \otimes \rho \longrightarrow \psi \otimes \tau}\ (\mathrm{t}) \qquad \frac{\tau \longrightarrow (\varphi \otimes \psi) \otimes \rho}{\overline{\tau \longrightarrow \varphi \otimes (\psi \otimes \rho)}}\ (\mathrm{a})$$

$$\frac{\varphi \longrightarrow 1 \otimes \psi}{\overline{\varphi \longrightarrow \psi}}\ (\mathrm{l}) \qquad \frac{\varphi \longrightarrow \psi \otimes 1}{\overline{\varphi \longrightarrow \psi}}\ (\mathrm{r})$$

In this case, a monoidal deductive system can be proven to be an instance of a monoidal category, hence the terminology. Following Mac Lane [52, pp. 161–2], a monoidal category is a category equipped with a tensor product $\otimes$ and a unit object I. The tensor product is associative, meaning that there is an arrow $a_{x,y,z}$ for each $\mathcal{D}$-objects x, y and z, and the unit object respects l_x and r_x for each $\mathcal{D}$-object x:

$$\begin{aligned} &a_{x,y,z} : (x \otimes y) \otimes z \longrightarrow x \otimes (y \otimes z) \\ &l_x : I \otimes x \longrightarrow x \\ &r_x : x \otimes I \longrightarrow x \end{aligned}$$

The arrows $\alpha_{x,y,z}$, l_x and r_x are natural isomorphisms.[6] Moreover, the tensor product and the unit object have to respect the triangle and pentagon identities, expressed by the two following commutative diagrams:

---

[5] A rigorous presentation needs a different notation for deductive systems and categories. We will use lower case letters of the end of the alphabet when speaking of categories and Greek letters when speaking of deductive systems.

[6] See Mac Lane [52, p. 16] for the definition.

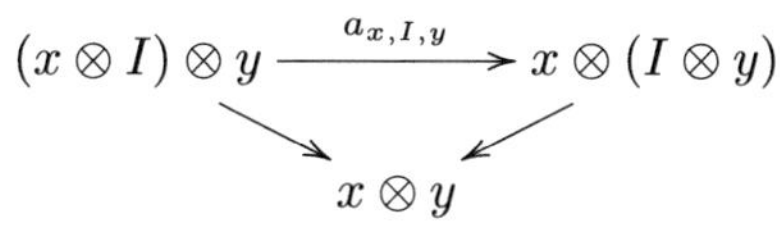

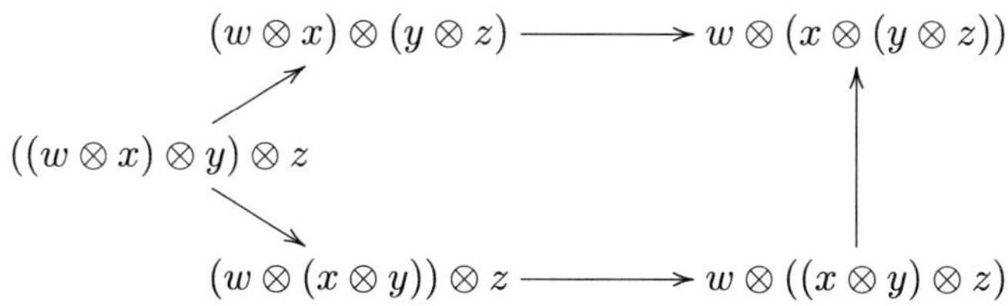

Having a conjunction at our disposition, we can do either one of two things. Either we add some structure to the conjunction or we add other logical connectives. Looking at Fig. 5.1 below, this means that either we go up in the structures or we go right and add some more material on the same categorical level. Let us examine these alternatives in turn.

First, it is possible to define a *symmetric deductive system* **S** as a monoidal deductive system satisfying a braiding rule (b).

$$\frac{\varphi \longrightarrow \psi \otimes \tau}{\varphi \longrightarrow \tau \otimes \psi} \text{ (b)}$$

In this case, the conjunction is commutative and we can show that a symmetric deductive system satisfies the definition of a symmetric category. Following Mac Lane [52, pp. 251–253], a symmetric monoidal category is a monoidal category obtained through the addition of a natural isomorphism $\beta_{x,y}$ which is its own inverse (i.e., $\beta_{y,x}\beta_{x,y} \cong 1_{x \otimes y}$). The tensor product also has to respect the two hexagon identities, represented by the commutative diagrams below. Since the braiding rule is its own inverse, one hexagon implies the other.

$$\beta_{x,y} : x \otimes y \longrightarrow y \otimes x$$

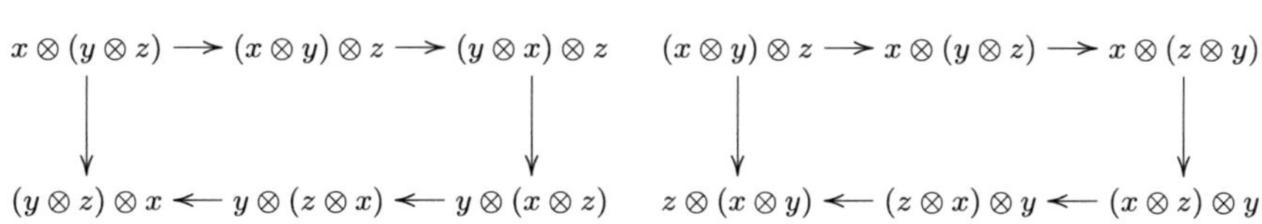

From a commutative tensor product, it is still possible to add some more structure by defining a Cartesian deductive system, which is characterized by two rules governing the introduction and the elimination of ⊗. A *Cartesian deductive system* **C** is defined as a monoidal deductive system satisfying the rules (⊗-in) and (⊗-out). Note that (⊗-out) abbreviates two rules.

$$\frac{\varphi \longrightarrow \psi \otimes \rho}{\varphi \longrightarrow \psi \qquad \varphi \longrightarrow \rho} \text{ (}\otimes\text{-out)} \qquad \frac{\varphi \longrightarrow \psi \qquad \varphi \longrightarrow \rho}{\varphi \longrightarrow \psi \otimes \rho} \text{ (}\otimes\text{-in)}$$

Note that a Cartesian deductive system can alternatively be defined as a deductive system satisfying ($\otimes$-in), ($\otimes$-out) and (!). Indeed, (!) is logically equivalent to ($\otimes$-out) in a monoidal deductive system, which in turn is logically equivalent to the axiom schemata $\varphi \otimes \psi \longrightarrow \varphi$ and $\varphi \otimes \psi \longrightarrow \psi$. Similarly, ($\otimes$-in) is logically equivalent to the copying axiom (Δ) within a monoidal deductive system. The conjunction in a Cartesian deductive system corresponds to the usual conjunction $\wedge$ in intuitionistic or classical propositional logics.

$$\frac{}{\varphi \longrightarrow 1}\ (!) \qquad\qquad \frac{}{\varphi \longrightarrow \varphi \otimes \varphi}\ (\Delta)$$

It can be proven that a Cartesian deductive system satisfies the definition of a Cartesian category, where the unit object is terminal and binary products are defined for each object. A monoidal category is Cartesian if the product respects the universal property defined by the following commutative diagram. This diagram must be interpreted as follows: for all f and g such that $f : z \longrightarrow x$ and $g : z \longrightarrow y$ there is one and only one arrow $\langle f, g \rangle : z \longrightarrow x \otimes y$ making the diagram commute.

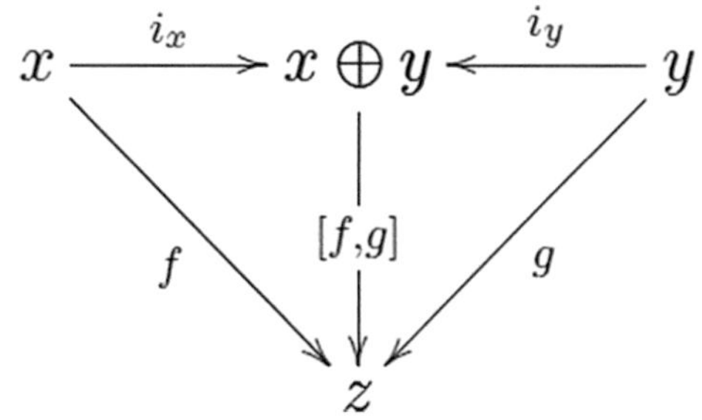

An interesting property of a Cartesian deductive system is that it satisfies idempotence of conjunction. An operation $\circ$ is said to be *idempotent* when $x \circ x \cong x$. Idempotence of conjunction follows from the rules governing the introduction and the elimination of $\otimes$ in a monoidal deductive system.

$$\frac{\dfrac{}{\varphi \longrightarrow \varphi}\ (1) \qquad \dfrac{}{\varphi \longrightarrow \varphi}\ (1)}{\varphi \longrightarrow \varphi \otimes \varphi}\ (\otimes\text{-in}) \qquad\qquad \frac{\dfrac{}{\varphi \otimes \varphi \longrightarrow \varphi \otimes \varphi}\ (1)}{\varphi \otimes \varphi \longrightarrow \varphi}\ (\otimes\text{-out})$$

So far, we have seen how one can build a logical system by adding some structure to the conjunction (i.e., by focusing on the properties of the tensor product). That being said, each of the aforementioned deductive systems can be augmented with either conditionals or a disjunction. On the one hand, we can define a closed deductive system by adding an adjoint to the tensor product. A *monoidal closed deductive system* **MC** is a deductive system satisfying either (cl) or (cl'). Note that there can be two conditionals given that $\otimes$ is not necessarily commutative.

$$\frac{\varphi \otimes \psi \longrightarrow \rho}{\varphi \longrightarrow \psi \multimap \rho}\ (\text{cl}) \qquad\qquad \frac{\varphi \otimes \psi \longrightarrow \rho}{\psi \longrightarrow \varphi \rhd \rho}\ (\text{cl'})$$

A monoidal closed deductive system satisfies the definition of a monoidal closed category, that is, a monoidal category in which the tensor product has a right adjoint.

Indeed, we can show using the aforementioned rules that there is an isomorphism between the class of morphisms from $\varphi \otimes \psi$ to ρ and the class of morphisms from φ to $\psi \multimap \rho$ (hence that it is left closed, $\mathbf{MC}_L$) and that there is an isomorphism between the class of morphisms from $\varphi \otimes \psi$ to ρ and the class of morphisms from ψ to $\varphi \rhd \rho$ (hence that it is right closed, $\mathbf{MC}_R$). Using the conditionals, we can introduce a special object 0 and define negations by $\sim \varphi =_{df} \varphi \multimap 0$ and $\neg\varphi =_{df} \varphi \rhd 0$.

From these definitions, we obtain negations that, when considered together, have an intuitionistic behavior. Double negations can be introduced via the following proofs. Note that if the deductive system is symmetric, then $\rhd$ and $\neg$ can be reduced to $\multimap$ and $\sim$ (and vice versa).

$$\dfrac{\dfrac{\dfrac{}{\varphi \multimap 0 \longrightarrow \varphi \multimap 0}\ (1)}{(\varphi \multimap 0) \otimes \varphi \longrightarrow 0}\ (\text{cl})}{\dfrac{\varphi \longrightarrow (\varphi \multimap 0) \rhd 0}{\varphi \longrightarrow \neg \sim \varphi}}\ (\text{cl'}) \qquad \dfrac{\dfrac{\dfrac{}{\varphi \rhd 0 \longrightarrow \varphi \rhd 0}\ (1)}{\varphi \otimes (\varphi \rhd 0) \longrightarrow 0}\ (\text{cl'})}{\dfrac{\varphi \longrightarrow (\varphi \rhd 0) \multimap 0}{\varphi \longrightarrow \sim \neg\varphi}}\ (\text{cl})$$

Then, one can require that the negation(s) behave classically (i.e., that double negations can be erased). A *monoidal closed deductive system with classical negations* **MCC** is a monoidal closed deductive system satisfying $(\neg \sim)$ and $(\sim \neg)$.

$$\dfrac{}{\sim \neg\varphi \longrightarrow \varphi}\ (\sim \neg) \qquad \dfrac{}{\neg \sim \varphi \longrightarrow \varphi}\ (\neg \sim)$$

If the deductive system is symmetric (and, further, if the co-tensor is defined by $\varphi \oplus \psi =_{df} \sim \varphi \multimap \psi$), we obtain a deductive system that satisfies the definition of a closed category with a dualizing object $*$ (with $x^* =_{df} x \rhd *$ and $\rhd$ the adjoint to $\otimes$), that is, an object such that x is isomorphic to x^{**} for each x.[7] A symmetric monoidal closed category with a dualizing object $*$ is a *-autonomous category—(cf. [8, 9]).

If one goes further and requires that the adjoint $x \rhd y$ to the tensor product be (naturally) isomorphic to $x^* \otimes y$ within a *-autonomous category, then one gets a compact closed category—(cf. [40]).

From the point of view of deductive systems, we can define a *compact deductive system* **CDS** as a symmetric monoidal closed deductive system satisfying (c1) and (c2). While **SCC** satisfies the definition of symmetric closed category and a **SCC** with the co-tensor defined by $\varphi \oplus \psi =_{df} \sim \varphi \multimap \psi$ satisfies the definition of a *-autonomous category, **CDS** satisfies the definition of a compact one, with an isomorphism $\varphi \multimap \psi \cong \sim \varphi \otimes \psi$. Vtrianglepair

$$\dfrac{}{\varphi \multimap \psi \longrightarrow \sim \varphi \otimes \psi}\ (\text{c1}) \qquad \dfrac{}{\sim \varphi \otimes \psi \longrightarrow \varphi \multimap \psi}\ (\text{c2})$$

Aside from the conditional connectives and the negations, one can also introduce a disjunction through co-deductive systems. It is noteworthy that the co-tensor does not need to be defined through the conditional. Indeed, we can define co-versions of the aforementioned deductive systems by reversing the arrows within the rules and axiom schemata, replacing $\otimes$ by $\oplus$, 1 by 0 and $\multimap$ /$\rhd$ by $\ltimes / \rtimes$. Hence, one can get co-monoidal co**M**, co-symmetric co**S**, co-closed deductive systems (e.g.,

[7] See Barr [9, p. 161] for the definition of a dualizing object.

co**SC**, co**SCC**), etc. Co-deductive systems can be proven to be instances of the dual (opposite) category $\mathcal{D}^{op}$ (see [52, p. 33]). In such co-deductive systems, the rules are used to define the co-tensor, which is a primitive connective. A **SCC** co**S** does *not* necessarily satisfy $\varphi \oplus \psi \cong\sim \varphi \multimap \psi$ [60].

Co-Cartesian deductive systems (co**C**) can be defined as co-monoidal deductive systems satisfying ($\oplus$-in) and ($\oplus$-out). Again, ($\oplus$-out) abbreviates two rules.

$$\frac{\varphi \oplus \psi \longrightarrow \rho}{\varphi \longrightarrow \rho \qquad \psi \longrightarrow \rho}\ (\oplus\text{-out}) \qquad \frac{\varphi \longrightarrow \rho \qquad \psi \longrightarrow \rho}{\varphi \oplus \psi \longrightarrow \rho}\ (\oplus\text{-in})$$

Note that similarly to Cartesian deductive systems, co-Cartesian deductive systems can alternatively be defined as deductive systems satisfying ($\oplus$-in), ($\oplus$-out) and (0).

$$\frac{}{0 \longrightarrow \varphi}\ (0)$$

From a categorical perspective, a co-Cartesian deductive system satisfies the definition of a co-Cartesian category, where there is a co-product $\oplus$ defined via the universal property expressed by the following commutative diagram together with an initial object.

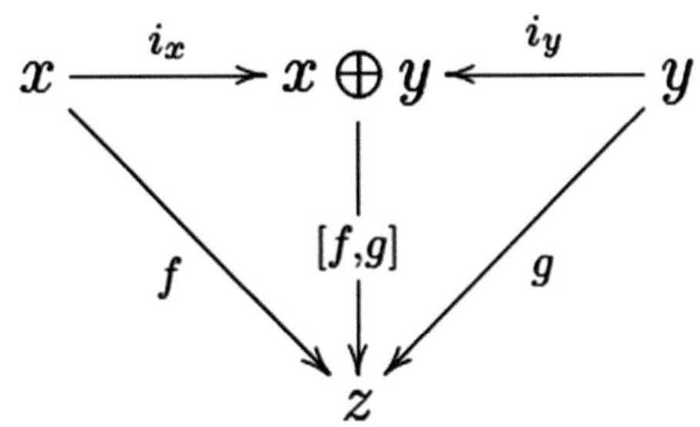

The universal property of the co-product means that for all $f : x \longrightarrow z$ and $g : y \longrightarrow z$ there is one and only one arrow $x \oplus y \longrightarrow z$ making the above diagram commute. As it was the case for the conjunction within a Cartesian deductive system, the disjunction of a co-Cartesian deductive system satisfies idempotence. The disjunction $\oplus$ in a co**C** corresponds to the usual $\vee$ in classical or intuitionistic logics.

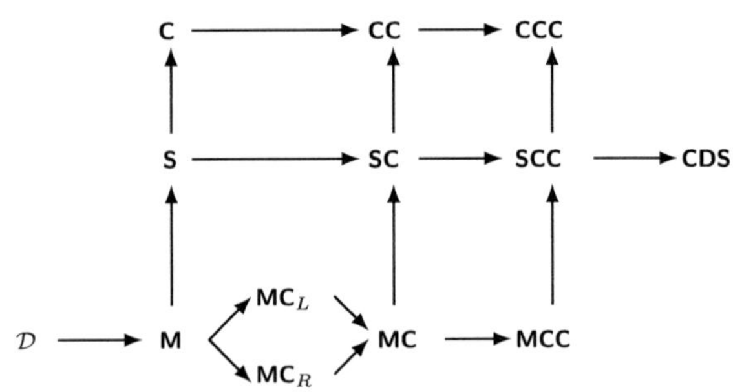

Fig. 5.1 Relations between deductive systems

5.2.3 *Summing Up*

We conclude this section with a brief recapitulation of what was presented so far. Figure 5.1 shows the relationships between the tensor's fragment of the deductive systems. From a proof-theoretical standpoint, logical systems can be defined via specific rules and axioms that make explicit the monoidal structure of the deductive systems and the properties of the connectives. Note that the proof-theory of a **CC**co**C** corresponds to intuitionistic logic and that the proof-theory of a **CCC**co**C** corresponds to classical logic.

The upshot of this mode of presentation is that the categorical framework enables us to see that different logical systems can be built upon weaker systems by focusing on specific properties of logical connectives. These definitions provide us with different types of logical systems. This allows for a proper classification where we can explicitly relate the properties of the logical connectives to the structural properties of the logical system. In what follows, we will show that most of the action logics that we find within the literature share a common Cartesian structure and we will argue that logical connectives that aim to model human actions do not have the properties that are related to this type of deductive system.

5.3 A Survey of Action Logics

Now turning our attention to the action logic literature, our aim is to illustrate how different approaches can be compared through the properties of their logical connectives. We will briefly present some of the main approaches one can find within the literature and show how each can be viewed as a variation of a monoidal deductive system. We begin by presenting dynamic logic and Kleene algebras, which will be followed by Pratt's action logic, linear logic, Boolean algebras of actions, STIT logics and Lucas's formalization of von Wright's conception of actions. We conclude this section with a comparison of these systems.

5.3.1 *Dynamic Logic*

5.3.1.1 Kleene Algebras

Dynamic logic was developed by Pratt [61, 62] and was meant to represent the evolution of a scenario after the execution of some *actions*, or, in Pratt's case, after the execution of some computer *program* (cf. [69]).[8] The main characteristic of dynamic logic is that it distinguishes between *propositions* (or *assertions*) and *actions* (or *programs*). It began with the introduction of a modal operator

[8] See [32] for an introduction and [7] for a historical sketch.

$[\alpha]\varphi$, which is interpreted as *after the execution of* α, φ *holds*.[9] Complex actions are composed from the connectives $\cup$, ; and *, representing respectively choice, sequence and iteration. The set of well-formed formulas is defined from the language $\mathcal{L} = \{Prop, Act, (,), \cup, ;, ^*, \neg, \supset, [\,]\}$ (with $Prop$ and Act denumerable sets of atomic propositions p_i and actions a_i) by:

$$\varphi := \; p_i \mid \neg\varphi \mid \varphi \supset \psi \mid [\alpha]\varphi$$

The set of complex actions $Act*$ is defined by [32, p. 165]:

$$\alpha := \; a_i \mid \alpha \cup \beta \mid \alpha; \beta \mid \alpha^*$$

The [ ]-modality is axiomatized as a normal K-modality. It is known that the action structure inherent to dynamic logic can be expressed by a Kleene algebra [32, p. 389]. A *Kleene algebra* $\mathcal{KA}* = (K, \cup, ;, 0, 1, *)$ is an idempotent semiring which satisfies the three following equations [32, p. 419]. The definition of $\alpha \leq \beta$ is $\alpha \cup \beta = \beta$.

$$1 \cup (\alpha; \alpha^*) = 1 \cup (\alpha^*; \alpha) = \alpha^* \tag{K*1}$$

$$\text{if } \beta \cup (\alpha; \gamma) \leq \gamma, \text{ then } \alpha^*; \beta \leq \gamma \tag{K*2}$$

$$\text{if } \beta \cup (\gamma; \alpha) \leq \gamma, \text{ then } \beta; \alpha^* \leq \gamma \tag{K*3}$$

Concentrating only on the fragment for choice and sequence, let $\mathcal{KA} = (K, \cup, ;, 0, 1)$. The logic of programs implicit to propositional dynamic logic can be expressed via an idempotent semiring. Indeed, a $\mathcal{KA}$ has to satisfy the following axioms. In short, 0 is the unit of $\cup$, which is associative (K1), commutative (K2) and idempotent (K3). Hence, $(K, \cup, 0)$ can be seen as an idempotent symmetric monoid.

$$\alpha \cup (\beta \cup \gamma) = (\alpha \cup \beta) \cup \gamma \tag{K1}$$

$$\alpha \cup \beta = \beta \cup \alpha \tag{K2}$$

$$\alpha \cup 0 = \alpha = \alpha \cup \alpha \tag{K3}$$

$$\alpha; (\beta; \gamma) = (\alpha; \beta); \gamma \tag{K4}$$

$$\alpha; 1 = \alpha = 1; \alpha \tag{K5}$$

[9] See [20] for an introduction to modal logics.

$$\alpha; 0 = 0 = 0; \alpha \tag{K6}$$

$$\alpha; (\beta \cup \gamma) = (\alpha; \beta) \cup (\alpha; \gamma) \tag{K7}$$

$$(\alpha \cup \beta); \gamma = (\alpha; \gamma) \cup (\beta; \gamma) \tag{K8}$$

A direct consequence of defining the pre-order using $\cup$ is that $(K, \cup, 0)$ satisfies the definition of a co**C**. First, note that given a pre-order we can recover the identity axiom since $\alpha \leq \alpha$ and, moreover, we can recover (cut) since if $\alpha \leq \beta$ and $\beta \leq \gamma$, then $\alpha \leq \gamma$. Second, the definition of the pre-order together with the axioms (K2) and (K3) yield the following injection maps. These maps follows from the fact that $\alpha \cup (\alpha \cup \beta) = \alpha \cup \beta$.

$$\alpha \leq \alpha \cup \beta \qquad \beta \leq \alpha \cup \beta$$

From these maps, we can show the rule governing the elimination of $\cup$. Assuming $\alpha \cup \beta \leq \gamma$, we can show the following equations using the transitivity of the pre-order.

$$\alpha \leq \alpha \cup \beta \leq \gamma \qquad \beta \leq \alpha \cup \beta \leq \gamma$$

Similarly, we can also show using the definition of $\leq$ that if $\alpha \leq \gamma$ and $\beta \leq \gamma$, then $\alpha \cup \beta \leq \gamma$. Indeed, assuming $\alpha \leq \gamma$ and $\beta \leq \gamma$, we have $\alpha \cup \gamma = \gamma$ and $\beta \cup \gamma = \gamma$, hence $\alpha \cup (\beta \cup \gamma) = (\alpha \cup \beta) \cup \gamma = \gamma$ and thus $\alpha \cup \beta \leq \gamma$. Therefore, the rule for the introduction of $\cup$ is satisfied.

Third, we can recover all the rules governing $\cup$ within a co-monoidal deductive system. While (K1), (K2) and (K3) enable us to prove (co-a), (co-b), (co-l) and (co-r), (co-t) can be recovered since assuming $\alpha \leq \beta$ and $\gamma \leq \delta$, we have $\alpha \leq \beta \cup \delta$ and $\gamma \leq \beta \cup \delta$, hence $\alpha \cup \gamma \leq \beta \cup \delta$ from the previous results.

Regarding the case of $(K, ;, 1)$, one can see that 1 is the unit of ; by (K5) and that ; is associative by (K4). From this, (a), (r) and (l) can be derived. Furthermore, it is possible to derive (t) given (K7) and (K8). Indeed, assuming $\alpha \leq \beta$ and $\gamma \leq \delta$ implies by definition $\alpha \cup \beta = \beta$ and $\gamma \cup \delta = \delta$. Hence, $\beta; \delta = (\alpha \cup \beta); \delta = (\alpha; \delta) \cup (\beta; \delta)$, which by definition implies that $\alpha; \delta \leq \beta; \delta$. Furthermore, $\alpha; \delta = \alpha; (\gamma \cup \delta) = (\alpha; \gamma) \cup (\alpha; \delta)$, which implies by definition that $\alpha; \gamma \leq \alpha; \delta$. Therefore, by the pre-order, it follows that $\alpha; \gamma \leq \beta; \delta$ and that (t) can be derived. The fragment $(K, ;, 1)$ thus satisfies the definition of a monoidal deductive system and, therefore, $\mathcal{KA}$ can be seen as an instance of a **M**co**C**.[10]

[10]Note that (K1)–(K8) can all be derived using the definition of a **MC**co**C**.

5.3.1.2 Deontic Dynamic Logic

After Pratt's developments, dynamic logic was first used in the context of deontic logic by Meyer [54, 55], who introduced the connectives $\bar{\alpha}$, $\alpha \cap \beta$ and $\varphi \rightarrow \alpha/\beta$, respectively representing action negation, joint action and the conditional action 'if φ, then α else β'. In addition to a denumerable set of atomic actions $Act = \{\varnothing, U, a_1, \ldots, a_n, \ldots\}$, Meyer includes the action which *fails* $\varnothing$ and the *universal action* U. The modality $[\alpha]$ is axiomatized as a normal K-modality. The set of well-formed formulas of propositional dynamic logic is defined recursively from the language $\mathcal{L} = \{Prop, Act, (,), \bar{}, \cup, \cap, ;, \neg, \supset, [\ ]\}$ (we leave aside the connective $\rightarrow$ /since it has not been considered further) as follows.

$$\varphi := \ p_i \mid \neg\varphi \mid \varphi \supset \psi \mid [\alpha]\varphi$$

The set of complex actions $Act*$ is defined by

$$\alpha := \ a_i \mid \bar{\alpha} \mid \alpha \cap \beta \mid \alpha \cup \beta \mid \alpha; \beta$$

The other logical connectives are defined as usual. The axioms for actions are as follows:

$$[\overline{\overline{\alpha}}]\varphi \equiv [\alpha]\varphi \tag{A1}$$

$$[\alpha; \beta]\varphi \equiv [\alpha]([\beta]\varphi) \tag{A2}$$

$$[\overline{\alpha; \beta}]\varphi \equiv [\overline{\alpha}]\varphi \wedge [\alpha]([\overline{\beta}]\varphi) \tag{A3}$$

$$[\alpha \cup \beta]\varphi \equiv [\alpha]\varphi \wedge [\beta]\varphi \tag{A4}$$

$$[\overline{\alpha}]\varphi \vee [\overline{\beta}]\varphi \supset [\overline{\alpha \cup \beta}]\varphi \tag{A5}$$

$$[\alpha]\varphi \vee [\beta]\varphi \supset [\alpha \cap \beta]\varphi \tag{A6}$$

$$[\overline{\alpha}]\varphi \wedge [\overline{\beta}]\varphi \equiv [\overline{\alpha \cap \beta}]\varphi \tag{A7}$$

$$[\varnothing]\varphi \tag{A9}$$

It is noteworthy that (A2) implies that sequence is not primitive. Hence, let us concentrate on the fragment constructed from $^{-}$, $\cap$ and $\cup$. Without explicitly defining the algebra of actions inherent to his approach, Meyer [55, p. 132] asserts that the following equations are satisfied.

$$\overline{\overline{\alpha}} = \alpha \tag{DL1}$$

$$\overline{\alpha \cup \beta} = \overline{\alpha} \cap \overline{\beta} \tag{DL2}$$

$$\overline{\alpha \cap \beta} = \overline{\alpha} \cup \overline{\beta} \tag{DL3}$$

$$\alpha \cup \varnothing = \alpha \tag{DL4}$$

$$\alpha \cup \alpha = \alpha \tag{DL5}$$

$$\alpha \cap \alpha = \alpha \tag{DL6}$$

$$\alpha \cap \overline{\alpha} = \varnothing \tag{DL7}$$

$$\alpha \cap (\beta \cup \gamma) = (\alpha \cap \beta) \cup (\alpha \cap \gamma) \tag{DL8}$$

$$\alpha \cup (\beta \cap \gamma) = (\alpha \cup \beta) \cap (\alpha \cup \gamma) \tag{DL9}$$

When writing on the algebraic properties of actions, Meyer did not refer to a Boolean algebra of actions. Hence, it is fair to assume that the algebra of actions he had in mind for his approach was somewhat weaker than a Boolean algebra. But he did not either refer explicitly to a Kleene algebra and so it is difficult to see exactly what he had in mind regarding the action logic inherent to his approach.

Let DDL refer to the (fragment of the) action logic inherent to the Dynamic Deontic Logic presented in [55]. In what follows, we will first expose without any extra assumption the properties of the action connectives we can infer from Meyer's presentation. We will compare his approach to a Kleene algebra and then we will examine what needs to be done if one wants to define explicitly DDL as a Boolean algebra. Finally, we will show that DDL actually *is* a Boolean algebra.

Although Meyer did not mention whether $\cap$ and $\cup$ were associative or commutative, associativity and commutativity of $\cup$ can be recovered from (A4). Idempotence of both $\cap$ and $\cup$ is given by (DL5) and (DL6), and distributivity of $\cap$ over $\cup$ is given by (DL8). Moreover, distributivity of $\cup$ over $\cap$ is given by (DL9). Finally, (DL1) gives us involutive negation.

Now, let us compare **DDL** and $\mathcal{KA}$. Associativity and commutativity of choice can be recovered in **DDL** from (A4), and hence (K1) and (K2) can also be recovered. (K3) is given by (DL4) and (DL5). (K7) is given by (DL8). There are no references to (K4), (K5), (K6) or (K8).

When one looks at (DL1), (DL2) and (DL3), one gets the feeling that Meyer wanted the structure of his action algebra to be classical. Anticipating the following sections, let us note that Meyer's algebra of actions must be stronger than the multiplicative fragment **MLL** of linear logic insofar as he assumes the distributivity of $\cap$ over $\cup$. Hence, since Meyer's action algebra cannot be reduced to **MLL**, it seems that the only way to obtain both De Morgan dualities and involutive negation is to assume a Boolean algebra.

The question we now want to examine is what extra assumptions are required to go from **DDL** to a Boolean algebra $\mathcal{BA}$. By definition, a $\mathcal{BA}$ is a distributive complemented lattice (see Sect. 5.3.4 below for the definition). From (DL8) and (DL9) we obtain the two conditions for distributivity, and from (DL7) we obtain one part of the condition for complementarity. Therefore, the extra assumptions one needs in order to obtain a $\mathcal{BA}$ from **DDL** are that it is a lattice and that $\alpha \cup \overline{\alpha} = U$.

Looking at Meyer's paper, it is hard to determine whether or not he assumed these two conditions. Meyer's approach takes place within the dynamic logic tradition and, therefore, it is reasonable to assume that he minimally considers $\cap$ to be associative and U to be its unit, which would enable him to recover (K4) and (K5). Moreover, (K6) can be recovered from distributivity and (DL7). Indeed, by (DL7) and (DL4) we have $(\alpha \cap \overline{\alpha}) \cup \varnothing = \varnothing$, hence by commutativity of $\cup$ and (DL9) we obtain $(\varnothing \cup \alpha) \cap (\varnothing \cup \overline{\alpha}) = \varnothing$. Since $\alpha \cup \varnothing = \alpha$, we get $\alpha \cap (\varnothing \cup \overline{\alpha}) = \varnothing$. By (DL8) we have $(\alpha \cap \varnothing) \cup (\alpha \cap \overline{\alpha}) = \varnothing$. Since $\alpha \cap \overline{\alpha} = \varnothing$, it gives us $(\alpha \cap \varnothing) \cup \varnothing = \varnothing$, hence from (DL4) we obtain $\alpha \cap \varnothing = \varnothing$.

Finally, the commutativity of $\cup$ and the De Morgan dualities suggest that Meyer assumed $\cap$ to be commutative. This would explain the absence of any reference to (K8), which would hence be derivable from (DL8) and commutativity. As a result, it is quite reasonable to assume that Meyer minimally presupposed the structure of a Kleene algebra for **DDL**.

We saw that a Kleene algebra is an idempotent semiring, hence a join-semilattice. Therefore, to explicitly define **DDL** as a $\mathcal{BA}$, one would need to assume that it is also a meet-semilattice (hence a lattice) and that it is complemented. It is noteworthy that a Kleene algebra is not a meet-semilattice insofar as the sequence operator is neither commutative nor idempotent. In Meyer's case, though, we saw that it would be safe to assume commutativity of $\cap$ and we already have idempotence. Consequently, assuming that **DDL** is a lattice would be consistent with his position. Moreover, the complementarity could easily be recovered. Indeed, from (DL4) and (DL7) we have $\alpha \cup (\alpha \cap \overline{\alpha}) = \alpha$, which by (DL9) implies $(\alpha \cup \alpha) \cap (\alpha \cup \overline{\alpha}) = \alpha$. From (DL5) we know that $\alpha \cup \alpha = \alpha$, and therefore we have $\alpha \cap (\alpha \cup \overline{\alpha}) = \alpha$. This suggests that $\alpha \cup \overline{\alpha}$ is the unit of $\cap$, and hence that $\alpha \cup \overline{\alpha} = U$ if we assume that U is the unit of $\cap$.

So far, we have seen how one could (explicitly) define DDL as a $\mathcal{BA}$ from Meyer's initial remarks. Now we show that DDL is actually implicitly a $\mathcal{BA}$. To accomplish this, we need to show that DDL satisfies the equations (L1)–(L14) that define a distributive complemented lattice (see Sect. 5.3.4 below). We already know that (L2) and (L4), which express the associativity and commutativity of $\cup$, can be recovered from (A4) and that (L5) and (L6) for idempotence are given by (DL5) and (DL6). The distributivity conditions (L11) and (L12) are given by (DL8) and (DL9) and the first part of complementarity (L13) is given by (DL7). (DL4) yields (L8). As such, we only need to show that DDL satisfies the following equations.

$$\alpha \frown (\beta \frown \gamma) = (\alpha \frown \beta) \frown \gamma \tag{L1}$$

$$\alpha \frown \beta = \beta \frown \alpha \tag{L3}$$

$$\alpha \frown 1 = \alpha \tag{L7}$$

$$\alpha \smile (\alpha \frown \beta) = \alpha \tag{L9}$$

$$\alpha \frown (\alpha \smile \beta) = \alpha \tag{L10}$$

$$\alpha \smile \bar{\alpha} = 1 \tag{L14}$$

Associativity and commutativity of $\cap$ expressed by (L1) and (L3) are recovered using the following equalities and (DL1).

$$\overline{\alpha \cap (\beta \cap \gamma)} = \overline{\alpha} \cup \overline{\beta \cap \gamma} \qquad \text{(by DL3)}$$

$$= \overline{\alpha} \cup (\overline{\beta} \cup \overline{\gamma}) \qquad \text{(by DL3)}$$

$$= (\overline{\alpha} \cup \overline{\beta}) \cup \overline{\gamma} \qquad \text{(by associativity)}$$

$$= \overline{\alpha \cap \beta} \cup \overline{\gamma} \qquad \text{(by DL3)}$$

$$= \overline{(\alpha \cap \beta) \cap \gamma} \qquad \text{(by DL3)}$$

$$\overline{\alpha \cap \beta} = \overline{\alpha} \cup \overline{\beta} \qquad \text{(by DL3)}$$

$$= \overline{\beta} \cup \overline{\alpha} \qquad \text{(by commutativity)}$$

$$= \overline{\beta \cap \alpha} \qquad \text{(by DL3)}$$

(L7) can be derived from the fact that:

$$\overline{\alpha} = \overline{\alpha} \cup \varnothing \qquad \text{(by DL4)}$$

Therefore, we can derive:

$$\alpha = \overline{\overline{\alpha}} \qquad \text{(by DL1)}$$

$$= \overline{\overline{\alpha} \cup \varnothing}$$

$$= \overline{\overline{\alpha}} \cap \overline{\varnothing} \qquad \text{(by DL2)}$$

$$= \alpha \cap \overline{\varnothing} \qquad \text{(by DL1)}$$

Hence, $\overline{\varnothing}$ is the unit of $\cap$. Moreover, since (K6) can be recovered in DDL, (DL3) gives us $\overline{\varnothing} = \alpha \cup \overline{\varnothing}$. Thus, it follows that the absorption laws (L9) and (L10) can be derived:

$$\alpha = \alpha \cap \overline{\varnothing} \qquad \text{(by L7)}$$

$$= \alpha \cap (\beta \cup \overline{\varnothing}) \qquad \text{(by K6 and DL3)}$$

$$= (\alpha \cap \beta) \cup (\alpha \cap \overline{\varnothing}) \qquad \text{(by DL8)}$$

$$= (\alpha \cap \beta) \cup \alpha \qquad \text{(by L7)}$$

$$= \alpha \cup (\alpha \cap \beta) \qquad \text{(by commutativity)}$$

$$\alpha = \alpha \cup \varnothing \qquad \text{(by DL4)}$$

$$= \alpha \cup (\varnothing \cap \beta) \qquad \text{(by K6)}$$

$$= (\alpha \cup \varnothing) \cap (\alpha \cup \beta) \qquad \text{(by DL9)}$$

$$= \alpha \cap (\alpha \cup \beta) \qquad \text{(by DL4)}$$

Finally, from (DL7) and (DL3) we have $\overline{\varnothing} = \overline{\alpha} \cup \overline{\overline{\alpha}}$. Thus, by commutativity and (DL1) we have (L14) $\overline{\varnothing} = \alpha \cup \overline{\alpha}$. As a result DDL satisfies the equations of a distributive complemented lattice.

The reader can compare Meyer's definition of DDL and the properties of a Boolean algebra in Table 5.1. When compared through the properties of their connectives, the only difference between them is that $\cap$ is not explicitly said to be commutative and associative within DDL. Otherwise, the result is the same and, as we saw, Meyer's definition yields implicitly a $\mathcal{BA}$. As such, it is noteworthy that if Meyer explicitly assumed that DDL is a lattice and that $\alpha \cup \overline{\alpha} = U$, then DDL would have been seen as a precursor of what Kozen [41] defined as a *Kleene algebra with test*, that is, a Kleene algebra embedded in a Boolean algebra to model negation of programs (see [32, p. 421]).

To sum up, a Kleene algebra $\mathcal{KA}$ (without the Kleene star *) can be classified as a **M**co**C** and, given that DDL satisfies the definition of a $\mathcal{BA}$, it can be classified as a **CCC**co**C** as we will see in the section below. Let us conclude this section by mentioning that in addition to $\mathcal{KA}$ and DDL, there are other approaches in dynamic logic which have explicitly used Boolean algebras for actions. For instance, the reader may look at the work of Castro and Maibaum [18], who develop a dynamic logic as an extension of a Boolean algebra of actions.[11]

5.3.2 Pratt

In addition to the introduction of dynamic logic, Pratt [63] also introduced another form of action logic, which looses the distinction between actions and assertions. Pratt's action logic is presented as a logic for programming that includes the three

[11]For more information on categories and dynamic logic, the reader can consult [73]. For other approaches in dynamic deontic logic, see [74] for a dynamic logic of permission, [66] for dynamic logic and agency and [14] for the analysis of action negation.

basic control elements *choice*, *sequence* and *iteration* [63, p. 100]. The choice $\alpha + \beta$ between α and β is introduced as a commutative and idempotent connective. Sequence $\alpha; \beta$ is introduced as a non-commutative conjunction where α can be performed before β or simultaneously (hence the non commutativity) and iteration α^* is the performance of α an arbitrary number of times [63, p. 100]. The constant 0 and 1 are introduced as the *empty* and the *skip* (or universal) actions respectively.

An Action algebra $\mathcal{A}* = \langle A, \leq, +, 0, ;, 1, \rightarrow, \leftarrow, ^* \rangle$ is defined by a pre-order and two monoids $(A, +, 0)$ and $(A, ;, 1)$, with $+$ commutative and idempotent, satisfying the following rules (cf. [63, p. 102]).

$$\frac{\alpha \leq \gamma \leftarrow \beta}{\alpha; \beta \leq \gamma} \qquad\qquad \frac{\beta \leq \alpha \rightarrow \gamma}{\alpha; \beta \leq \gamma}$$

Note that these rules are sometimes called the *residuation laws* in the literature—(see [27]). An action algebra also has to satisfy the following equations.

$$\frac{1 + \beta; \beta + \alpha \leq \beta}{\alpha^* \leq \beta} \qquad\qquad 1 + \alpha^*; \alpha^* + \alpha \leq \alpha^*$$

The symbols $\rightarrow$ and $\leftarrow$ stand respectively for pre-implication (had α, then β) and post-implication (β if ever α). We concentrate only on the fragment for conjunctive action and disjunctive action without iteration. Hence, let $\mathcal{A} = \langle A, \leq, +, 0, ;, 1, \rightarrow, \leftarrow \rangle$.

While the pre-order relation gives us (1) and (cut), the fact that $(A, ;, 1)$ is a monoid implies that the operation ; is associative and 1 is its unit. Thus, (a), (r) and (l) can be derived. Furthermore, the aforementioned rules imply that (cl) and (cl') are satisfied, which implies that (t) also is. Indeed, assume that $\alpha \leq \beta$ and $\gamma \leq \delta$:

$$\begin{aligned}
&\beta; \delta \leq \beta; \delta \\
&\quad \delta \leq \beta \rightarrow (\beta; \delta) \\
&\quad \gamma \leq \beta \rightarrow (\beta; \delta) \\
&\beta; \gamma \leq \beta; \delta \\
&\quad \beta \leq (\beta; \delta) \leftarrow \gamma \\
&\quad \alpha \leq (\beta; \delta) \leftarrow \gamma \\
&\alpha; \gamma \leq \beta; \delta
\end{aligned}$$

Accordingly, $(A, ;, 1, \rightarrow, \leftarrow)$ satisfies the definition of a **MC**. Furthermore, given the definition of $(A, +, 0)$ as a commutative monoid and the definition of the pre-order via $\alpha \leq \beta =_{df} \alpha + \beta = \beta$, we can show as in Sect. 5.3.1.1 that $(A, +, 0)$ satisfies the definition of a co**C**. Therefore, it follows that $\mathcal{A} = \langle A, \leq, +, 0, ;, 1, \rightarrow, \leftarrow \rangle$ satisfies the definition of a **MC**co**C**.[12]

[12]See also [12] for Kleene algebras and Pratt's action logic.

5.3.3 Linear Logic

Linear logic was introduced by Girard [28] and was meant to provide a constructive framework different than intuitionistic logic. The main objective of linear logic is to be able to deal with limited resources. This led to the introduction of two types of conjunction (with respect to their dual disjunction). The syntax of linear logic is built from the language $\mathcal{L} = \{Prop, \ ^{\perp}, \top, \bot, \mathbf{1}, \mathbf{0}, \otimes, \parr, \&, \oplus, !, ?, \forall, \exists\}$, with $Prop$ a denumerable set of atoms, $!/?$ operators to add and subtract resources and $\forall/\exists$ the usual quantifiers. For the sake of our analysis, we will only present the **MALL** fragment of linear logic, that is, the multiplicative and additive fragment. It is constructed from $\mathcal{L} = \{Prop, \ ^{\perp}, \top, \bot, \mathbf{1}, \mathbf{0}, \otimes, \parr, \&, \oplus\}$. The main feature of linear logic is that it incorporates two pairs of dual connectives, namely $\otimes/\parr$ and $\&$ $/\oplus$, with a linear implication $\multimap$ defined as the adjoint of $\otimes$.

Seely [67] showed how the multiplicative fragment **MLL** can be obtained through a *-autonomous category and the **MALL** fragment results by adding products and coproducts. As such, the proof-theory of **MALL** can be reproduced by a *-autonomus category with products and coproducts.[13] The **MALL** fragment of linear logic is constructed from the multiplicative $\{\otimes, \parr, \multimap, \mathbf{1}, \bot\}$ and additive $\{\&, \oplus, \mathbf{0}, \top\}$ fragments (see [29, p. 10] for the definitions). Recall the definition of a *-autonomous category: it is a symmetric closed monoidal category with a dualizing object. From a proof-theoretical perspective, the multiplicative fragment **MLL** satisfies the definition of a **SCC** satisfying $\varphi \parr \psi =_{df} \varphi^{\perp} \multimap \psi$, with $\parr$, $\multimap$ and $\mathbf{1}$ being respectively the dual, adjoint and unit of $\otimes$. Negation is defined using the dualizing object $\bot$, satisfying $\varphi \cong \varphi^{\perp\perp}$. The additive fragment **ALL** satisfies the definition of a **C**co**C**, with $\mathbf{0}$ the initial object and $\top$ the terminal.

The multiplicative fragment of linear logic **MLL** is thus comparable to a symmetric closed monoidal deductive system with with classical negation satisfying $\varphi \oplus \psi =_{df} \sim \varphi \multimap \psi$. Hence, the tensor and the co-tensor are both associative and commutative, but neither are idempotent. Note that in **MLL** neither $\otimes$ nor $\parr$ distribute over the other. In this respect, **MLL** is not distributive. However, **MLL** is *weakly* or *linearly* distributive (cf. [21]). Indeed, there are natural transformations $\varphi \otimes (\psi \parr \rho) \longrightarrow (\varphi \otimes \psi) \parr \rho$ and $(\varphi \parr \psi) \otimes \rho \longrightarrow \varphi \parr (\psi \otimes \rho)$ (cf. [13]). This is a consequence of the definition of $\parr$ using $\multimap$.

It is noteworthy that there are other variations of linear logic that involve different properties for the multiplicative connectives. First, following Casadio [16, p. 167], there is the multiplicative (and exponential free) fragment of the non-commutative intuitionistic linear logic **NILL** introduced in [2], which corresponds to Lambek's [42] Syntactic Calculus with negations, (see also [1]). The multiplicative fragment of **NILL** built from $\{\otimes, \multimap, \multimapinv, \mathbf{1}\}$ corresponds to what Lambek [43] defined as a residuated category and can be compared to a **MC**. From a proof-theoretical perspective, **NILL** without negations (i.e., the Syntactic Calculus) shares the same structure than the fragment $(A, ;, 1)$ of Pratt's Action algebra $\mathcal{A}$. Furthermore, there is the multiplicative fragment **MNLL** of the non-commutative linear logic **NLL** introduced in

[13] On this subject, see also Blute and Scott [13].

[3]. It can be obtained from a syntactical point of view when one introduces involutive negations, (see also [4]), and thus it can be compared to a **MCC**.

5.3.4 Boolean Algebras of Actions

We now turn our attention to Boolean algebras of actions. The classical structure of action propositions can be fully expressed when one uses a Boolean algebra of actions rather than the usual propositional dynamic logic. Segerberg [68], for instance, proposes to define a deontic action algebra on the grounds of a Boolean algebra of actions. Also distinguishing between *actions* and *propositions*, he assumes a language $\mathcal{L} = \{Act, \smile, \frown, {}^{-}\}$ with Act a denumerable set of atomic actions together with the impossible action $\mathbf{0}$ and the universal action $\mathbf{1}$. Segerberg [68, p.269] uses $\smile$, $\frown$ and $^{-}$ to denote the union, the conjunction and the negation of actions. The set A of actions is recursively defined as follows.

$$\alpha := \ \mathbf{0} \mid \mathbf{1} \mid a_i \mid \bar{\alpha} \mid \alpha \frown \beta \mid \alpha \smile \beta$$

A logic of action is understood as a Boolean algebra. Following Goldblatt [30, p. 134], a Boolean algebra $\mathcal{BA} = (A, \frown, \smile, {}^{-}, \mathbf{0}, \mathbf{1})$ is by definition a distributive complemented lattice. A lattice can algebraically be defined by the following equations.

$$\alpha \frown (\beta \frown \gamma) = (\alpha \frown \beta) \frown \gamma \tag{L1}$$

$$\alpha \smile (\beta \smile \gamma) = (\alpha \smile \beta) \smile \gamma \tag{L2}$$

$$\alpha \frown \beta = \beta \frown \alpha \tag{L3}$$

$$\alpha \smile \beta = \beta \smile \alpha \tag{L4}$$

$$\alpha \frown \alpha = \alpha \tag{L5}$$

$$\alpha \smile \alpha = \alpha \tag{L6}$$

$$\alpha \frown \mathbf{1} = \alpha \tag{L7}$$

$$\alpha \smile \mathbf{0} = \alpha \tag{L8}$$

$$\alpha \smile (\alpha \frown \beta) = \alpha \tag{L9}$$

$$\alpha \frown (\alpha \smile \beta) = \alpha \tag{L10}$$

(L1) and (L2) express associativity of $\smile$ and $\frown$, as (L3)-(L4) and (L5)-(L6) represent commutativity and idempotence. (L7) and (L8) make **1** and **0** into the respective unit of $\frown$ and $\smile$, and (L9)-(L10) are the absorption laws.

A lattice is said to be *distributive* when it satisfies the distribution laws represented by (L11) and (L12).

$$\alpha \smile (\beta \frown \gamma) = (\alpha \smile \beta) \frown (\alpha \smile \gamma) \tag{L11}$$

$$\alpha \frown (\beta \smile \gamma) = (\alpha \frown \beta) \smile (\alpha \frown \gamma) \tag{L12}$$

Equivalently, a distributive lattice can be defined as a closed poset $\mathcal{P} = (A, \leq)$ which possesses meets and joins, that is, for each $\alpha, \beta, \gamma \in A$:

$$\alpha \frown \beta \leq \beta \text{ and } \alpha \frown \beta \leq \beta \tag{meets}$$

$$\text{if } \gamma \leq \alpha \text{ and } \gamma \leq \beta, \text{ then } \gamma \leq \alpha \frown \beta$$

$$\alpha \leq \alpha \smile \beta \text{ and } \beta \leq \alpha \smile \beta \tag{joins}$$

$$\text{if } \alpha \leq \gamma \text{ and } \beta \leq \gamma, \text{ then } \alpha \smile \beta \leq \gamma$$

Moreover, for each $\alpha \in A$:

$$\alpha \leq \mathbf{1}$$
$$\mathbf{0} \leq \alpha$$

The closed poset is obtained by requiring the satisfaction of $\alpha \frown \beta \leq \gamma$ iff $\alpha \leq \beta \rhd \gamma$, (see [5, p. 129]). Here, $\alpha \frown \beta$ is a greatest lower bound and $\alpha \smile \beta$ a least upper bound.

The definition of a distributive lattice corresponds to the notion of a Heyting algebra. Furthermore, a distributive lattice can easily be shown to be an instance of a **CC**co**C**. While the partial-order gives us (1) and (cut), the aforementioned definition trivially yields the rules for introduction and elimination of $\smile$ and $\frown$ together with (!) and (0).[14] As a closed poset, (cl) is implicitly satisfied.

A Boolean algebra is a Heyting algebra satisfying $\alpha \smile \bar{\alpha} = \mathbf{1}$. In a Boolean algebra, $\alpha \rhd \beta$ can be defined by $\bar{\alpha} \smile \beta$. A Boolean algebra can also be defined as a distributive lattice that is complemented, i.e. which satisfies (L13) and (L14).

$$\alpha \frown \bar{\alpha} = \mathbf{0} \tag{L13}$$

$$\alpha \smile \bar{\alpha} = \mathbf{1} \tag{L14}$$

This enables us to derive the axiom for classical negation. Indeed, we have $\bar{\bar{\bar{\alpha}}} = \bar{\bar{\alpha}} \rhd 0 = \bar{\alpha} \smile 0 = \bar{\alpha}$, hence $\mathbf{1} = \bar{\alpha} \smile \alpha = \bar{\bar{\bar{\alpha}}} \smile \alpha = \bar{\bar{\alpha}} \rhd \alpha$, which implies that $\mathbf{1} \frown \bar{\bar{\alpha}} \leq \alpha$ and thus $\bar{\bar{\alpha}} \leq \alpha$.

As a result, while a Heyting algebra can be seen as an instance of a **CC**co**C**, a Boolean algebra can be seen as an instance of a **CCC**co**C**. For approaches that consider the logic of actions as a Boolean algebra of actions, see for instance Lindahl and Odelstad [48], Trypuz and Kulicki [72] or Castro and Maibaum [18].

5.3.5 Stit Logics

Although the origin of STIT logic is usually attributed to the work of Belnap and Perloff [11],[15] who introduced the modal operator $[i\ stit : \varphi]$, this variant of action logic can actually be traced back to the work of Kanger [36].[16] The first formal semantics for STIT logics was proposed by Chellas [19] (cf. [69, p. 369]) and the modal operator E_i (originally written D_i), which is another version of the STIT operator and was meant to be read '*agent i brings it about that*', was introduced in the work of Pörn [64]. The work of Belnap and Perloff [11] was further developed by Xu [79] and, currently, one can find STIT logics for instance in the work of Horty [33], Horty and Belnap [34], Pacheco and Carmo [57] and Broersen [15].

STIT logics are modal logics that use a normal K-modality indexed to an agent (or a set of agents). The STIT modality is usually axiomatized by a KT-modality, as in the work of [64] or Pacheco and Carmo [57], but it can also be axiomatized via an $S5$-modality (cf. [33]) or by a KD-modality as in the XSTIT framework of Broersen [15]. Assuming a language $\mathcal{L} = \{Prop, (,), \neg, \supset, [i\ stit]\}$ (with $Prop$ a denumerable set of atomic sentences), one can recursively define the set of well-formed formulas as follows.

$$\varphi := \ p_i \mid \neg\varphi \mid \varphi \supset \psi \mid [i\ stit]\varphi$$

[14] Note that when a poset is seen as a category, $\leq$ must be viewed as an arrow between equivalence classes of actions rather than actions per se insofar as $\leq$ is an antisymmetric relation.

[15] Recall that STIT stands for *seeing to it that*.

[16] This was reprinted in [37] and further developed in [38, 39].

The other connectives are defined as usual. The $[i\ stit]$ modality is axiomatized through the usual axioms and rules of modal logic together with suitable axiom(s), depending on the framework adopted. If $[i\ stit]$ behaves according to the axiom schema (T), then one can infer from $[i\ stit]\varphi$ that φ is true, which will be accompanied by a reflexive relation R on the semantical model. Likewise, truth conditions for the $[i\ stit]$ modality will be usually defined via the accessibility relation on the model $\mathcal{M}$, the only difference with the usual model of modal logic being that $\mathcal{M}$ is augmented with a set of agent Ag and the truth condition for $[i\ stit]$ is indexed to a relation R_i, and thus one obtains a class of accessibility relations indexed to some agents.

Although STIT logics do not distinguish between actions and propositions, one interesting feature is that they enable one to distinguish between doing a negative action and simply not doing an action. Indeed, they enable one to distinguish between not doing φ, represented by $\neg[i\ stit]\varphi$, and consciously doing $\neg\varphi$, expressed by $[i\ stit]\neg\varphi$. As the distinction between actions and assertions can be seen as an argument in favour of dynamic logic over STIT logics, the fact that the latter allow one to distinguish between '*not doing* φ' and '*doing not* φ' can be seen as an argument in their favour.

In a STIT framework one considers descriptions of states rather than actions in themselves. The modal operator $[i\ stit]\varphi$ means that the agent i brings about (makes true) some description of the world φ. Therefore, the emphasis is not put on the action in itself but is rather put upon the *result* of the action (i.e., its outcome), that is, some specific description of the world. That being said, since STIT logics are modal logics, it follows that their structure is fundamentally classical (Boolean). As with Boolean algebras, the logic of the connective $\wedge$, $\vee$, $\neg$ and $\supset$ satisfies the definition of a **CCC**co**C**.

5.3.6 *Thierry Lucas*

Lucas's work stands among the few approaches that explicitly applied category theory to deontic logic. His logic of action, inspired by the work of von Wright, is meant to express the distinction between the algebraic structure of action sentences and the propositional structure of normative sentences. In the first part of his paper, Lucas [49] reconstructs von Wright's logic of actions using action operators and a modality for necessity. However, for the sake of our analysis, we will only concentrate on the second part of the 2006 paper, where Lucas emphasizes the algebraic structure of an action logic, which was further analyzed in [50, 51].

The originality of Lucas's [49, p. 101] approach is that he considers actions as *morphisms* in a category-theoretic framework. Indeed, he understands actions as mapping from a set of initial conditions to a set of results. In [51, p. 370], he defines an algebra of action as a triple $\langle B, \mathcal{ACT}, C\rangle$ with B and C two Boolean algebras (that may be different from each other) and $\mathcal{ACT} = \langle A, 0, 1, \neg, \wedge, \vee, C_0, \rangle$ a poset that behaves as a bi-intuitionistic logic [49, p. 86] (the notion of cosupport C_0 is introduced to represent the conditions under which the negation of the action is obtained).[17]

Assuming that $\mathcal{F}_1$ and $\mathcal{F}_2$ are respectively the set of well-formed formulas of B and C, Lucas [49, p. 102] views an action as a morphism $a : \Sigma \longrightarrow \mathcal{F}_2$, with $\Sigma \subseteq \mathcal{F}_1$. Hence, an action is considered as a mapping from a set of conditions to a set of results [49, p. 101]. Actions have to satisfy a coherence condition such that $a(\sigma) = a(\sigma')$ for $\sigma, \sigma' \in \Sigma$, meaning that the action a yields the same result for every sub-condition $\sigma \in \Sigma$. In other words, the sub-conditions are consistent with each other, that is, the action regarding two sub-conditions of a set Σ of conditions yields the same result. Actions are pre-ordered by $\leq$ [51, p. 369], 1 is the *empty* action and 0 the *total* (or *null*) action. The empty action is the equivalent of doing nothing (the trivial action) while the null action is impossible to be performed.

On the grounds of $(A, 0, 1, \wedge, \vee)$, Lucas [49, p. 106] defines $(A, 0, 1, \wedge, \vee, \sim, \rightarrow)$ as an intuitionistic logic (hence a **CC**co**C**), with $\rightarrow$ the adjoint of $\wedge$ and $\sim$ the intuitionistic negation defined by $\sim \alpha =_{df} \alpha \rightarrow 0$ (see also [51, p. 383]). However, there is more to it: he also defines an adjoint to disjunction, together with another negation υ. The adjoint to disjunction is governed by the rule (co-cl), and negation is defined by $\upsilon\alpha =_{df} 1\backslash\alpha$.

$$\frac{\gamma \leq \alpha \vee \beta}{\gamma\backslash\beta \leq \alpha} \qquad \text{(co-cl)}$$

This is why $\mathcal{ACT}$ is defined as a bi-intuitionistic logic. A bi-intuitionistic logic is understood as a bi-Heyting algebra, that is, a bounded distributive lattice which is a Heyting and a co-Heyting algebra. From this definition, the fragment $(A, 0, 1, \wedge, \vee, \sim, \rightarrow, \backslash)$ can be compared to a **CC** co**CC**.

5.3.7 *Summing Up*

All of the aforementioned approaches can be compared on the grounds that they are all variations of monoidal deductive systems. Table 5.1 provides a comparison between the systems discussed so far, concentrating only upon the fragments including tensor $\otimes$, co-tensor $\oplus$, adjoint(s) $\multimap$, $\triangleright$, and negation $\sim$ when they are defined (with only the multiplicative fragments for variations of linear logic). The notation has been made uniform according to the analysis provided in Sect. 5.2. We included a Heyting algebra $\mathcal{HA}$ and Lambek's Syntactic Calculus $\mathcal{SC}$ for comparison.[18] Although we will not make explicit each consequence of the comparison made in Table 5.1, let us note that:

1. $\otimes$ is non-commutative for **NILL**, **MNLL**, $\mathcal{A}$ and $\mathcal{KA}$;

[17] Note that 'bi-intuitionistic' here refers to an intuitionistic logic with two conditionals. The relationship between bi-Heyting algebras and modal logic has been discussed by Reyes and Zolfaghari [65].

[18] The abbreviations in Table 5.1 stand for associativity, commutativity, idempotence and distributivity.

Table 5.1 Properties of action connectives

	Ass.	Comm.	Idem.	Dist.	Adjoint(s)	Negation
$\mathcal{SC}$	$\otimes$				$\multimap$ $\triangleright$	
NILL	$\otimes$ $\oplus$				$\multimap$ $\triangleright$	$\alpha \leq \sim\sim \alpha$
MNLL	$\otimes$ $\oplus$				$\multimap$ $\triangleright$	$\alpha \cong \sim\sim \alpha$
$\mathcal{A}$	$\otimes$ $\oplus$	$\oplus$	$\oplus$	$\otimes$ over $\oplus$	$\multimap$ $\triangleright$	
$\mathcal{KA}$	$\otimes$ $\oplus$	$\oplus$	$\oplus$	$\otimes$ over $\oplus$		
DDL	$\oplus$	$\oplus$	$\otimes$ $\oplus$	$\otimes$ over $\oplus$ $\oplus$ over $\otimes$		$\alpha \cong \sim\sim \alpha$
MLL	$\otimes$ $\oplus$	$\otimes$ $\oplus$			$\multimap$	$\alpha \cong \sim\sim \alpha$
$\mathcal{HA}$	$\otimes$ $\oplus$	$\otimes$ $\oplus$	$\otimes$ $\oplus$	$\otimes$ over $\oplus$ $\oplus$ over $\otimes$	$\multimap$	$\alpha \leq \sim\sim \alpha$
$\mathcal{ACT}$	$\otimes$ $\oplus$	$\otimes$ $\oplus$	$\otimes$ $\oplus$	$\otimes$ over $\oplus$ $\oplus$ over $\otimes$	$\multimap$ $\backslash$	$\alpha \leq \sim\sim \alpha$
$\mathcal{BA}$	$\otimes$ $\oplus$	$\otimes$ $\oplus$	$\otimes$ $\oplus$	$\otimes$ over $\oplus$ $\oplus$ over $\otimes$	$\multimap$	$\alpha \cong \sim\sim \alpha$

2. in every case except the linear fragments, $\otimes$ distributes over $\oplus$;
3. negation is involutive in **MNLL**, DDL, **MLL** and $\mathcal{BA}$;
4. $\oplus$ is commutative and idempotent in $\mathcal{A}$, $\mathcal{KA}$, DDL, $\mathcal{HA}$, $\mathcal{ACT}$ and $\mathcal{BA}$;
5. $\otimes$ is idempotent in DDL, $\mathcal{HA}$, $\mathcal{ACT}$ and $\mathcal{BA}$;
6. only **NILL**, $\mathcal{HA}$ and $\mathcal{ACT}$ provide a non-involutive negation.

As we said at the beginning of this paper, our aim is to provide a logic able to model *human* actions. So far, our analysis enabled us to classify the different action logics that can be found within the literature according to the categorical structure of their proof-theory. As such, we were able to correlate the properties of their logical connectives with their categorical structure. In what follows, we will provide a philosophical analysis of action connectives in order to expose which properties should be satisfied. In light of our framework, we will thus be able to see what type of deductive system should be used to adequately model human actions and why the current approaches fail to do so.

5.4 Philosophical Analysis of Action Connectives

5.4.1 Preliminaries

The starting point of our investigation is the distinction between *actions* and *assertions*. It was already pointed out by Goldman [31, p. 47] that it is unclear whether or not propositional connectives could or should be applied to actions. It is our view that *actions* and *propositions* have different structures and, as such, they do not share the same connectives. Following the dynamic logic tradition but also Lucas [49], we assume that the algebraic structure of actions is fundamentally different from the structure of propositions.[19] In what follows, we propose to define an action logic as a logic that can represent the formal structure of actions, while a propositional action logic is a logic that expresses the structure of the language we use to talk about actions. Hence, actions are not understood as declarative sentences in an action logic. What we mean by an *action* will be discussed further below in Sect. 5.4.4.[20]

The first thing that needs to be done is to determine which types of actions we wish to consider for our action logic. Assuming that there are such things as *atomic actions*, we must determine the action connectives that can be used to construct complex actions (cf. [22–24]). It was mentioned at the beginning of this paper that the aim is to provide an analysis of action logics from a philosophical perspective rather than a computational one. Although some action connectives are relevant for programming, we wish to develop an action logic that is relevant to human actions in deontic contexts instead of a logic that is useful for computer science.

Following Pratt [63, p. 99], the three basic control elements of imperative programming languages are choice, sequence, and iteration. The first consequence of our philosophical standpoint is that we will not consider iteration of action insofar as it is not relevant to human actions. Iteration and recursion are fundamental for programming languages and algorithms, but they are not part of *human* actions. Of course, there are actions that require some form of repetition (e.g., a work chain or a serial killer), but these actions can be seen as finite sequences rather than pure recursive actions. In this respect, iteration of human actions can be modeled using sequence, which is taken as a primitive action connective. A second consequence of our approach is that *choice* will not be considered as a connective of our action logic but will rather be considered as a connective of the propositional action logic. From the perspective of human actions, *choice* is not an action in itself. The choice between two actions can be expressed within the language that we use to *talk* about actions. However, there is no such thing as an action *choice*. Despite the fact that one can face a choice between two actions, *choice* is not an action that will be performed: one can choose between two actions, but one is not performing *choice*. The act of *choosing* can be represented by an atomic action.

[19] Segerberg [68] also distinguishes between actions and assertions but uses a Boolean algebra of actions and Boolean propositional connectives.

[20] At this point, one might be tempted to object that 'logic' is concerned with declarative sentences (i.e., with sentences that have the potential to be true or false) and, in this respect, the notion of an action 'logic' where actions are not understood as declarative sentences violates the current use of the term. It should be noted, however, that this view of 'logic' is too restrictive. Indeed, the notion of 'truth' is only a mere tool that can be discarded. For instance, in proof-theory, formulas are not understood as declarative sentences but are taken as having the potential to be 'provable' (rather than 'true').

5.4.2 Action Conjunction and Sequence

The primary action connective we consider is the conjunction of action $\otimes$, where $\alpha \otimes \beta$ means the action α *together with* β. The conjunction of actions expressed by $\otimes$ implies that both actions are done simultaneously. It implies that α is done in conjunction with β (hence the use of $\otimes$ to suggest simultaneity). Action conjunction is introduced similarly to the multiplicative conjunction of linear logic. Since the 'with' operator implies simultaneity and sequence implies a temporal order, it follows that action conjunction and action sequence are two different action connectives. While $\otimes$ is commutative and associative, the sequence connective $\curvearrowright$ is only associative (and not commutative). A sequence action $\alpha \curvearrowright \beta$ means α *and then* β. We assume that the special action $*$ is the unit of both $\otimes$ and $\curvearrowright$. The special action $*$ is interpreted as the *no change* or the *nothing* action. Doing α together with *no change* is the same as doing α, and vice versa. Similarly, doing α and then *no change* is the same as doing α, as it is the case for doing *no change* and then α. Considering that we do not want *choice* to be included within our action logic, we do not define the co-tensor.

Now, the question that needs to be asked is whether $\otimes$ possesses an adjoint and, if so, then what is its philosophical interpretation. Since implication is usually the adjoint to conjunction, it seems natural to first ask if $\supset$ is the adjoint of $\otimes$. While we answer positively the first question ($\otimes$ possesses an adjoint), its adjoint is not implication. A conditional 'if α, then β' can have one of two meanings within natural language when talking about actions. In order to illustrate this, consider the two following conditionals:

(1) If Jones murders Smith gently, then Jones murders Smith.
(2) If Jones murders Smith, then he will try to hide the body.

The first conditional expresses a semantical entailment. However, this kind of entailment does not produce an action per se. Representing (1) by $\alpha \supset \beta$, '$\supset$' is a propositional connective (i.e., a connective that can be applied to declarative statements) rather than an action connective. Hence, (1) expresses a semantical entailment between propositions rather than actions. The conditional 'If Jones murders Smith gently, then Jones murders Smith' is not an action in itself but is rather a semantical entailment between 'murdering Smith gently' and 'murdering Smith'. Rather than pertaining to our action logic, the entailment expressed by (1) will be represented within the propositional action logic.

The second conditional expresses that an action is conditional to another one. That Jones tries to hide the body of Smith requires that he murdered Smith in the first place. A conditional action happens when an action cannot be done unless some specific requirements have been met. In order to checkmate one's opponent, white must first open, then black must do the same, and then one will move another piece, etc. When building a house, building the roof is conditional to building the foundations and the walls. Similarly, do not try to frost a cupcake if it has not been cooked yet. Conditional actions are not, however, actions per se. Whether an action is conditional or not can be expressed when *talking* about actions. We can speak of conditional actions using

$\supset$ but it does not produce a complex action. The fact that an action α is conditional to other actions implies that α can only be accomplished when some requirements have been met within a specific order. Hence, when saying that an action is conditional, we are saying that this action can only be realized through some specific sequences.

The connective $\supset$ is a propositional one and hence does not pertain to an action logic. Thus, it is not the adjoint to action conjunction. Another question that can be asked is whether or not the sequence connective can be defined as an adjoint to conjunctive action. If so, the following rule would be satisfied.

$$\frac{\alpha \otimes \beta \longrightarrow \gamma}{\alpha \longrightarrow \beta \curvearrowright \gamma} \quad \text{(cl)}$$

However, the interpretation of this rule would not be philosophically accurate. Indeed, an arrow between two actions represents logical entailment. Thus, (cl) would imply that if γ is deducible from $\alpha \otimes \beta$, then the sequence $\beta \curvearrowright \gamma$ is deducible from α. But γ is not something which would be done subsequently after β: it is only a logical consequence of the conjunctive action. The distinction between sequence and conjunction is meant to distinguish between simultaneity and succession. Assuming that sequence is the adjoint of conjunction would blur that distinction. It is not because the action α together with β logically implies the action γ that α logically implies that γ will be done after β.

5.4.3 *The Without Connective*

There is, however, an action connective than can be seen as the adjoint of $\otimes$. Indeed, it is our view that the appropriate adjoint for the *together with* connective is the *without* connective $\ominus$. To the best of our knowledge, the action connective 'without' has never been considered within the action logic literature. The behaviour of $\ominus$ is represented by (cl).

$$\frac{\alpha \otimes \beta \longrightarrow \gamma}{\beta \longrightarrow \gamma \ominus \alpha} \quad \text{(cl)}$$

We wrote $\gamma \ominus \alpha$ instead of $\alpha \ominus \gamma$ to keep the intuitive reading of γ without α, hence the notation reminding us of the subtraction γ minus α. In other words, if γ is a consequence of α together with β, then γ without α is a consequence of β (and γ without β is a consequence of α since we assumed that conjunction is commutative). That is to say, if adding β to α gives us the action γ, then subtracting β from γ gives us α back. The following derivation is a particular instance of (cl):

$$\frac{\alpha \otimes \beta \longrightarrow \alpha \otimes \beta}{\beta \longrightarrow (\alpha \otimes \beta) \ominus \alpha} \quad \text{(cl)}$$

For example, from the action of *walking together with singing* one can conclude that the action of *walking* yields the action of *walking and singing but without singing*.

Having introduced the conjunction and its adjoint, we can now see that we do not want $\otimes$ to be idempotent. First, if $\otimes$ is idempotent, then we obtain both $\alpha \longrightarrow \alpha \otimes \alpha$ and $\alpha \otimes \alpha \longrightarrow \alpha$. Although the latter might seem desirable at first glance (i.e., if one does α together with α it logically follows that one does α), both should be rejected. One of the main concern of Girard when developing linear logic was to be able to deal with limited resources. That said, $\alpha \longrightarrow \alpha \otimes \alpha$ violates this principle. Indeed, doing an action once is not the same thing as doing an action twice. For instance, giving one dollar is not logically equivalent to giving one dollar *and* giving one dollar.

But moreover, since $\ominus$ is the adjoint of $\otimes$, idempotence would give us that from $\alpha \otimes \alpha \longrightarrow \alpha$ (which seemed a desirable consequence) one can infer $\alpha \longrightarrow \alpha \ominus \alpha$, which is clearly undesirable from a philosophical perspective. Indeed, it would mean that doing the action α logically implies doing α without doing α. As a result, an action logic that aims to model human actions should not satisfy idempotence of conjunction and, therefore, it should not satisfy the definition of a **C**.[21]

5.4.4 Action Negation

The last action connective we wish to investigate is also the most intriguing from a philosophical perspective. When applied to deontic logic, action logics usually include an action connective for negative actions. But what does it mean, if it means anything, to say that one is *doing* a *negative action*? It might make sense to use negation when talking about actions,[22] but as Walton [77, p. 321] pointed out one must be careful when trying to interpret action negation ontologically. Although action negation has a meaning when we are *talking* about actions, it is unclear whether or not it makes any sense to speak of an *action negation* within an action logic per se.

Negation of actions (or programs) are often introduced within the logic literature without further ado. Wiseman [78], for example, introduced a connective for action negation but did not explain its meaning. He sees a problem with action logics but only because they presuppose the existence of *acts*, which he finds dubious. In the computer science literature, Meyer [55] introduced action negation without explaining its meaning. Further, negation of programs were introduced by Kozen [41] to model tests, but the significance of a *program negation* was left unexplained. Similarly, when thinking about programs, Solin [71] introduces the negation of a program as a mere useful fictional technical tool, without saying more about what it means for a program to be *negated*.

All of these interrogations regarding action negation rely upon a much more fundamental question: what is an *action*? What is it to *act*? Von Wright [76, p. 38] answers the latter by saying that acting is to intentionally bring about or prevent a change in the world. This understanding, however, is not accepted unanimously. Davidson [25, p. 50], for instance, rather says that acting is to bring about a state of affair but is "neutral with respect to the question whether the action was intentional"

[21]By the same reasoning, if we want to be able to deal with limited resources, sequence must not be idempotent.

[22]Walton [77, p. 320], for example, distinguished between fourteen different locutions of action negation within the natural language.

or not. The main characteristic of an action is that it (causally) brings change in the world. This change, however, need not be voluntary: to act is not to be responsible for acting. Goldman [31, p. 46] would agree with this understanding. These conceptions are also in accordance with Pörn [64, p. 5], for whom acting is to bring about a state of affair, although this state of affair can either be the end of the action, its result or one of its consequences.

Philosophers agree that there are such things as *actions*. These *actions* are characterized by the fact that they have causal repercussions in the world. As such, actions are often associated with the description of the state they produce. This is not to say that an action is a mere change in the world [75, p. 36], or that it can be reduced to a description. Rather, it is to say that there *are* actions, and one way to see them is to look at the changes they produce. In the STIT literature, actions are associated with the description of the state they bring about. It is presupposed that there are such things as *actions*, and these actions can be performed to obtain results in the world. For Lucas, actions are considered as mappings from the description of a state to another description. An action is thus something which can influence the evolution of the world. Even though dynamic logic tends to distinguish actions from assertions, its ontology of actions is understood similarly. When looking at the semantics of [68] or [55], one sees that actions are considered as sequences of events (or points in time). Actions are seen, in these cases, as sequences of descriptions.

One can thus safely assume that an action has causal power and that it brings change in the world. This change may be done actively or passively and it might also result from *not doing*. As Goldman [31, p. 10] distinguishes between *action-types* and *action-tokens*, von Wright [75, p. 36] speaks of *act-categories* and *act-individuals*. Act-categories are classes of particular actions. For Goldman [31, p. 10], an act-token is an instance of an act-type. Since the negation of an act-type is also an act-type [31, p. 18], it follows that an instance of the negation of an act-type is an act-token. Hence, Goldman [31, pp. 47-8] admits that there are such things as *negative acts* but he leaves the matter open since it brings difficult problems. Von Wright, on the other hand, always sees an action as *positive*. He distinguishes between *productive* and *preventive* actions [76, p. 38]. One can either produce by bringing about that the state stays the same or by bringing about a change and one can prevent by leaving the state unchanged or by letting a change happen.

This brings us to the notions of *omission*, *refraining* and *forbearance*. Although this might seem paradoxical, the doing of a negative action is usually understood as not doing. One *performs* a negative action by not performing said action. This non-performance can be intentional or not. If it is, we speak of a forbearance and, if not, we speak of an omission. An omission is thus understood as an intentional or unintentional not-doing (i.e., consciously not-doing or unconsciously not-doing). Mossel [56] presented a nice review of the literature on negative actions. He argued that there are such things as *negative acts* and they are characterized by the fact that they have causal power. Moreover, he argues that we are *responsible* for our negative actions [56, p. 330]. While an *omission* is not necessarily voluntary, *forbearance* (*refraining*) is. Hence, a negative action in Mossel's view is a form of forbearance, which is intentional. According to his understanding, omissions are not negative acts.

For von Wright [75, p. 36], *acting* is to bring about 'at will' a change in the world. On Davidson's [25, p. 50] and Goldman's [31, p. 46] accounts, *acting* should be understood independently from *intention* and *responsibility*. It is peculiar, though, that while Goldman asserts that the notion of action should be analyzed independently from *intentionality*, he also says [31, p. 48] that *negative actions* are characterized by the fact that they are *intentional* omissions (i.e., forbearances). Similarly, although von Wright asserts that to *act* is to bring about a state of affair intentionally, he also says that forbearing to do something does not presuppose awareness [75, pp. 45-6]. In both cases, the authors seem to assume some form of duality between *acting* and *not-acting*. One considers that *acting* presupposes intentionality but *not-acting* does not, while the other assumes that *acting* does not presuppose intentionality but *not-acting* does.

It is noteworthy that Davidson's and Goldman's accounts are consistent with the legal understanding of *actions*. In (Canadian) criminal law, liability is determined on two grounds: the *mens rea* and the *actus reus*. The *actus reus* corresponds to the action *per se* while the *mens rea* refers to the agent's mind, to his intention before and while acting. A convict can only be held criminally responsible if he *did* indeed commit the crime and it was done *voluntarily*.[23] Hence, from a legal point of view, *action* and *voluntariness* are two notions that should be treated separately.

But if *acting* does not presuppose any form of intentionality and negative actions are a form of acting, how can one be justified in assuming that negative actions are intentional *per se*? These dualities are quite puzzling. An action does not presuppose intentionality, yet not-acting, which is also a form of action, does? Are these duality inherent to the structure of actions?

In our view, no. Even though there are cases within natural language where these dualities are at play, there are also cases where they are not. In criminal law, for instance, if an agent *did* indeed commit a crime, then he did it (*actus reus*) intentionally (*mens rea*). The *actus reus* is an action that does not presuppose intentionality but the action of committing the crime does. If the agent *did not* commit the crime in question, then either he did it unintentionally or he did not do it (intentionally or not). Thus, when one commits a crime it is intentional but when one does not it may be intentional or not. As such, assuming that acting is intentional does not necessarily entail that not-acting is not. Similarly, if an agent respects the law, it needs not be intentional. Consider 'it is forbidden to steal'. One respects this norm either by omitting or forbearing to steal (i.e., one respects the norm by *not stealing*). But if an agent breaks the law (does not respect the law), it can be with or without intention: it is not because an agent broke the law (*actus reus*) that he will be held responsible for it. In a nutshell, one can respect the law intentionally or not, as one can *not* respect

[23]To be precise, the *mens rea* refers to the agent's state of mind. Acting *volontarily* does not presuppose an explicit intention but only means that the agent is in a state of mind such that he has the capacity to understand what he is doing, even though he cannot anticipate every possile outcome of his actions.

the law intentionally or not. Hence, it is not a necessary assumption that *acting* (or *not-acting*) presupposes intentionality, while *not-acting* (or *acting*) does not.[24]

It would be a mistake to presuppose that negative actions are intentional *per se*. A negative action can either be a forbearance or an omission. Otherwise, if omission is neither an act nor a negative act, how could we say that someone committed a crime by omission? There are situations where one can break the law by omission and, thus, *omission* must be considered as some sort of negative action. We think that Goldman's and Davidson's conceptions of action without intentionality should be preferred to von Wright's insofar as it is more faithful to our understanding of action in criminal law, where one can act without being responsible for one's actions. This, however, also implies that negative actions are not necessarily intentional.

Actions are thus understood minimally as things that have causal powers. An action is characterized by the fact that it affects the world. This is the minimal understanding of *action* we adopt. To *act* is to affect the world, either by changing it or by not changing it, intentionally or not. To act is to play a role within the evolution of the world. It might seem paradoxical to say that one can play a role by not playing a role, but it actually is not. By not acting, either intentionally or not, one influences (directly or not) how the world changes. This understanding allows us to speak of both positive and negative actions, the latter including omissions and forbearances.

As we mentioned earlier, it is sometimes argued that STIT logics should be preferred to dynamic logics insofar as they enable the distinction between doing the negation of an action ($[i\ stit]\neg\varphi$) and not doing an action ($\neg[i\ stit]\varphi$), which are not (and should not) be equivalent. In the STIT frameworks, we usually have $[i\ stit]\neg\varphi \supset \neg[i\ stit]\varphi$ (unless the STIT modality is not axiomatized by KD or KT), but the converse does not hold. Omission does not imply forbearance but forbearance does imply omission. If it is true that the agent saw to it that not-α was done, then it is true that he did not see to it that α was. Note that the aforementioned duality is at play in this conditional. While the $[i\ stit]$ modality presupposes intentionality, its negation does not. It is noteworthy that this duality appears when we are *talking* about actions (i.e., when we use declarative statements). As such, this duality must be taken care of within a propositional action logic.

Lenk [47] was, to the best of our knowledge, the first to explore the connections between different lattice structures and action negation. Lenk distinguishes between three types of negations: non-execution, omission and forbearance. He argued that Boolean negation roughly represents non-execution [47, p. 258] and that the negation within a Heyting algebra is better suited to represent forbearance [47, p. 260]. He thus concludes that a logic of actions should include at least two negations, a classical and an intuitionistic one, to represent respectively non-execution (not-doing) and

[24]Although criminal liability usually requires both an illegal act and a criminal mind, it should be noted that there are exceptions within Canadian law. First, some cases involve the notion of *absolute* liability, which depends on the act alone (see *R. v. Canning, 2004 SKPC 13* for the definition). Second, there are specific cases where murders are straightaway considered as first degree murders (e.g., the killing of a police officer or a warden acting in the course of their duties), notwithstanding the intention of the person perpetrating the act (see for instance the *Criminal Code, R.S.C., 1985, c. C-46, art. 231.4-231.6.*).

forbearance. The negation for non-execution is classical insofar as it respects the law of excluded middle: either an action is performed or it is not. However, it is not because an action is not performed that its negation is. Hence, negation representing forbearance is intuitionistic since to refrain from forbearing to do α is different from doing α. Doing α implies not refraining from doing it, but it is not because one is not refraining from doing α that one is doing it (since one might simply not do α).

Lenk [47, p. 262] then suggests that if one wants to represent *omission*, then one should look at a structure which is neither a Boolean nor a Heyting algebra. The distinction between an *omission* and *non-execution* is meant to distinguish between actions that are in one's power to *not do* from actions that are not. For example, Paul cannot *omit* or *refrain* from flying into the Sun, but Superman can. What is of interest to our approach is that Lenk [47, p. 263] actually suggests that an action logic could be inspired by the logic of quantum mechanics. Indeed, he hinted at the parallel that can be made between quantum mechanics and action logics, where action operators are not distributive but are linearly (or weakly) distributive and where actions have some form of complement.

Neither the logical tools nor the terminology were available for him at the time, but what Lenk hinted at is that one should look at a logic that has the structure of a compact closed deductive system to model the structure of actions. This is the basic idea that guides our conception of an action logic.

5.5 A Logic for Human Actions

Although we came by this idea independently, Lenk's suggestion that an action logic could perhaps be constructed in the light of quantum mechanics and the properties of particles is at the core of our approach. As Lenk suggested, we will use a structure that is neither a Boolean nor a Heyting algebra to model negative actions that can be omissions. In what follows, we understand an omission as an action which is not done, whether it is intentional or not. Hence, an omission encompasses forbearance and is seen as the complement of an action. Moreover, following Lenk's suggestion, we will use an intuitionistic negation to model forbearance in a propositional action logic.

It is our view that an action possesses a *dual* or a *complement*, which can be referred to as a *negative action*. Since an action does not presuppose any intentionality, its dual also does not. Hence, a negative action is understood as either an omission or a forbearance. Let α^* be the *dual* (or the *complement*) of an action α. It can be understood as its *negation*, although *dual* might be a proper unambiguous term. Dual actions are defined by $\alpha^* =_{df} * \ominus \alpha$. The dual of α is understood as *no change without* α. Consider the 'no change' situation, subtract α from it and one gets its dual. Four consequences of this definition are:

$$\alpha \otimes \alpha^* \longrightarrow * \tag{5.1}$$

$$\alpha \longrightarrow \alpha^{**} \tag{5.2}$$

$$\beta \ominus \alpha \cong \alpha^* \ominus \beta^* \tag{5.3}$$

$$\alpha \otimes \beta \cong (\beta^* \ominus \alpha)^* \tag{5.4}$$

Since neither α nor α^* presuppose intentionality, α can be seen as an action which is *present* (intentionally or not) while α^* refers to the absence of α (intentionally or not). The first consequence means that the action α together with its dual leads to *no change*. That is, they cancel each other. This is consistent with Lenk's suggestion. The second consequence means that if an action is present, then its absence is absent. The third consequence implies that β without α is isomorphic to α^* without β^*. Equation 5.4 follows from the fact that $\ominus$ is the adjoint of $\otimes$ (see [9, p. 161]).

Let us pursue the analogy between quantum mechanics (Hilbert spaces) and an action logic by considering the following property. An action and its dual in an action logic can be compared to a particle and its corresponding antiparticle in physics.

$$\alpha^* \otimes \beta \cong \beta \ominus \alpha \tag{5.5}$$

This isomorphism means that β without α is logically equivalent to β together with α^*. This principle is actually quite plausible. It means, for example, that the action *driving without drinking a bottle of scotch* is isomorphic to *driving together with not drinking a bottle of scotch*.

There is thus a strong connection between the structure of an action logic and quantum mechanics considering that, as a Hilbert space, an action logic can be seen as an instance of a compact closed category in virtue of Eq. 5.5. Here are some consequences of adopting Eq. 5.5.

$$* \longrightarrow \alpha \otimes \alpha^* \tag{5.6}$$

$$\alpha \cong \alpha^{**} \tag{5.7}$$

$$(\alpha \otimes \beta)^* \cong \alpha^* \otimes \beta^* \tag{5.8}$$

$$\alpha \otimes \beta \cong (\beta^* \otimes \alpha^*)^* \tag{5.9}$$

$$* \cong *^* \tag{5.10}$$

Equations 5.6 and 5.7 follow from the fact that the unit dualizes, Eq. 5.8 is a consequence of Eq. 5.5 (see [40, p.194]) and Eq. 5.9 follows from 5.7, 5.8 and the fact that if $\alpha \cong \beta$, then $\alpha^* \cong \beta^*$.[25]

To sum up, our proposal is to define an action logic where:

1. action conjunction $\otimes$ is commutative, associative, and has a unit $*$;
2. $\otimes$ possesses an adjoint $\ominus$;

---

[25] This is quite straightforward assuming that $\alpha \cong \beta$ since $\alpha \cong \beta \cong \beta^{**} \cong * \ominus \beta^*$, and hence $\alpha \otimes \beta^* \cong *$ by the fact that $\ominus$ is the adjoint of $\otimes$, and by the same reasoning one obtains $\beta^* \cong * \ominus \alpha \cong \alpha^*$.

3. action negation is defined by $\alpha^* =_{df} * \ominus \alpha$;
4. action negation is involutive (from Eq. 5.7);
5. $\alpha^* \otimes \beta \cong \beta \ominus \alpha$;
6. sequence is associative and has a unit $*$.

The special action $*$ is the *no change* or *nothing* action but it is also an action that cannot be accomplished. Put differently, $*$ is both trivial (no change) and impossible (it cannot be performed). This might seem paradoxical at first glance but it actually is not. Indeed, $*$ is logically equivalent to $\alpha \otimes \alpha^*$. Hence, $*$ is an action that cannot be accomplished: it would be impossible to do both an action and its complement at the same time. The action $*$ can thus informally be understood as something that does not have any causal powers. Strictly speaking, we assumed that an action is something that has causal powers and, in this respect, $*$ is not an action per se. Nonetheless, we can see why $*$ is both trivial and impossible: it is trivial since it won't have any impact on the world (i.e., it will make no change) and it is impossible to perform given that it has no causal powers (while both acting and not acting do). Put differently, since either acting or not acting will have causal repercussions in the world, it follows that one cannot accomplish an action that will not have causal repercussions in the world. Consequently, $*$ is trivial but impossible to perform, (cf. [58, pp. 423-4]).

As a result of our analysis, we propose to define an action logic as a compact deductive system with sequence. More specifically, assume a collection of atomic actions a_i together with a collection of equivalence classes of deductions (proofs). Let Act stands for the collection of all atomic actions. The language of an action logic $\mathcal{AL}$ is $\mathcal{L}_{\mathcal{AL}} = \{Act, \otimes, \ominus, \curvearrowright, *, (,)\}$. Well-formed formulas ($WFF_{\mathcal{AL}}$) are defined recursively as follows:

$$\alpha := a_i \mid * \mid \alpha \otimes \beta \mid \alpha \ominus \beta \mid \alpha \curvearrowright \beta$$

Definition 1 *Sequence* is axiomatized by a **M** from $\mathcal{L}_{\curvearrowright} = \{Act, \curvearrowright, *, (,)\}$ and $WFF_{\mathcal{AL}}$.

Definition 2 An *action logic* $\mathcal{AL}$ is defined on the grounds of $\mathcal{L}_{\mathcal{AL}}$ and $WFF_{\mathcal{AL}}$ by:

1. Sequence;
2. The fragment $\{Act, \otimes, \ominus, *, (,)\}$ is axiomatized by a **CDS** with
   a. $*$ the unit of $\otimes$ (which also dualizes);
   b. $\alpha^* =_{df} * \ominus \alpha$.

Hence, an action logic is understood as a compact deductive system with sequence. Having an action logic at our disposal, we can now define a logic for action propositions. The idea is to develop a two-sorted language that includes both descriptive propositions and action propositions. To avoid confusion, let us define the set AP of action propositions as follows: if $\alpha \in WFF_{AL}$, then $\boldsymbol{\alpha} \in$ AP.

The bold notation $\boldsymbol{\alpha}$ is used when actions are understood as declarative statements. Hence, $\boldsymbol{\alpha}$ is the proposition one uses to refer to the action α. A propositional action logic is understood as a two sorted deductive system where the collection of objects is composed of members of AP and of a collection $Prop$ of descriptive atoms p_i (and where arrows are equivalence classes of proofs). Let $\mathcal{L}_{\mathcal{PAL}} = \{\mathsf{AP}, Prop, \wedge, \vee, \supset, \top, \bot, (,)\}$ be the language of a propositional action logic. Well-formed formulas ($WFF_{\mathcal{PAL}}$) are defined recursively as follows:

$$\varphi := \boldsymbol{\alpha} \mid p_i \mid \top \mid \bot \mid \varphi \wedge \psi \mid \varphi \vee \psi \mid \varphi \supset \psi$$

Definition 3 A *propositional action logic* $\mathcal{PAL}$ is a **CC**co**C** built from $\mathcal{L}_{\mathcal{PAL}}$ and $WFF_{\mathcal{PAL}}$. Negation is defined by $\neg\varphi =_{df} \varphi \supset \bot$. It has to satisfy the rule (Ψ).

$$\frac{\alpha \longrightarrow_{\mathsf{AL}} \beta}{\boldsymbol{\alpha} \longrightarrow_{\mathsf{PAL}} \boldsymbol{\beta}}\ (\Psi) \qquad (\text{with } \Psi(*) = \bot)$$

Thus defined, action propositions have an intuitionistic behavior. We use indexed arrows to avoid confusion between the different deductive systems. It goes without justification that propositional conjunction, disjunction and implication can be axiomatized within a **CC**co**C**. The only thing we need to show is that negation, when applied to action propositions, is not involutive. The underlying issue is whether or not we want $\neg\neg\boldsymbol{\alpha} \longrightarrow \boldsymbol{\alpha}$.

Consider the action α 'Paul drives his car' and the action proposition $\boldsymbol{\alpha}$. Saying that $\boldsymbol{\alpha}$ is false can have one of two meanings: either (a) Paul forbears to drive his car or (b) Paul omits to drive his car. Each time one adds a propositional negation before an action proposition, one will obtain these two possibilities. As Lenk suggested, saying that it is false that an agent forbears does not necessarily imply that the action was done since it might be false in virtue of a simple omission. Hence, the negation of the negation of $\boldsymbol{\alpha}$ will not necessarily give us $\boldsymbol{\alpha}$ back.

Furthermore, from a normative point of view, an action proposition can be considered as *true* only if we have a proof of its truthfulness. For instance, an agent will be liable for a crime only if we can prove that the agent did indeed commit the crime, in which case we will have sufficient reason to believe in the truthfulness of the action proposition. The negation of an action proposition thus has an intuitionistic behavior: not having a proof that an action was not done is not a proof that it was done.

Although we won't be going into the details of the definition, it should be noted that the rule (Ψ) is defined as a fibration from the atomic fragment of $\mathcal{AL}$ to the atomic fragment of $\mathcal{PAL}$, (cf. [58, pp. 428–435]). It enables us to import the structure of $\mathcal{AL}$ into $\mathcal{PAL}$. As such, if there is some relation between two actions α and β (i.e., if there is an arrow $\alpha \longrightarrow \beta$), then there will also be an arrow between the atomic sentences that are used to refer to these actions within $\mathcal{PAL}$. For instance, if we assume that there is a relation between the action 'stealing a car' and the more general action 'stealing', as for example if we assume that 'if Tom steals a car, then Tom steals', then we obtain in $\mathcal{PAL}$ that if it is true that Tom steals a car, then it is also true that Tom steals.

An interesting consequence of this construction is that we obtain the following result.

$$\Psi(\alpha \longrightarrow_{\mathsf{AL}} *) = \boldsymbol{\alpha} \longrightarrow_{\mathsf{PAL}} \bot$$

Hence, whenever there is a deduction arrow from α to $*$ in $\mathcal{AL}$, it follows that $\boldsymbol{\alpha}$ is necessarily false within $\mathcal{PAL}$. For instance, since 'Signing while not signing' is isomorphic to $*$, we obtain that it is necessarily false that 'Paul sings while not signing'.

Another consequence of this definition is that it allows us to import the logical equivalences of $\mathcal{AL}$ into $\mathcal{PAL}$. If there are $\alpha \longrightarrow \beta$ and $\beta \longrightarrow \alpha$ in $\mathcal{AL}$, then we obtain $\boldsymbol{\alpha} \cong \boldsymbol{\beta}$ in $\mathcal{PAL}$. As a result, logically equivalent actions will be true in the same conditions in the language we use to talk about actions.

5.6 Closing Remarks

Summing up, we proposed analyzing the proof-theory of action logics from the perspective of category theory. Using a framework inspired by Lawvere's and Lambek's work, we showed how different logical systems can be compared on the grounds that they share the same categorical structure. One upshot of this mode of analysis is that the properties of the logical connectives can be correlated with the categorical structure of the logic. In this respect, we provided a philosophical analysis of action connectives and showed which type of deductive systems are adequate to model human actions. One important property that human action conjunction should not satisfy is idempotence. Accordingly, we showed that human action should not be modeled using an instance of a Cartesian deductive system.

We proposed to distinguish between an action logic and a propositional action logic. The former represents the logical structure of actions while the latter expresses the propositional structure of the sentences we use to talk about actions. We argued that a logic able to adequately model human action should be defined as a compact deductive system with sequence while a propositional action logic should be defined as a **CC**co**C** (i.e., as a deductive system logically equivalent to intuitionistic logic). This difference in structures is consistent with our assumption that the structure of actions is fundamentally different from the structure of the language we use to talk about actions.

The closer system to our approach would be the tensor's fragment of **MLL** but without linear implication and with a radically different philosophical interpretation. Our approach contributes to the literature insofar as the action logic $\mathcal{AL}$ provides a new system of logic that has, to the best of our knowledge, never been explored by philosophers.[26]

One interest of this approach is that it provides an action logic that can be used to define a deontic deductive system. For such work, see [58]. For future research, we intend to explore how this framework can be incorporated with dynamic logic.

[26] It should be noted that logics that are analogues to compact deductive systems are used in quantum physics and quantum logic. See for instance Duncan's [26] Ph.D. thesis and further work.

Acknowledgments I would like to thank Jean-Pierre Marquis for many discussions on this subject. I am also grateful to Andrew Irvine, François Lepage and Yvon Gauthier for helpful comments and discussions. Thanks to an anonymous referee for comments on a previous draft of this paper. This research was financially supported by the Social Sciences and Humanities Research Council of Canada.

References

1. Abrusci, V. M. (1990a). A comparison between Lambek syntactic calculus and intuitionistic linear propositional logic. *Mathematical Logic Quarterly*, *36*(1), 11–15.
2. Abrusci, V. M. (1990b). Non-commutative intuitionistic linear logic. *Mathematical Logic Quarterly*, *36*(4), 297–318.
3. Abrusci, V. M. (1991). Phase semantics and sequent calculus for pure non-commutative classical linear propositional logic. *The Journal of Symbolic Logic*, *56*(4), 1403–1451.
4. Abrusci, V. M., & Ruet, P. (2000). Non-commutative logic I: The multiplicative fragment. *Annals of Pure and Applied Logic*, *101*, 29–64.
5. Awodey, S. (2006). *Category theory* (2nd ed.). Oxford University Press.
6. Baez, J. C., & Stay, M. (2011). Physics, topology, logic and computation: A Rosetta stone. In B. Coecke (Ed.), *New structures for physics, Lecture Notes in Physics* (Vol. 813, pp. 95–174). New York: Springer.
7. Balbiani, P. (2008). Propositional dynamic logic. In E. N. Zalta (Ed.), *The stanford encyclopedia of philosophy*.
8. Barr, M. (1979). *∗-autonomous categories, Lecture Notes in Mathematics* (Vol. 752). Springer.
9. Barr, M. (1991). ∗-autonomous categories and linear logic. *Mathematical Structures in Computer Science*, *1*(2), 159–178.
10. Barr, M., & Wells, C. (1990). *Category theory for computing science* (Vol. 10). Prentice Hall.
11. Belnap, N., & Perloff, M. (1988). Seeing to it that: A canonical form for agentives. *Theoria*, *54*(3), 175–199.
12. Bimbó, K., & Dunn, J. M. (2005). Relational semantics for Kleene logic and action logic. *Notre Dame Journal of Formal Logic*, *46*(4), 461–490.
13. Blute, R., & Scott, P. (2004). Category theory for linear logicians. In T. Ehrhard, J. -Y. Girard, P. Ruet & P. Scott (Eds.), *Linear Logic in Computer Science* (Vol. 316, pp. 3–64). Cambridge University Press.
14. Broersen, J. (2004). Action negation and alternative reductions for dynamic deontic logics. *Journal of Applied Logic*, *2*(1), 153–168.
15. Broersen, J. (2011). Deontic epistemic stit logic distinguishing modes of mens rea. *Journal of Applied Logic*, *9*(2), 137–152.
16. Casadio, C. (2001). Non-commutative linear logic in linguistics. *Grammars*, *4*(3), 167–185.
17. Casadio, C., Scott, P. J., & Seely, R. A. G. (2004). Introduction: The Lambek program. In C. Casadio, P. J. Scott & R. A. G. Seely (Eds.), *Language and grammar, studies in mathematical linguistics and natural language, pages xi-xxxi*. Center for the Study of Language and Information.
18. Castro, P. F., & Maibaum, T. S. E. (2009). Deontic action logic, atomic Boolean algebras and fault-tolerance. *Journal of Applied Logic*, *7*(4), 441–466.
19. Chellas, B. F. (1969). *The logical form of imperatives*. Perry Lane Press.
20. Chellas, B. F. (1980). *Modal logic: An introduction*. Cambridge University Press.
21. Cockett, R., & Seely, R. A. G. (1997). Weakly distributive categories. *Journal of Pure and Applied Algebra*, *114*(2), 133–173.
22. Czelakowski, J. (1996). Elements of formal action theory. In A. Fuhrmann & H. Rott (Eds.), *Logic, action, and information* (pp. 3–62). de Gruyter.

23. Czelakowski, J. (1997). Action and deontology. In *Logic, action and cognition*, *Trends in Logic* (Vol. 2, pp. 47–87). Springer.
24. Czelakowski, J. (2015). Action and deontology. In *Freedom and enforcement in action*, *Trends in Logic* (Vol. 42, pp. 143–193). Springer.
25. Davidson, D. (1967). The logical form of action sentences. In *The essential davidson (2006)*. Oxford University Press.
26. Duncan, R. (2006). *Types for quantum computing*. Ph. D. thesis, Oxford University.
27. Galatos, N., Jipsen, P., Kowalski, T., & Ono, H., (Eds.). (2007). *Residuated lattices: An algebraic glimpse at substructural logics*, *Studies in Logic and the Foundations of Mathematics* (Vol. 151). Elsevier.
28. Girard, J.-Y. (1987). Linear logic. *Theoretical Computer Science*, *50*(1), 1–102.
29. Girard, J. -Y. (1995). Linear logic: Its syntax and its semantic. In J. -Y. Girard, Y. Lafont & L. Regnier (Eds.), *Advances in linear logic* (Vol. 222, pp. 1–42). Cambridge University Press.
30. Goldblatt, R. (2006). *Topoi: The categorical analysis of logic*. Dover Publications.
31. Goldman, A. I. (1970). *A theory of human action*. Princeton University Press.
32. Harel, D., Kozen, D., & Tiuryn, J. (2000). *Dynamic logic*. MIT Press.
33. Horty, J. (2001). *Agency and deontic logic*. Oxford University Press.
34. Horty, J., & Belnap, N. (1995). The deliberative stit: A study of action, omission, ability and obligation. *Journal of Philosophical Logic*, *24*(6), 583–644.
35. Jacobs, B. (1999). *Categorical logic and type theory*, *Studies in Logic and the Foundations of Mathematics* (Vol. 141). Elsevier.
36. Kanger, S. (1957). *New foundations for ethical theory*. Stockholm.
37. Kanger, S. (1971). New foundations for ethical theory. In R. Hilpinen (Ed.), *Deontic logic: Introductory and systematic readings* (pp. 36–58). D. Reidel Publishing Company
38. Kanger, S. (1972). Law and logic. *Theoria*, *38*(3), 105–132.
39. Kanger, S., & Kanger, H. (1966). Rights and parliamentarism. *Theoria*, *32*(2), 85–115.
40. Kelly, G. M., & Laplaza, M. L. (1980). Coherence for compact closed categories. *Journal of Pure and Applied Algebra*, *19*, 193–213.
41. Kozen, D. (1997). Kleene algebra with tests. *ACM Transactions on Programming Languages and Systems*, *19*(3), 427–443.
42. Lambek, J. (1958). The mathematics of sentence structure. *The American Mathematical Monthly*, *65*(3), 154–170.
43. Lambek, J. (1968). Deductive systems and categories I. *Mathematical Systems Theory*, *2*(4), 287–318.
44. Lambek, J. (1969). Deductive systems and categories II. standard constructions and closed categories. In P. J. Hilton (Ed.), *Category theory, homology theory and their applications I*, *Lecture Notes in Mathematics* (Vol. 86, pp. 76–122). Springer.
45. Lambek, J., & cott, P. (1986). *Introduction to higher order categorical logic*. Cambridge University Press.
46. Lawvere, F. W. (1963). *Functorial semantics of algebraic theories and some algebraic problems in the context of functorial semantics of algebraic theories*. Ph.D. thesis, Columbia University.
47. Lenk, H. (1977). Complements and different lattice structures in a logic of action. *Erkenntnis*, *11*(2), 251–268.
48. Lindahl, L., & Odelstad, J. (2011). Stratification of normative systems with intermediaries. *Journal of Applied Logic*, *9*(2), 113–136.
49. Lucas, T. (2006). Von Wright's action revisited: Actions as morphisms. *Logique et Analyse*, *49*(193), 85–115.
50. Lucas, T. (2007). Axioms for action. *Logique et Analyse*, *50*(200), 103–123.
51. Lucas, T. (2008). Deontic algebras of actions. *Logique et Analyse*, *51*(202), 367–389.
52. Mac Lane, S. (1971). *Categories for the working mathematician* (2nd ed.). Springer.
53. Marquis, J. -P. (2009). *From a geometrical point of view: A study of the history and philosophy of category theory*. Springer.
54. Meyer, J.-J. C. (1987). A simple solution to the "deepest" paradox in deontic logic. *Logique et Analyse*, *30*(117–118), 81–90.

55. Meyer, J.-J. C. (1988). A different approach to deontic logic: Deontic logic viewed as a variant of dynamic logic. *Notre Dame Journal of Formal Logic*, *29*(1), 109–136.
56. Mossel, B. (2009). Negative actions. *Philosophia*, *37*(2), 307–333.
57. Pacheco, O., & Carmo, J. (2003). A role based model for the normative specification of organized collective agency and agents interaction. *Autonomous Agents and Multi-Agent Systems*, *6*(2), 145–184.
58. Peterson, C. (2014). The categorical imperative: Category theory as a foundation for deontic logic. *Journal of Applied Logic*, *12*(4), 417–461.
59. Peterson, C. (2015). Contrary-to-duty reasoning: A categorical approach. *Logica Universalis*, *9*(1), 47–92.
60. Peterson, C. (2016) A comparison between monoidal and substructural logics. *Journal of Applied Non-Classical Logics, 26*(2), 126–159.
61. Pratt, V. (1976). Semantical considerations of Floyd-Hoare Logic. Technical Report MIT/LCS/TR-168.
62. Pratt, V. (1980). Application of modal logic to programming. *Studia Logica*, *39*(2–3), 257–274.
63. Pratt, V. (1991). Action logic and pure induction. In J. Eijck (Ed.), *Logics in AI*, *Lecture Notes in Computer Science* (Vol. 478, pp. 97–120). Springer.
64. Pörn, I. (1970). *The logic of power*. Basil Blackwell.
65. Reyes, G. E., & Zolfaghari, H. (1996). Bi-Heyting algebras, toposes and modalities. *Journal of Philosophical Logic*, *25*(1), 25–43.
66. Royakkers, L. (1998). *Extending deontic logic for the formalisation of legal rules*. Kluwer Academic Publishers.
67. Seely, R. A. G. (1989). Linear logic, $*$-autonomous categories and cofree coalgebras. In J. W. Gray & A. Scedrov (Eds.), *In categories in computer science and logic*, *Contemporary Mathematics* (Vol. 92, pp. 371–382).
68. Segerberg, K. (1982). A deontic logic of action. *Studia Logica*, *41*(2), 269–282.
69. Segerberg, K. (1992). Getting started: Beginnings in the logic of action. *Studia Logica*, *51*(3), 347–378.
70. Segerberg, K., Meyer, J. -J., & Kracht, M. (2009). The logic of action. In E. N. Zalta (Ed.), *The stanford encyclopedia of philosophy*.
71. Solin, K. (2012). Dual choice and iteration in an abstract algebra of action. *Studia Logica*, *100*(3), 607–630.
72. Trypuz, R., & Kulicki, P. (2010). Towards metalogical systematisation of deontic action logics based on Boolean algebra. In G. Governatori & G. Sartor (Eds.), *DEON 2010*, *Lecture Notes in Computer Science* (Vol. 6181, pp. 132–147). Springer.
73. van Benthem, J. (1991). *Language in action: Categories, lambdas and dynamic Logic*. Elsevier.
74. van der Meyden, R. (1996). The dynamic logic of permission. *Journal of Logic and Computation*, *6*(3), 465–479.
75. von Wright, G. H. (1963). *Norm and action*. Routledge & Kegan Paul.
76. von Wright, G. H. (1968). *An essay in deontic logic and the general theory of action*, *Acta Philosophica Fennica* (Vol. XXI). North-Holland Publishing Company.
77. Walton, D. (1980). Omitting, refraining and letting happen. *American Philosophical Quarterly*, *17*(4), 319–326.
78. Wiseman, C. (1970). The theory of modal groups. *The Journal of Philosophy*, *67*(11), 367–376.
79. Xu, M. (1995). On the basic logic of stit with a single agent. *The Journal of Symbolic Logic*, *60*(2), 459–483.

Chapter 6
Reasoning with Comparative Moral Judgements: An Argument for Moral Bayesianism

Ittay Nissan-Rozen

Abstract The paper discusses the notion of reasoning with comparative moral judgements (i.e. judgements of the form "act a is morally superior to act b") from the point of view of several meta-ethical positions. Using a simple formal result, it is argued that only a version of moral cognitivism that is committed to the claim that moral beliefs come in degrees can give a normatively plausible account of such reasoning. Some implications of accepting such a version of moral cognitivism are discussed.

Keywords Moral bayesianism · Moral uncertainty · Moral reasoning · Conditionalization · Commutativity · The lottery paradox

6.1 Introduction

Moral judgements[1] are an object of reasoning. People consider them, try to justify them, change them in light of new arguments and so on. This is particularly true regarding comparative moral judgements (CMJs), i.e. judgements of the form 'a is morally superior to b', when 'a' and 'b' are either acts or states of affairs.

Comparative moral judgements, many (but not all) philosophers believe, ought to be consistent in the decision theoretic sense. Minimally, they ought to be transitive. In this paper, I argue that accepting the latter claim while also accepting the claim that agents can reason with comparative moral judgements commits one to accepting a meta-ethical position that I call moral Bayesianism (MB). Moral Bayesianism is a form of moral cognitivism, the thesis that moral judgements are beliefs. It is the

[1]The term 'moral judgements' is used in our everyday discourse, in philosophical discourse and in discourses in other disciplines (such as psychology) in two different ways. Sometimes it is used in order to refer to an act (typically a verbal act) of judging. E.g. when I say 'it is wrong to lie' I am performing a verbal act and this act is a moral judgement. Other times, however, it is used in order to refer to the mental attitude that is expressed by such an act. E.g. the reason for my saying 'it is wrong to lie' is my mental judgement that it is wrong to lie. In this paper I am going to use the term in the second way.

I. Nissan-Rozen (✉)
Philosophy and PPE, The Hebrew University of Jerusalem, Jerusalem, Israel
e-mail: ittay.nissan@mail.huji.ac.il

R. Urbaniak and G. Payette (eds.), *Applications of Formal Philosophy*,
Logic, Argumentation & Reasoning 14, DOI 10.1007/978-3-319-58507-9_6

version of moral cognitivism according to which moral beliefs come (just like non-moral beliefs) in degrees of credence.

The structure of the argument is as follows. I will first present a characterization of reasoning with CMJs and will argue that all those who deny MB must accept this characterization (those who accept MB, however, must reject this characterization). I will then show that, under this characterization, reasoning with CMJs is impossible if we accept the transitivity of CMJs. In the last section I will show that moral Bayesians do have a good way to both characterize and conduct reasoning with CMJs.[2]

The rest of this paper will be organized in the following way. In Sect. 6.1, I present MB in more detail and try to identify motivations for accepting it. The discussion in this section also serves as a good introduction to the argument that follows as it leads, in a natural way, to the problem of moral reasoning. In Sect. 6.2, I present a very general formal characterization of reasoning with CMJs and argue that all non-Bayesians should accept it, while Bayesians can (and should) reject it. In Sect. 6.3, within the framework of the characterization introduced in Sect. 6.2, I present three axioms that whoever accepts the characterization must accept, and show that they are inconsistent with each other. In Sect. 6.4, I argue that moral Bayesians have a good alternative way of characterizing reasoning with CMJs that does not lead to similar problems. This last discussion also explains what it is about non-Bayesian approaches that makes them unfit to explain reasoning with CMJs.

6.2 Moral Bayesianism

In this paper I take moral cognitivism to be the thesis that moral judgements are beliefs.[3] MB is the thesis that these beliefs come (just like non-moral beliefs) in degrees. Although the term Moral Bayesianism is mine, the position it denotes is not new. Many philosophers explicitly accept it and many others would be ready to accept

[2]The formal result presented in Sect. 6.3 applies, of course, not only to CMJs, but also to comparative judgements regarding any other type of linear ordering (such as rational preferences for example). In fact a version of it can be applied to any set of propositions which is governed by a semantics that rules out as inconsistent at least one possible distribution of truth values over any three propositions in the set. However, my justification for the claim that the characterization of reasoning with comparative judgements that I present should be accepted by all non-Bayesian (with regards to a given type of comparative judgement) does not apply to reasoning with comparative judgements about all types of linear orderings. Specifically, I do not try to suggest (and I think it is not the case that) it holds with regards to a rational agent's personal preference ordering.

[3]In any case, it takes moral judgements to have truth values. For a discussion of different types of moral cognitivist positions see [27]. The discussion here is independent of the issues discussed there.

it if it were presented to them.[4] There are, however, philosophers that explicitly deny the thesis contained in this position.

It will be useful to distinguish two types of opponents to MB. On the one hand are non-cognitivists. Non-cognitivists deny that moral judgements are beliefs and so they deny that moral judgements are quantitative beliefs.[5] On the other hand are non-Bayesian moral cognitivists. Such moral cognitivists accept that moral judgements are beliefs, but deny that they come in degrees. Moral judgements, they argue, are beliefs of a binary form: you can either hold them or not, you can either accept them or deny them.[6] The negative part of my argument is directed against both types of non-Bayesians. Neither of these positions can account for moral reasoning with CMJs, or so I will argue.

Two clarifications are in order. First, the degrees in question here are degrees of belief, not the degrees of moral value of the objects of these beliefs or judgements. Both non-cognitivists and non-Bayesian cognitivists can (and probably do) accept that the objects of moral judgements usually do come with degrees of moral value. What they reject is that the judgements themselves come with degrees.

Moral Bayesians insist that the judgements themselves also come with degrees. On the one hand, I can be equally certain that it is wrong to lie and that it is wrong to murder, while still judging murder to be much worse than lying. On the other hand, I can believe that in case an early-stage foetus has the same moral status as an adult human being, having an abortion is as bad as murdering, but still be uncertain—to different degrees—whether an early-stage foetus has the moral status of an adult human being.

Second, up to this point I have discussed MB as a conceptual (possibly with some descriptive elements) thesis: a family of attitudes, which we call moral judgements, exists and MB is the thesis that these attitudes are beliefs. Now, however, I want to add a normative component to MB, which is that degrees of moral judgements obey the laws of probability. I am not going to argue for this claim. In the literature there are several justifications of this demand for the case of non-moral beliefs (for example see [12]) and I see no special reason to believe that these justifications fail specifically when it comes to moral beliefs. I will not discuss this possibility here. However, most (but not all) of what I argue is independent of this assumption.

Why be a moral Bayesian? Well, the first reason is that moral judgements do intuitively come with degrees. Notice that the feeling of moral uncertainty usually arises in the following type of situations. When an agent discovers through reflection that she holds moral judgements that are inconsistent according to some criterion of consistency that she herself takes to be valid, she usually feels that she must give

[4]Two examples of discussions in which at least some of the participants explicitly accept MB are the literature discussing moral uncertainty (for example see [10, 16, 22, 25]) and the literature discussing the desire as belief thesis (for example see [1, 2, 19–21]). I use the term 'moral Bayesianism' to highlight the relevance—rarely discussed—of this thesis to moral reasoning.

[5]Van Roojen [27] offers a good overview of different non-cognitivist views.

[6]I do not know of any published work that explicitly argues against MB from a cognitivist point of view. However, in private conversations and other types of informal communication I have met the non-Bayesian cognitivist position quite often.

up on at least one of the inconsistent judgements. They are inconsistent, after all, so how can she hold all of them? However, typically, she also feels uncertain regarding which of the inconsistent judgements she must reject. All of them seem initially plausible (otherwise she wouldn't have held them in the first place), but they cannot all be held consistently at the same time. Which one(s), then, should she reject?

The last question feels like a question to which a good answer can be given. At least, it is clear that not all answers are good to the same degree. MB explains this phenomenon in a straightforward way: the feeling of uncertainty is just that, uncertainty. Non-Bayesian approaches would seem to face a much greater difficulty in explaining the phenomenon. Now, I agree that this argument[7] is a good yet inconclusive one, but I also think that its strength is frequently underappreciated. In Sects. 6.2 and 6.3 I will present a stronger argument for MB, according to which degrees of certainty are not only something we intuitively attach to moral judgements, but also an attribute of moral judgements that is necessary in order for us to account for the phenomenon of moral reasoning. Before I do so, however, I want to reflect a little more on the intuition that moral judgements come with degrees.

Consider the two versions of the trolley problem: the side-track version, in which it is possible to save 5 people from death by pulling a switch that moves a train to a side track on which it is going to kill only 1 person; and the fat man version, in which it is possible to save the 5 people by pushing a fat man from a bridge on the tracks thereby killing the fat man while stopping the train from hitting the 5 people. Now let us consider the following three claims regarding these two versions of the trolley problem.

- S—it is permissible to pull the switch in the side-track version.
- F—it is impermissible to push the man in the fat-man version.
- N—there is no morally significant difference between the two versions.

As is well-known (for example see [17]), about 70% of the population judge both S and F to be true. I know of no study that has explicitly examined people's attitudes toward N, but one of the central disagreements in the vast philosophical literature on the trolley problem concerns N. Those who argue in favour of accepting both S and F try to point to some differences between the two versions and argue that these differences are morally significant (significant enough to justify accepting both S and F). Those who argue against try to argue that these differences are not morally significant (or, if significant, not strong enough to justify accepting both S and F). In any case, it seems that everybody agrees that it is irrational to hold all three judgements at the same time.

When I come to evaluate the three claims, however, I tend to accept all three of them. Note that this is not just my intuitive reaction to the three claims being presented to me. Rather, this is a true description of my attitudes toward the three claims after deep reflection: after I read many papers on the subject, sat in many lectures on it, discussed the issue with many people whose opinions I appreciate and thought about it for many hours. At the same time, I take the three claims to be

---

[7] Introduced in a much more detailed way by [25].

Table 6.1 Moral uncertainty in the trolley problem

	Credence		1/3	1/3		1/3			
State		S	S	S	S	−S	−S	−S	−S
		F	F	−F	−F	F	F	−F	−F
		N	−N	N	−N	N	−N	N	−N

inconsistent with one another: if there is no morally relevant difference between the two versions it cannot be the case that in one of them it is permissible to save the 5 people and in the other not.

How can this be the case? Maybe, it cannot. Maybe, the problem is with my attitudes. I suspect, however, that many other people also share these judgements and, if this is so, many people would surely welcome an explanation of this pattern of judgments that does not attribute irrationality to all of us. MB allows us to do this. For a Bayesian, to accept that the three claims are inconsistent is to accept that the probability of the intersection of all three of them should get probability 0 in the credence function of a rational agent. In other words, for a Bayesian it follows from the inconsistency of the three claims that $c(F|SN) = 0$, where $c(.)$ is a rational agent's credence function. It does not follow from this, however, that if a rational agent believes to a high degree that S and N are true she should believe to a high degree that F is false.

There are infinitely many credence functions that respect the former condition while violating the latter. Table 6.1 shows one:

Here, the agent believes in each one of the three propositions, S, F and N, with a relatively high probability of 2/3, while giving a probability of 0 to each state in which all three are true.

One way to view this type of situation is as a version of the famous Lottery Paradox (LP), introduced by Henry [13]. In the LP, an agent finds herself accepting each one of the propositions included in some set (the proposition that says that ticket 1 will not win the lottery, the proposition that says that ticket 2 will not win the lottery, and so on), while rejecting their conjunction (i.e. that no ticket will win the lottery).

This is exactly what happens here. I tend to accept each one of the three claims, F, S and N, while rejecting their conjunction seems to me unavoidable.[8] Now, one of the main lessons from the LP is that there is nothing irrational in holding such degrees of belief. Given the evidence, it is not only not irrational but, arguably, the only rational

[8] An anonymous referee commented that, while in the trolley problem we can intuitively accept any conjunction of two out of the three propositions involved, the example presented in Table 6.1 assigns to each one of these conjunctions a low probability (of 1/3). I am not sure that I share the referee's intuition (when I think about a conjunction of two out of the three propositions I become immediately aware of the fact that they entail the negation of the third), but in any case it is easy to construct an example in which each one of the three conjunctions (i.e. SF, SN and FN) gets a high probability while the conjunction of all three of them gets a low probability (though not 0 and see footnote 10). For example, this is the case when the agent assigns a credence value of 9/24 to SFN, and a credence value of 5/24 to SF-N, S-FN and -SFN.

thing to believe with high probability regarding each ticket that it will not win the lottery while at the same time believing with probability 1 that one of the tickets will win it. In other words, sometimes the evidence is such that it supports holding credence functions that constitute a LP. Sometimes the evidence is misleading.

Now, in the trolley problem case, as also in the case of many other moral debates,[9] most people seem to take our moral intuitions as an important type of moral evidence: having a strong intuition that it is impermissible to push the fat man is evidence for the claim that it is impermissible to do so. In the literature one can find different explanations for why moral intuitions will tend to mislead us in many cases: they are emotionally loaded (see for example [7]), they are the product of an evolutionary process which took place in a very different environment from ours (see for example [24]), they are the product of heuristics that work well most of the time but give the wrong prescriptions in some cases (see [26]) and so on.

Thus, we have good reasons to suspect that our moral evidence will mislead us. So it is not surprising that we find ourselves in some cases in the kind of moral lottery paradox described above. There is nothing irrational about this. It is the rational reaction to misleading evidence.

So while non-Bayesian approaches must treat my attitudes to the two versions of the trolley problem and to other moral debates as irrational,[10] MB sees them as perfectly rational. Moreover, it seems that the explanation that MB gives to the apparent inconsistency among my attitudes succeeds in capturing exactly what is going on in my mind. This I take to be strong evidence for MB.

Notice now that while all the accounts mentioned above explain why we should expect our moral intuitions to mislead us in some cases, they do not tell us which intuition it is that is misleading us in a specific case. When we find ourselves in a moral LP of the sort described above we know that at least one of our intuitions misleads us, but we do not know which one it is. Neither our psychological knowledge nor our biological knowledge can tell us that. Thus, it seems perfectly reasonable to give a positive answer to the question 'do you accept this claim?' when asked regarding

[9]For example, consider Peter Singer's argument from "Famine, Affluence, and Morality" [23], which has a structure almost identical to the one discussed here, as it is based on the apparent inconsistency between the following three claims: 1. It is obligatory to save a drowning child when the child is in a pond next to you, even if by doing this you will ruin your new pair of shoes. 2. It is permissible not to save dying children in far away countries, even if it is possible to do so for the cost of a new pair of shoes. 3. There are no morally significant differences between the case of the child drowning in a pond next to you and the case of dying children in far away countries.

[10]However, see [5] for an argument against the demand that a rational agent must hold a deductively consistent set of full beliefs at all times (Kyborg rejected this demand too). As noted, I do not take the discussion in this section to be a knock-down argument for MB. The aim of the discussion is merely to motivate MB. The real argument for MB is presented in the following sections. Still, it is worth mentioning that in [4] an extension of the position presented in [5] is used in order to suggest a new subjective interpretation for probability. Thus, it seems (but I cannot say this with complete certainty) that, to the extent that the non-Bayesian can give a satisfactory account for the phenomenon described in the main text in terms of full beliefs, it is possible to view this account as a version of MB.

each one of the claims, S, F and N. At the same time, it seems reasonable to give a negative answer when asked this question regarding the conjunction of S, F and N.[11]

'How can you accept all three claims while you deny their conjunction?' the psychologist might ask his subject. The subject, then, would simply reply 'you asked me for a binary judgement, but my attitudes here are not binary so I just went with the higher probability'. This seems to me a reasonable reply.

Moreover, there seems to be nothing irrational in acting according to the same strategy (i.e., the strategy of 'going with the higher probability'). I believe it is more likely than not that morality requires me to pull the switch in the side-track version, so I do. I believe it is more likely than not that morality requires me not to push the fat man in the fat-man version, so I don't. It is true that I believe with high probability that there is nothing that can justify this pattern of choices, but so what? Can you point to a principle of rationality that I have violated?

When do the problems with the 'going with the higher probability' strategy start? They start when, by following it, an agent must violate a principle of consistency of choice. For example, when by following it an agent must choose intransitively. How can this happen?

The first thing to notice is that transitivity is a condition of rationality that applies both to choices and to comparative moral judgements (and other betterness judgements). Rationality requires us both to choose in a transitive way and to hold transitive comparative moral judgements. Although some philosophers deny one or both of these requirements, in this essay I take at least the second, i.e., transitivity of CMJs, as an assumption. By doing so I do not, however, exclude from my audience all those who deny this condition.

First, I take it that most scholars that deny that transitivity is a condition of rationality would still accept that it should be possible for agents to hold transitive CMJs. It is surely not irrational to do so, after all. Thus, meta-ethical or ethical accounts that do not allow for transitivity of CMJs must be seen as problematic in some way: they do not allow for patterns of attitudes that seem perfectly reasonable.

Second, the debate between MB and non-Bayesian approaches seems independent of the debate regarding the status of the transitivity condition. Most non-Bayesians, even ones who are willing to bite the bullet and accept that full transitivity of CMJs is impossible, will hesitate, or so I presume, to take this claim as one of the building blocks of their meta-ethical position. It should be possible to be a non-Bayesian even if you do believe in transitivity, shouldn't it?

So I am not going to justify either the transitivity of CMJs or the transitivity of choices conditions. Rather, I take them as assumptions (I only need the transitivity of choices condition for purposes of exposition, though. The argument does not depend on it. It does depend on transitivity of CMJs). But what exactly does the transitivity condition say?

Well, for a non-Bayesian the answer is obvious: it says that if one holds both the judgements that one act, A, is morally superior to another act, B, and that B is morally

[11] And as mentioned in footnote 8, I am not sure about the case of the conjunction of only two out of the three propositions.

Table 6.2 A condorcet-like profile

1/3	1/3	1/3
A	C	B
B	A	C
C	B	A

superior to a third act, C, then one ought to hold the judgement that A is morally superior to C.[12] But for a Bayesian the requirement says something else. It says that one ought to give credence 0 to any intransitive ranking of the three acts, A, B and C. It does not say that if one believes with high probability that A is morally superior to B and that B is morally superior to C then one ought to believe with high probability that A is morally superior to C. There are infinitely many credence functions that respect the former condition and violate the second. Table 6.2 shows an example:

Here, the agent believes with probability 2/3 in any one of the three CMJs, A > B, B > C, and C > A, but still gives credence 0 to the possibility that the moral superiority relation is intransitive.

However, when one holds such degrees of belief then, by following the 'going with the higher probability' strategy, one is led to choose in an intransitive way. This is something one wants to avoid, I assume. What should an agent that finds herself in such an epistemic condition do?

Well, there are two different questions here. The first question is how should she choose? I am not going to discuss this question here (I discuss it in [18]). Most of the moral uncertainty literature deals with exactly this question. However, there is another question that does not receive enough attention in the literature and arises in this context. This is the question, not of choice, but of reasoning: when the agent finds herself in such an epistemic condition she knows for sure that at least one of the claims in which she believes with high probability is false. Thus, if there is no need for an immediate choice, it is clear that the agent ought to try and reason herself out of the unfortunate epistemic position she has found herself in. She can, for example, try to collect more moral evidence (consult with people whose opinions she values, think of more arguments, read some books, etc.), or revaluate the evidence she already has.

Notice that the same holds also for non-Bayesians. When a non-Bayesian agent finds himself accepting inconsistent CMJs, and when there is no need for an immediate decision, he too ought to try and reason himself out of the inconsistency. In the next section I am going to step out of the Bayesian point of view, at least for

---

[12]There are actually three possible readings of the requirement. Let 'i > j' stand for 'i is morally superior to j' with appropriate prefixes which make it a readable that-clause. Then, the three readings are the following ones. 1. If one accepts A > B and B > C then one ought to accept A > C. 2. If one accepts A > B then one ought to accept 'if B > C then A > C'. 3. One ought to accept 'if A > B and B > C then A > C. I think one should accept the requirement under all three readings. In any case, nothing in my argument depends on this.

the time being, and will attack the problem of reasoning with CMJs from the non-Bayesian point of view. I will return to the Bayesian point of view only after we reach a dead-end by applying non-Bayesian thinking to the problem.

6.3 Reasoning with CMJs

At the most general level, reasoning with CMJs is a process in which an agent changes some of the CMJs he holds in a non-arbitrary way. It is true, of course, that reasoning might not lead an agent to change any one of the CMJs he holds, but if we want to discuss a non-trivial concept of reasoning, we must also allow for cases in which reasoning does lead the agent to change some of the CMJs he holds. Moreover, in the kind of cases that motivates my discussion of reasoning, cases in which an agent realizes through reflection that he holds inconsistent CMJs (according to a criterion of consistency he accepts), the reasoning must lead the agent to change at least one of the CMJs he originally held. Otherwise, he would end the process of reasoning holding inconsistent CMJs, which we can assume cannot be the case for a rational agent.

Of course, not any process in which an agent changes some of the CMJs he holds constitutes a process of reasoning. In order for an instance of a change in one's CMJs to constitute an instance of reasoning with CMJs the change should be due to a reason. An arbitrary change is not a matter of reasoning.[13] I will exploit this last distinction shortly in order to present a necessary condition for an instance of a change in one's CMJs to constitute an instance of reasoning with CMJs. Before I do so, however, one comment must be made.

There might be (and probably are) some non-cognitivists who will not mind accepting that reasoning with CMJs is impossible (some Humean non cognitivists will probably even be happy to accept this claim). Such non-cognitivists take a change in one's CMJs to be a similar matter to a change in one's preferences between, say, vanilla ice cream and chocolate ice cream. My argument is not directed at such non-cognitivists. My argument takes as an assumption that reasoning with CMJs is possible and, using this assumption, argues for MB. It does not assume, however, that reasoning must be made with quantitative beliefs. In my starting point I allow for both reasoning with qualitative beliefs and with other attitudes (for discussions regarding the idea of reasoning with preferences see, for example, [3]).

---

[13] I am not trying to perform here a conceptual analysis of the concept of reasoning. Maybe an arbitrary change in one's CMJs does constitute an instance of reasoning with CMJs. Broome [3] discussion of reasoning with preferences seems to assume that this might be the case. Broome, then, limits his discussion to what he calls 'correct reasoning'. I do not mind, of course, accepting such terminology. The important point is that there is a sense in which we do demand from a rational moral agent, who is involved in a process in which he changes some of the CMJs he holds, to do so in a non-arbitrary way. Whether this demand is a necessary condition in order for the process to count as reasoning or only a necessary condition in order for the process to count as correct reasoning does not concern me. For convenience, then, I will assume that the former condition holds.

Although some non-cognitivists will be willing to accept that reasoning with CMJs is impossible, most non-cognitivists, I believe, will not. Most non-cognitivists do want to accommodate within their meta-ethical framework most of the features people intuitively attribute to morality. The idea that it is possible to reason with moral judgements—in the minimal sense that I have presented—is surely one of these features.

Those who do believe that reasoning with CMJs is possible must accept the very general characterization I have presented here for such reasoning. I am now going to use this characterization in order to present a necessary condition for an instance of a change in one's CMJs to constitute an instance of reasoning with CMJs. In order to do so we have to distinguish, first, between two stages involved in a process of (at least some instances of) reasoning with CMJs.

The first stage is the stage in which the agent—after being exposed to some information he takes to be morally significant (I will call this 'moral evidence')—decides to change at least one of the CMJs he holds. The second stage is the stage in which, following this change, and as a result of it, the agent changes other CMJs he holds in order to keep the set of judgements he holds consistent.

For example, consider an agent who judges one act, A, to be morally superior to another act, B. Since he judges A to be superior to B, he also judges, we can assume, that a lottery that brings A with probability 0.9 and B with probability 0.1 is morally superior to a lottery that brings B with probability 0.9 and A with probability 0.1. The agent is then exposed to a piece of moral evidence that makes him change his judgement regarding the moral superiority relation that holds between A and B, and judges now B to be morally superior to A. Maybe, for example, he thinks of an argument he had never thought of before, or maybe he reads a book with such an argument, or maybe he does not think of any argument but simply learns that a person, whose opinion he values, judges B to be superior to A, or maybe he has a dream in which God tells him that B is superior to A. I do not assume anything, at this stage, about what can justifiably constitute moral evidence (later on I will present one weak assumption about what cannot constitute moral evidence).

In any case, the agent changes his judgement regarding the moral superiority relation between A and B, and this change brings with it other changes in the set of CMJs the agent holds. For example, since the agent now judges B to be superior to A, he should also change his judgement regarding the moral superiority relation that holds between the lottery that brings B with probability 0.9 and A with probability 0.1 and the lottery that brings A with probability 0.9 and B with probability 0.1. He should now judge the former to be superior to the latter.

The point of this example is not to argue that the agent should normatively change his judgements in such a way. I believe he should, but even if he shouldn't there must be other cases in which an initial change in one's CMJs carries with it further changes. The point is rather to make clear the distinction between the two stages involved in the process of reasoning with CMJs: the stage in which an agent—after being exposed to some moral information—performs one change in the set of CMJs he holds, and the stage in which the initial change brings some further changes.

The condition I will shortly present applies only to the second stage, not to the first stage. There might be specific instances of reasoning with CMJs in which the initial change does not bring with it any further changes because it conflicts with none of the other CMJs the agent holds, but if we want to allow for non-trivial reasoning we must also allow for cases in which the initial change does bring with it some other changes.

In the same way, there might be some instances of reasoning with CMJs in which the second change I was referring to is absent because the agent's immediate response to the moral evidence he is exposed to is a change in each one of the CMJs he holds (and thus there is no need for the second stage in which the agent adjusts the set of his CMJs to the initial change). However, arguing that there are no cases of reasoning with CMJs in which the second stage plays a significant role trivializes the concept of reasoning used. Intuitively, most instances of reasoning with CMJs are instances in which the evidence the agent is being exposed to (e.g. an intuition triggered by an 'intuition pump') supports a change in a set of CMJs that is a proper subset of the set of all CMJs the agent has.

Thus, accounting for the second stage I have described is necessary for anyone who wants to allow for a non-trivial concept of reasoning with CMJs.

Here is the idea: when one reasons with one's CMJs one changes them in a way that is only sensitive to what one takes to be relevant considerations. To be clear, the demand is not that the way one changes one's CMJs will always be sensitive to what one takes to be relevant considerations. This demand is too strong. Sometimes agents judge some considerations to be relevant to their CMJs but, nevertheless, do not take them to be strong enough to trigger a change in the CMJs they accept. The demand I have in mind is a much weaker one. It is the demand that the way one changes one's CMJs is not sensitive to what one takes to be irrelevant considerations.

Consider an agent who changes her CMJs and is then asked to supply a reason for the change. The agent can either deny that there was any reason for the change or supply a reason. In the first case, the change is surely not an instance of reasoning with CMJs as the agent herself claims there was no reason for the change. In the second case, when the agent does supply a reason for the change, if she does so in a way that violates the condition then the reason she supplies must be a reason that the agent herself takes to be irrelevant to the change she has made. This just means that it is not a genuine reason for the change.[14]

We are thus looking on a process in which, following an initial change in one's CMJs, one changes other CMJs one previously held and, while doing so, one takes the initial change to be relevant to the other changes one makes. If you are a non-Bayesian, you can think of such a process in the following way. There is a function, f(x, y), which takes as input ordered pairs, in which the first element is a ranking of

[14]You can read the last two paragraphs while interpreting the phrases 'takes to be a reason' and 'takes to be relevant' anyway you like. I think of them as referring to beliefs: 'takes X to be a reason' means 'believes that X is a reason'; but if you prefer to understand the agent's attitudes toward the status of the information he gets differently this is okay too. In any case, if you do want to talk about reasoning in the way I have characterized here, you must make a distinction between information that the agent takes to be relevant and information that he does not.

alternatives that belong to some set of alternatives, X, and the second element is an order of all the alternatives in some subset of X, and produces as output an order of all the alternatives in X.

The interpretation is quite simple. An agent starts with some set of CMJs regarding some of the alternatives available to her. These need not be complete or transitive, but we can assume that they are as this does not matter for the discussion that follows and it will make the presentation simpler. Then, she receives some input. This can be the result of either the agent thinking about the matter, or discussing the issue with other people, or reading a book, or receiving a message from God, or what have you. The input, however, must be in the form of an order (i.e. a complete and transitive relation) over some of the alternatives available to the agent. In the simplest case—which is sufficient for us to consider (since problems begin even with it)—the input is just a moral superiority relation between two of the alternatives.

Now, the agent must decide how to incorporate the input she has received into her set of CMJs. I assume that the agent will do so in such a way that the result will be an ordering of the alternatives (this represents the agent's commitment to rationality). For example, consider an agent that holds the following three judgements: act a is morally superior to act b, act b is morally superior to act c, act a is morally superior to act c. Suppose the agent—following a discussion with her philosophy professor—changes her judgement regarding the moral superiority relation that holds between a and b, and now judges b to be morally superior to a. The function that we are considering tells the agent how this should affect her other judgements.

This is a very general characterization of the process of reasoning with CMJs.[15] It assumes nothing about what triggered the change; it allows the agent to start with any CMJs and is very pluralistic regarding the richness of the input. What is not allowed by this characterization is that either the input or the output will be in the form of a probability distribution over the set of all possible orders of the alternatives, or of a new probability value for a specific moral superiority relation between two alternatives (as by allowing that, one becomes a Bayesian).

Maybe it will be useful to make explicit a distinction that I have implicitly used in this characterization. This is the distinction between the information that the agent has received and the effect of this information on the agent's mental states. I assume nothing about the information itself. Different people can take different things as constituting for them moral information. Assuming that an agent did receive information that she takes to be morally relevant, and assuming that this information was strong enough to change at least one of the agent's judgements, the question of how this change should effect the agent's other judgements arises.

As explained, the characterization presented here tries to capture only this last stage—the stage in which one change in an agent's judgements brings other changes. Now, all non-Bayesians must accept that the change must take a binary form, that is, the agent must incorporate a new judgement into her set of judgements. It is true that both non-Bayesian cognitivists and non-cognitivists can argue that the information

[15]See [8, 9] (for example) for a similar characterization of reasoning with preferences and for preference change generally.

itself might support a specific change in the agent's judgements to different degrees. However, once a piece of information is powerful enough to trigger a change, the change is binary; it does not come in degrees.

Only the Bayesian can reject that. For the Bayesian, the input itself (not only the strength of the information) can come in degrees. A piece of morally relevant information can, for the Bayesian, make an agent raise the probability she attaches to a specific judgement, without accepting it. More importantly, the judgements themselves—the outputs of the function—are not binary; they always come with a probability value attached to them. This is exactly the difference between Bayesians and non-Bayesian: Bayesians believes that beliefs—not only the evidential support for a given belief, but the beliefs themselves—come in degrees; non-Bayesians deny that. Thus, while Bayesians can (and should) reject the characterization presented here for reasoning (or merely revising) CMJs, non-Baycsians have no reason to do so.

In the next section I will introduce three axioms on f and will argue that all non-Bayesians must accept them. I will then show that these axioms are inconsistent with each other.

6.4 Non-Bayesian Reasoning with CMJs Is Impossible

In order to get our result it is enough for us to consider only the simplest case in which there are only three alternatives and the agent starts with an order over them and gets an input that is a relation over two of the three. Even in this simplest case, we get an impossibility result.

Let f be our function, let $X = \{a, b, c\}$ be a set of alternatives, let $>$ be a complete and transitive relation over X. We will use the notation $i > j > k$ to say that the CMJ ranks i over j, j over k and i over k, and in the same way, the notation $i > j$ will denote that the CMJ ranks i over j.[16] Consider now the following three restrictions on f:

1. Stability: if the input $l > m$ is consistent with $i > j > k$, then $f(i > j > k, l > m) = i > j > k$.
2. Respectfulness: $f(i > j > k, l > m)$ is consistent with $l > m$.
3. Commutativity: for any two consistent inputs, $l > m$ and $z > x$, $f(f(i > j > k, l > m), z > x) = f(f(i > j > k, z > x), l > m)$.

The first two axioms are necessary in order for us to treat the function as a function that describes a CMJs change (not specifically reasoning with CMJs). They are, in a sense, pre-conditions for any discussion regarding CMJs change.[17] Stability requires

[16]There is an implicit semantics in the background that—for the sake of simplicity—I choose not to make explicit. It is easy to see what it should look like, though, in our simple case. Hansson [8, 9] provide a much more general framework.

[17]Stability is a special case of Hansson's vacuity, and respectfulness is equivalent to Hansson's success.

that if the agent gets no new information, i.e., if he gets an input that is consistent with his CMJs, he does not change his CMJs. Respectfulness requires that if the agent learns something, i.e., if he receives an input that is inconsistent with his CMJs, he incorporates it into his new set of CMJs.

The commutativity axiom is the one that characterizes the function as a function that describes reasoning with CMJs and not just any kind of CMJs change. I discuss the axiom at length in Sect. 6.4. For now, I only want to point to the general motivation for accepting it. When one reasons with one's CMJs one changes one's CMJs in a way that is only sensitive to what one judges to be relevant to the question at hand. One thing that most people in most cases take to be irrelevant to the question of which CMJs they ought to have is the order in which they receive different consistent pieces of relevant information.

It does not seem to matter whether I first thought of one argument, A, that made me change one of my judgements, and then of another argument, B, which made me change another judgement (consistent with the judgement produced by A), or first thought of B and then of A. Of course, if the judgement produced by A is inconsistent with the judgement produced by B, the order does matter (for example, if I first learn A and then not-A, I should believe not-A, but if the order is reversed, I should believe A), but commutativity refers only to consistent inputs. When it comes to consistent inputs, the order does not matter, unless one holds very peculiar meta-ethical or epistemological convictions.

For example, one might insist that the order in which an agent gets different pieces of information has a constitutive role for some moral facts, or one might insist that this order—in itself—reveals to the agent something about morality (maybe God communicates with us through the order in which She sends us different pieces of information, for instance). It is hard to think of any example in which the order does intuitively seem to matter. However, even if there are some contexts like that, surely in most cases the order does not matter; yet the result that will be presented now shows that it should always matter!

Theorem 1 *There is no function that respects stability, respectfulness and commutativity.*[18]

Proof Consider WLOG, an agent that starts with a set of CMJs according to which $a > b > c$. Consider first the two consistent inputs, $a > b$ and $c > a$. From respectfulness and stability we immediately get that $f(a > b > c, c > a) = f(f(a > b > c, a > b), c > a)$ is equal to either $c > a > b$, or to $c > b > a$ or to $b > c > a$. From applying respectfulness and commutativity again we get $f(f(a > b > c, c > a), a > b) = c > a > b$ and from commutativity we get $f(f(a > b > c, a > b), c > a) = c > a > b$. Thus, $f(a > b > c, c > a) = c > a > b$.

[18]I was not able to locate an earlier proof of this (very simple) result. However, there are obvious formal connections between this result and the literature dealing with judgement aggregation on the one hand (for a good introduction see [15]) and the literature dealing with preference change on the other hand (such as [8, 9]). The literature that explores the relation between Bayesian updating and the AGM approach to belief revision is also of obvious relevancy here. See for example [14].

Table 6.3 Demonstration

		A	B	C	D	E	F
		a > b	b > a	a > c	c > a	b > c	c > b
1	a > b > c				c > a > b (b > c > a)		
2	a > c > b						
3	b > a > c						
4	b > c > a						
5	c > a > b						
6	c > b > a						

However, consider now the two consistent inputs, b > c and c > a. Similar considerations to the ones we used in the previous paragraph show that f(a > b > c, c > a) = f(f(a > b > c, b > c), c > a) = f(f(a > b > c, c > a), b > c) = b > c > a, contrary to our conclusion that, f(a > b > c, c > a) = c > a > b.[19]

Table 6.3 demonstrates more clearly what is going on:

In this table columns A to F represent all the possible inputs the agent in our simple case can get. Rows 1–6 represent all the possible orders the agent can start from. The function we are looking for assigns to every cell in the table one of the possible orders.

We start from row 1 and consider first the (consistent) inputs a > b and c > a. If the agent gets the input a > b first, he stays in row 1 (by stability), and then when he gets the input c > a he moves to one of the rows in which c is preferred to a, i.e., to either 6 or 5 or 4. If the agent gets the input c > a first, he moves to either row 6 or 5 or 4, and then, when he gets the input a > b, he must move to a row in which a is ranked above b. However, from commutativity we know that he must end up in either row 6 or 5 or 4. The only row out of these three in which a is preferred to b is 5. This is the reason why I wrote c > a > b in cell D1. However, following exactly the same line of reasoning regarding the inputs b > c and c > a leads to the conclusion that the value of cell D1 should be b > c > a. However, the function cannot give two different values to the same argument.

The result shows that, after she receives some information, a non-Bayesian (cognitivist or not) cannot change her CMJs in a commutative way. The non-Bayesian now has two options. She can either argue that the order in which an agent receives the information always matters (either epistemically or metaphysically) or accept that indeed reasoning with CMJs is impossible. There is no other option. If she refuses

[19]It is easy to see that the result can be generalized to any system of propositions that includes at least three pairwise inconsistent propositions. The inconsistency need not be always in the form of intransitivity. I thank Christian List for pointing this out to me. However, as indicated in footnote 2, I do not try to suggest that the result is of any philosophical interest when considered under different interpretations. Maybe it is under some interpretations, but I could not think of any other than the CMJ interpretation presented here.

to accept that the order in which different pieces of information are revealed to an agent matters, she must admit that the agent has no way to change her CMJs in a way that is not sensitive to irrelevant features of reality. If she insists that there must be such a way, she must accept that the order matters. But, of course, the order does not matter. Thus, a non-Bayesian must admit that reasoning with CMJs is impossible.

Bayesians, however, reject the claim—implicit in the formalization—that one's attitudes regarding comparative moral claims can always be represented in the form of a relation between some alternatives. For the Bayesian, the information can lead the agent to change his degrees of belief in some of his judgements, without making him change the judgements he accepts (or believes, as the word is used in a qualitative way).

It would be misleading to argue that the Bayesian is free to reject both respectfulness and stability, as what the Bayesian really rejects is the mere characterization of reasoning presented. For a Bayesian, respectfulness and stability do not make any sense: they assume that one's attitudes regarding comparative moral claims are binary, when these attitudes are not. For a Bayesian, an argument that supports a judgement an agent already holds (however you choose to cash out the term 'hold', maybe in terms of high degree of belief, for example) can make the agent raise his degree of belief in this judgement, and an argument that supports a judgement an agent rejects can make him lower his degree of belief in the judgement. This does not mean, however, that the agent changes the judgements he accepts. When it comes to reasoning, acceptance has no normative role for the Bayesian.

Again, this move is unavailable both to the non-cognitivist and to the non-Bayesian cognitivist. Non-Bayesians can argue that a piece of information is not strong enough to change the judgements an agent accepts, but once a piece of information is strong enough to do so, the change is binary.

Some non-cognitivists might consider adding a measure of confidence to the judgements one accepts, but still argue that this measure is not a measure of strength of beliefs. If this measure is probabilistic, then the difference between such non-cognitivists and Bayesians becomes merely terminological: comparative moral judgements behave, according to both accounts, in the same way; they just get different tags attached to them. If this measure is not probabilistic, then I am tempted to look at this move as an unorthodox version of Bayesianism, which accepts that beliefs come in degrees but rejects that these degrees should (normatively) obey the probability axioms. One can find in the literature some arguments against such a move but also some arguments in favour of it (see [12] for a discussion).

The lesson, therefore, seems to be the following. Moral reasoning requires degrees of confidence in one's comparative moral judgements. Bayesianism allows for this. It takes degrees of confidence in comparative moral judgements to be degrees of belief in the propositions that are the objects of the judgements. All versions of non-Bayesianism do not allow for this and thus they cannot give a normatively plausible account of reasoning with comparative moral judgements.

The fact that Bayesianism is not vulnerable to the result presented in this section does not mean, however, that it is on safe ground. A plausible account of Bayesian reasoning with comparative moral judgements should still be presented. In the next section I will argue that such an account is, in fact, available. The discussion will also make clear what it is about non-Bayesianism that makes it unfit to account for moral reasoning.

6.5 What Can the Bayesian do?

The first thing to notice is that, for the case in which the input itself is learned with certainty (although the agent can still be uncertain regarding other judgements), classical Bayesian updating is a commutative method of reasoning that has many other virtues that are well discussed in the literature.

However, especially when it comes to moral judgements, it seems that we must allow for probabilistic inputs: in most cases the morally relevant information we get does not make us absolutely certain in a specific moral judgement; rather it makes us raise the probability we attach to it. In such cases, the standard Bayesian move is to use Jeffrey's conditionalization (JC), which allows for belief updating when nothing is learnt with certainty (see [11]).

Bayesian updating can take place only if the agent has become completely certain that some proposition is true; that is; if he has raised the probability he assigns to some proposition to 1. Jeffrey [11] argued that not all cases of learning can be characterized in such a way.

In cases where this happens, however, we can still demand that the way the agent updates his beliefs will obey some consistency conditions. Jeffrey claimed that the main consistency condition the agent should obey is that, after raising the probability of some proposition, A, the conditional probability of all other propositions given A should stay the same. He showed that this condition is satisfied iff the following condition holds for all propositions A and B[20].

Jeffrey's Conditionalization (for the simple case): if $p(A) \neq O, 1$

$$q(B) = q(A)p(B \mid A) + q(\neg A)p(B|\neg A)$$

Here, p(.) is the agent's initial probability distribution and q(.) is his new probability distribution after raising the probability of A. It is easy to see that classical Bayesian updating is just a special case of JC.

Jeffrey's conditionalization is commutative, however, only in special cases. It is commutative for two inputs when the two inputs are probabilistically independent. When two inputs are probabilistically dependent, the order in which an agent considers the two inputs does matter for the final probability distribution the agent holds.

Does this feature of JC constitute a problem for MB? The answer is, at least on the face of it, yes. To see this, consider the case in which there are 6 states and each state is one of the possible orders over three alternatives, A, B and C.

---

[20] For simplicity I will deal here only with the simple case in which the agent has only raised the probability he assigns to one proposition. Nothing in the discussion depends on this.

Table 6.4 A case with two independent pairs

A	A	B	B	C	C
B	C	A	C	A	B
C	B	C	A	B	A
	1/16	3/16	9/16	3/16	

Question 1: is there a probability distribution over the 6 states such that each state gets a positive probability and all the events of the form (‘$\alpha > \beta$’ with $\alpha, \beta \in \{A, B, C\}$) are probabilistically independent of each other?

Answer: no.

Question 2: is there a probability distribution over the 6 states such that every state gets a positive probability and two pairs of such events (e.g., $A > B$ and $A > C$ and $B > C$ and $A > C$) are probabilistically independent?

Answer: no.[21]

There is, however, a probability distribution such that only 4 states get positive probability and there are two pairs of such events which are probabilistically independent. Here is one (Table 6.4):

Here both ($B > C$ and $C > A$) and ($A > B$ and $C > A$) are probabilistically independent:

$$p(C > A) = 12/16$$

$$p(A > B) = 4/16 = 1/4$$

$$p(A > B \mid C > A) = p(C > A > B)/\, p(C > A) = 3/12 = 1/4, \text{ and}$$

$$p(C > A) = 12/16$$

$$p(B > C) = 12/16 = 3/4$$

$$p(B > C \mid C > A) = p(B > C > A)/\, p(C > A) = 9/12 = 3/4$$

Since both $B > C$ and $A > B$ are probabilistically independent of $A > C$, Jeffrey’s conditioning on both pairs ($B > C$ and $C > A$) and ($A > B$ and $C > A$) is commutative.

This happens, however, only when the order $A > B > C$ gets probability 0. If $A > B > C$ gets a positive probability, then Jeffrey’s conditioning is not commutative over at least one of these pairs.

[21]The proofs are straightforward but still they require some space. As the questioner’s purpose is only to highlight how severe the threat is that JC’s incommutativity poses to MB (which is the position I argue for), my argument in this paper does not depend on the truth of the answers I gave in the main text and so, for the sake of simplicity, I chose to omit the proofs. The answers are correct, though.

Notice that this phenomenon is analogous to the one that caused the problems in the non-Bayesian framework: when one starts from $A > B > C$, f cannot be commutative (if the two other axioms are respected) over both ($A > B$ and $C > A$) and ($B > C$ and $C > A$). In the Bayesian case our conclusion is that Jeffrey's conditioning is commutative with regards to both ($A > B$ and $C > A$) and ($B > C$ and $C > A$) only when $p(A > B > C) = 0$.

So, on the face of it, it seems that Bayesians face an analogous problem to the one non-Bayesians face in accounting for moral reasoning. I will argue shortly that this is not the case, but before I do so, it might be useful to discuss why it is that JC is not commutative (in those cases that it is not).

One way to understand the incommutativity of JC is the following.[22] When two inputs are probabilistically dependent then, when an agent learns one of them, she does not only learn something about the proposition that is the object of the judgement in question, but also about how she ought to evaluate the other input in case she receives it.

For example, consider an agent that starts with a probability distribution according to which the probability of $a > b$ depends on the probability of $c > a$, so $p(a > b \mid c > a) \neq p(a > b)$. Now the agent gets some information that makes her raise $p(c > a)$, but still keep it below 1. If the agent uses Jeffrey's conditionalization, one effect of receiving this input is that the agent will change $p(a > b)$, but this is not the only effect. The agent must also now change $p(c > a \mid a > b)$. In other words, the agent must change the way she evaluates $c > a$ under the supposition that $a > b$ is true. Thus, she learns something not only about the moral superiority relation that holds between a and c, and not only about the moral superiority relation that holds between a and b, but also about what is the right way to evaluate inputs she might receive in the future.[23]

In such a case, it can be argued, commutativity ought to be violated because the method of reasoning has changed: after the agent learned the new input she evaluates moral information in a different way and thus different orders should lead to different outcomes. But if this is so, then surely there is a sense in which the information that the agent gets when she learns, first the input that makes her change the probability she attaches to $a > b$, and then the input that makes her change the probability she attaches to $c > a$, is different from the information the agent gets when she gets the inputs in the opposite order.

This, then, should lead us to question the justifiability of the commutativity axiom: what we just saw is that there is an intuitive sense in which the order in which an agent gets two consistent pieces of information can matter epistemically. It matters epistemically when it teaches the agent something about the way she would update her beliefs in a possible future. We also saw that in the kind of cases we are interested in, this is always the case!

---

[22] I thank Richard Bardley for explaining this point to me.

[23] Now we can see that the case of two inconsistent inputs is just a limiting case of two probabilistically dependent inputs.

Can this be true? Can it be true that the order in which an agent gets any two possible consistent pieces of moral evidence always teaches the agent something about the way she would update her beliefs in the future? Of course not. For example, if the agent gets two inputs that make her raise the probability she attaches to both $a > b$ and $b > c$ to 1, the order does not teach her anything about possible future belief updating: no matter which of the two inputs the agent gets first, she would end up believing with probability 1 in $a > b > c$, and this belief will not change after gaining any new piece of evidence.

So what is going on here? Why is it the case that when dealing with learning certainties Bayesian updating (which is a special case of JC) is commutative while when dealing with learning without certainty JC is not commutative? The answer is simple and is not new to Bayesians. It will, however, be helpful to consider it here in the context of our inquiry.

When it comes to learning certainties, the evidence the agent learns can be identified with one of the propositions in his algebra. To say that an agent raised the probability he attaches to $a > b$ to 1 is to say that the agent has learnt $a > b$. However, when it comes to learning without certainty, this is not the case. The agent was exposed to a new piece of evidence that cannot be captured neatly by any one of the propositions in his algebra. This is exactly why he does not raise the probability of any proposition in his algebra to 1. So we must make a distinction between the evidence to which the agent was exposed and the effect of this evidence on the agent's degrees of belief in the propositions in his algebra (in our case, the propositions, $a > b$, $b > c$, $a > c$, their conjunctions, disjunctions, and negations). It will be convenient to use the term 'input' (which I have already used in this way) in order to refer to this effect.

Using the above distinction we can make another distinction, a distinction between two conditions of commutativity. The first condition is commutativity with respect to inputs (i.e., the effect of the evidence on the agent's degrees of belief). The second condition is commutativity with respect to experiences (i.e., the different pieces of information to which the agent was exposed).

Now, it is clear that the incommutatitivty of JC we have discussed is incommutativity of the first type, not the second. But of course, this second kind of incommutativity should not be ruled out: raising the probability of a proposition from 0.2 to 0.4, and then from 0.4 to 0.6, must, of course, lead to a different probability distribution than raising the probability of the same proposition from 0.2 to 0.6 and then lowering it from 0.6 to 0.4.

We want the way in which we update our beliefs to be commutative with respect to experiences, not with respect to their effects. The rationale for commutativity is the following. Holding your initial belief state constant, if you are exposed to the same pieces of evidence you should end up in the same belief state, but the order in which you are exposed to different pieces of evidence is not a piece of evidence in itself. This does not mean, however, that the way we update our beliefs should be commutative with respect to the effect of the evidence.

To make things clearer consider again the example above while keeping in mind that there is an underlying stream of evidence that is responsible for the changes in the agent's degrees of belief. In the example an agent raises the probability of a proposition from 0.2 to 0.4 after being exposed to an experience, x, and then raises it from 0.4 to 0.6 after being exposed to another experience, y. The same agent would raise the probability of the same proposition from 0.2 to 0.6 after being exposed to an experience, y', and would lower the probability of this proposition from 0.6 to 0.4, after being exposed to an experience, x'.

In order for the incommutativity with respect to the inputs to be also incommutativity with respect to the experiences it must be the case that x = x' and y = y', but this is surely not the case. While x made the agent raise the probability of the proposition, x' made the agent lower this probability. While y made the agent raise the probability of the proposition from 0.4 to 0.6, y' made him raise this probability much more, from 0.2 to 0.6.

So, generally speaking, incommutativity with respect to inputs does not automatically translate into incommutativity with respect to experiences. Is JC commutative with respect to experiences? Well, the answer depends on the type of experiences the agent is being exposed to. Hartry Field [6] showed that there is a large family of experiences regarding which JC is commutative. This family is the family of all experiences that we can identify using Bayes factors, i.e., using the expression

$$q(A)p(\neg A)/q(\neg A)p(A).$$

For example, an agent who starts with degrees of beliefs that can be represented by a probability distribution p(.) is exposed to a piece of evidence, X, that makes him change his degree of belief in a proposition, A, in such a way that his new degrees of beliefs can be represented by the probability distribution q(.). We can identify X using the Bayes factor we calculate using the change in the agent's degrees of belief in A. In the second stage, we can deduce what will be the effect of learning X on the agent's degrees of belief if he had started from a different degree of belief in A.

Here is a short demonstration that might help to show things more clearly. Assume the agent starts from a probability distribution $p'(.)$ and is now exposed to X, what should his new degree of belief in A be (in his new probability distribution, $q'(.)$)?

The answer is simple. By the assumption that we can identify X with the Bayes factor, we know that:

$$X = q'(A)p'(\neg A)/q'(\neg A)p'(A).$$

Since we know the value of X (calculated using p(A) and q(A)) and the value of $p'(A)$, we can find out the value of $q'(A)$.

So we can point to a well-defined family of experiences regarding which JC is commutative. It turns out (it is easy to see this by looking at q(A) as a function of p(A) while keeping the Bayes factor constant) that these experiences are experiences that have a decreasing marginal contribution to the agent's degree of belief, i.e., the same experience makes the agent raise the probability of a proposition less the higher his initial degree of belief in the proposition (this feature does not fully characterize the family but it is a feature of all its members). Since our intuitions regarding these matters are not very vivid, I leave it to readers to think of what they take to be paradigmatic cases of moral evidence and see if these cases have this property. I believe they do.

There is one last worry that should be dealt with. I have just argued that commutativity with respect to the effect of experiences is not the kind of commutativity we should demand from an updating method. Instead we should demand commutativity with respect to the experiences themselves. While JC is commutative (for a large family of experiences) in the latter sense, it is not commutative in the former sense. Can the same type of argument work for the non-Bayesian?

The answer is no, and it will be revealing to show why this is the case. Notice that by replacing my original commutativity condition for the non-Bayesian characterization of reasoning with a commutativity condition with respect to experiences we can only exchange the impossibility result for a triviality result.

To see this recall that the result discussed in Sect. 6.2 shows that f cannot be commutative at the same time regarding both (C > A and A > B) and (C > A and B > C). Let us assume it is not commutative regarding (C > A and A > B). Now suppose the agent's reasoning is commutative with respect to experiences. Thus, there can be no two experiences, x, y, such that x makes the agent change B > A to A > B and stay on A > B if the agent is there and y makes him change A > C to C > A and stay on C > A if the agent is there. If there were two such experiences, f would be commutative with regards to (C > A and A > B), which we assumed is not the case. Thus, it is either the case that:

(1) No experience is powerful enough to change one of these judgements. Or
(2) The same experience can change B > A to A > B and A > B to B > A.

Option (1) trivializes reasoning. Accepting (1) means accepting that out of every possible three alternatives, there is at least one pair of alternatives such that the agent's judgement regarding the moral superiority relation that holds between them is such that nothing can make the agent change it.

Whether option (2) also trivializes reasoning is a matter of interpretation, but it surely significantly limits the scope of instances of reasoning that remain open for the non-Bayesian. Accepting (2) means that out of every possible three alternatives there is at least one pair of alternatives such that the agent's judgement regarding the moral superiority relation that holds between them is such that only experiences that behave in a very special way can make the agent change them: experiences that make the agent reject the judgement that he holds (regarding the superiority relation between these two alternatives), no matter what it is.

Arguably some types of high-order evidence, such as pieces of evidence that undermine the evidential relevancy of pieces of evidence the agent already has, behave like this. However, intuitively, most instances of reasoning with CMJs are not made as a result of exposure to this kind of evidence. Usually, when we reason with CMJs we do so on the basis of first order evidence, e.g. after being exposed to an argument that makes us change a specific CMJ we hold in a specific way. The result shows that the non-Bayesian must accept that out of any three alternatives there are at least two alternatives such that the agent's CMJ regarding the moral superiority relation that holds between them is insensitive to such simple moral evidence.

I think it is clear that we want to allow for cases of reasoning that are based on this type of evidence. More generally, we want to allow for instances of reasoning that do not fall into either one of the two categories indicated above, and I think it is clear that anyone who is committed to the idea of reasoning with CMJs wants to allow for such cases too.

Notice that what made the transformation from impossibility to triviality so easy is the close connection, in the non-Bayesian framework, between experiences and their effects on agent's attitudes. For the non-Bayesian, granted that an experience is powerful enough to change one of the agent's CMJs, the change is binary. Thus, many different experiences (which constitute pieces of evidence with varying degrees of strength) that push in the same direction have exactly the same effect on the agent's attitudes. This is not the case for the Bayesian. For the Bayesian, stronger evidence leads to greater increase in degrees of belief.

So Bayesians use all the information to which they are exposed. Stronger evidence brings greater effect. Non-Bayesians, however, do not use all the information available to them. All pieces of evidence that do trigger a change bring the same change. Some information is lost. It is no wonder, then, that non-Bayesians get into trouble when it comes to reasoning.

6.6 Conclusion

The conclusion of this paper is quite strong: moral Bayesianism, the thesis according to which moral judgements are quantitative beliefs, is the only meta-ethical view that can account for moral reasoning. Nevertheless, the argument is sound.

Although the conclusion is strong, it should—I believe—be welcomed. As is demonstrated in Sect. 6.1, it is not only very intuitive, but also carries with it many exciting and yet unexplored opportunities.

Acknowledgments This research has been supported by the Israeli Science Foundation (grant number:1042/13). I thank Richard Bradley, Christian List and two anonymous referees for their useful suggestions.

References

1. Bradley, R., & List, C. (2009). Desire-as-belief revisited. *Analysis*, *69*(1), 31–37.
2. Broome, J. (1991). Desire, belief and expectation. *Mind*, *100*(2), 265–267.
3. Broome, J. (2006). Reasoning with preferences? *Royal Institute of Philosophy Supplements*, *59*, 183–208.
4. Easwaran, K. Dr. truthlove, or how i learned to stop worrying and love bayesian probabilities. forthcoming.
5. Easwaran, K., & Fitelson, B. (2015). Accuracy, coherence, and evidence. *Oxford Studies in Epistemology*, *5*, 61.
6. Field, H. (1978). A note on Jeffrey conditionalization. *Philosophy of Science*, *45*(3), 361–367.

7. Greene, J.D. (2007). The secret joke of kants soul. *Moral Psychology: Historical and Contemporary Readings*, 359–372.
8. Hansson, S. O. (1995). Changes in preference. *Theory and Decision*, *38*(1), 1–28.
9. Hansson, S. O. (2001). *The structure of values and norms*. Cambridge: Cambridge University Press.
10. Jackson, F., & Smith, M. (2006). Absolutist moral theories and uncertainty. *The Journal of philosophy*, 267–283.
11. Jeffrey, R. (1992). *Probability and the art of judgment*. Cambridge: Cambridge University Press.
12. Joyce, J. M. (1998). A nonpragmatic vindication of probabilism. *Philosophy of Science*, *65*(4), 575–603.
13. Kyburg, H.E. (1961). Probability and the logic of rational belief.
14. Lepage, F., & Morgan, C. (2011). Revision with conditional probability functions: Two impossibility results. In *Dynamic formal epistemology* (pp. 161–172). Berlin: Springer.
15. List, C. (2012). The theory of judgment aggregation: An introductory review. *Synthese*, *187*(1), 179–207.
16. Lockhart, T. (2000). Moral uncertainty and its consequences.
17. Mikhail, J. (2007). Universal moral grammar: Theory, evidence and the future. *Trends in cognitive sciences*, *11*(4), 143–152.
18. Nissan-Rozen, I. (2012). Doing the best one can: A new justification for the use of lotteries. *Erasmus Journal for Philosophy and Economics*, *5*(1), 45–72.
19. Oddie, G. (1994). Harmony, purity, truth. *Mind*, *103*(412), 451–472.
20. Piller, C. (2000). Doing what is best. *The Philosophical Quarterly*, *50*(199), 208–226.
21. Price, H. (1989). Defending desire-as-belief. *Mind*, *98*(389), 119–127.
22. Sepielli, A. (2009). What to do when you don't know what to do. *Oxford studies in Metaethics*, (4). Oxford University Press.
23. Singer, P. (1972). Famine, affluence, and morality. *Philosophy and Public Affairs*, 229–243.
24. Singer, P. (2005). Intuitions, heuristics, and utilitarianism. *Behavioral and Brain Sciences*, *28*, 560–1.
25. Smith, M. (2002). Evaluation, uncertainty and motivation. *Ethical Theory and Moral Practice*, *5*(3), 305–320.
26. Sunstein, C. R. (2005). Moral heuristics. *Behavioral and brain sciences*, *28*(4), 531–541.
27. Van Roojen, M. (2009). Moral cognitivism vs. non-cognitivism. *Stanford Encyclopaedia of Philosophy*.

Part II
Epistemology

Chapter 7
Is Theory Choice Using Epistemic Virtues Possible?

Kit Patrick and Kate Hodesdon

Abstract According to the popular 'epistemic virtue account' (EVA) of theory choice, we should choose between scientific theories on the basis of their epistemic virtues: empirical fit, simplicity, unity etc. More specifically, we should use a rule that aggregates theories' virtues into a ranking of the overall goodness of the theories. However, an application of Arrow's impossibility theorem shows that, given plausible premises, there is no rule that can aggregate theories' virtues into a theory ranking. The EVA-supporter might try to avoid the impossibility result of Arrow's theorem by asserting that we have more fine-grained distinctions between theories' epistemic virtues than initially supposed. We show that implausibly fine-grained distinctions between virtue quantities are necessary to escape the impossibility result. This is shown via novel proofs of Arrow's theorem for cases in which the quantities to be aggregated are measured on *any* combination of different scales of information, as is likely the case when aggregating epistemic virtues.

Keywords Theory choice · Arrow's theorem · Epistemic virtues · Simplicity · Unity

7.1 Introduction

According to the epistemic virtue account of scientific theory change (EVA) we ought to choose between scientific theories on the basis of their epistemic virtues. One theory is preferable to another if it is better with respect to some collection of epistemic virtues. An important virtue is 'empirical fit': how well the theory fits the observable data. But there are also other considerations important in theory choice

K. Patrick (✉)
Department of Philosophy, University of Bristol, Bristol, UK
e-mail: kit.patrick@bristol.ac.uk

K. Hodesdon
Laboratoire d'Histoire des Sciences et de Philosophie,
University of Lorraine, Nancy, France
e-mail: Kate@hodesdon.com

R. Urbaniak and G. Payette (eds.), *Applications of Formal Philosophy*,
Logic, Argumentation & Reasoning 14, DOI 10.1007/978-3-319-58507-9_7

such as simplicity, unifying power and capacity to make novel predictions. Given information about the extent to which each theory possesses each of the epistemic virtues, EVA claims that we can apply a rule to a collection of theories that ranks them in order of overall goodness and allows us to select the best one.

This article presents a powerful argument against EVA. Given certain plausible assumptions, an application of Arrow's theorem shows that there is *no rule* that an EVA-supporter could use to aggregate the information about how virtuous each theory is into a ranking of theories' overall goodness. This is the 'impossibility result'. The application of Arrow's theorem to theory choice takes theories to be ordered linearly according to how much of each of a number of epistemic virtues they possess. The proof of Arrow's theorem shows that there is no aggregation of theories on the basis of these virtue orderings which satisfies certain plausible assumptions. Howard Darmstadter [9] has used Arrow's theorem to argue that there is no way to aggregate scientists' personal preferences. In a recent paper, Samir Okasha [18] argues that the problem of theory choice on the basis of epistemic virtues is "formally identical" to the social choice problem that Arrow's impossibility theorem is typically applied to. Okasha argues, furthermore, for the plausibility of the premises of Arrow's theorem when applied to theory choice on the basis of epistemic virtues. After outlining the core commitments of EVA (Sect. 7.2), we show here that the EVA-supporter is committed to all of the plausible assumptions required to reach the impossibility results, except for two (Sect. 7.3). The first of these exceptions is the unrestricted domain assumption: that theories with any combination of epistemic virtues can be ranked. The second of these exceptions is the informational assumption: that for any two theories and a given virtue we have (only) ordinal information about the theories' possession of that virtue. This is to say, we know which one has more of that virtue, or that they have an equal amount of it.

In his paper, Okasha [18] argues the best hope for the scientific realist to avoid the impossibility result is to deny the informational assumption. Since this assumption is not implied by an EVA supporter's core commitments, denying it is one possible way to save EVA. Moreover, the EVA-ist has independent reason to deny the informational assumption of Arrow's theorem, for it does seem too strong: it seems that we can know *more* than simply the relative orderings of theories for each epistemic virtue. Thus the informational assumption is perhaps the least plausible of the plausible assumptions, and some weakening of it is independently motivated. This article demonstrates that every plausible weakening of the informational assumption fails to avoid the impossibility result, given EVA's core commitments (Sects. 7.4, 7.5, 7.6 and 7.7). This, we take it, renders problematic Okasha's conclusion that the EVA-ist had better deny the informational assumption of Arrow's theorem.

In particular, this article proves for the first time that the impossibility result holds for a range of ways of weakening Arrow's theorem that are very likely to obtain in the theory choice case. Hitherto, proofs of Arrow's theorem have dealt exclusively with cases where all the choices to be aggregated are measured with the same strength of information. Such proofs have been widely discussed and are well-known to game-theorists. It's easy to understand why these proofs have always dealt with cases where we aggregate choices with the same strength of information. After all, the original, paradigmatic application of Arrow's theorem is the aggregation of voters' preferences. In this application, the capacity to carry out aggregation over choices measured with mutually different scales is not an obvious desideratum. For, if we have a way of measuring one voter's choice using a cardinal scale, then in a fair election we ought to be able to measure every voter's preference in this way. However, in the new application of Arrow's theorem to the aggregation of theories' virtues, there is no reason to suppose that every epistemic virtue is measured with the same scale, or that it is meaningful to assign the same richness of information to measurements of each virtue. Therefore, aggregation over input orderings measured on a number of different scales is precisely what is needed for an application of Arrow's theorem to the problem of scientific theory choice. This article examines for the first time what aggregation rules, if any, are permissible when the choices to be aggregated are measured on mutually different information scales—some giving merely ordinal-scale information, others ratio-scale, and so on. We prove that Arrow's theorem holds for the aggregation of epistemic virtues measured on any plausible combination of informational scales. As a consequence, the EVA-ist can only avoid the impossibility result either by denying the unrestricted domain assumption, or by giving up the EVA account altogether.

7.2 The Epistemic Virtue Account

The view that scientific theory choice is determined by the epistemic virtues of the contending theories has been a dominant view in philosophy of science at least as far back as the scientific revolution. McMullin [16] gives a historical survey of philosophers and scientists who have endorsed this account. The details of the method by which theories' virtues are aggregated to determine overall theory choice has seldom been explicitly considered. Instead, debate has focused almost exclusively on what the virtues are. However, where virtue aggregation has been considered, a view much like EVA has been endorsed. Kuhn assumes that scientists aggregate theories' virtues as EVA describes [13, 18]. Graham Priest (forthcoming) asserts that EVA is standardly assumed in philosophy of science. Furthermore, formal methods of scientific theory choice in economics also assume that virtues are aggregated according to an algorithm much as EVA claims (see, e.g., Brock and Durlauf [5]; Bonilla [4]).

EVA has been used to defend realism from the underdetermination of theories by evidence [20]. But even anti-realists may endorse EVA: for example, constructive empiricists choose which empirically equivalent theory to accept on the basis of pragmatic virtues such as simplicity and explanatory power [25, p. 88]. If a constructive empiricist requires there to be a rule for aggregating these virtues, then she is endorsing EVA.[1]

Many details of EVA remain unsettled. For instance, there's widespread disagreement on what the epistemic virtues are.[2] This oversight is understandable: it's natural to think that until we know what the epistemic virtues are we won't know how to combine them. However, our arguments apply no matter what the epistemic virtues are and no matter how numerous they are. We restrict our consideration to virtues that are independent of one another in the sense that it is possible for theories to possess any combination of either.

According to EVA, theory choice is a two-stage process. First, we evaluate the extent to which each theory possesses each epistemic virtue. Second, we apply a rule that aggregates this information into an overall 'ranking' of theories, which lists every theory in order of preference. This ranking allows us to select the best theory. Thus EVA-supporters are committed to the following claims:

EVA-1: There is more than one epistemic virtue.

For if there were only one epistemic virtue, then the 'aggregation' stage would be redundant.

EVA-2: The ranking of theories is determined by their epistemic virtues alone.

We intend this in a strong sense: two theories' relative ranking depends only on those two theories' virtues, and not on the virtues that other theories have. EVA-2 distinguishes EVA from Kuhn's ([13]; see also [18]) and Duhem's ([7], see also [10]) views. They both endorse accounts of theory choice that involve epistemic virtues, but deny that there is a single rule or 'algorithm' that could be used to settle theory choice.

We can see EVA in use in Kepler's preference for Copernicus' theory over both Ptolemaic astronomy and Tycho Brahe's geocentric model. While empirical fit is an important consideration when scientists make such theory choices, empirical fit alone doesn't suffice to establish Copernicus' theory as best, for the fit of Copernicus' theory was worse than that of both Brahe and Ptolemy's theories. Copernicus' theory predicted unobserved stellar parallax and contradicted observed changes in the size of Venus; nevertheless, Kepler preferred Copernicus' theory because it was simpler. This historical case-study suggests a third commitment:

EVA-3: A small enough deficit in one epistemic virtue can always be outweighed by a big enough surplus in other epistemic virtues.

---

[1] Van Fraassen [25] doesn't subscribe to EVA since his voluntarism denies that we should follow any rule at all. Changes in one's degrees of belief can be rational even if one is following no hard and fast epistemic rule in making those changes.

[2] See e.g. [2, 14, 20].

7.3 Arrow's Theorem

The EVA-supporter needs to find a permissible aggregation rule that will tell them which theory is best. For this, the rule must give a total ordering of the theories; i.e. for every pair of theories considered, either they are ranked equally—in which case we will say the rule is indifferent between them—or one is ranked above the other. We reserve the term 'ranking' to describe a list of theories outputted by the aggregation rule, and denote it by '>'. The ranking must be transitive, so a cyclical ranking where A>B>C>A is no good. We will now give a deductive argument from premises describing constraints on permissible aggregation rules to the 'impossibility result' which says that no rules jointly satisfy them. Unlike many proofs in the social choice literature where the theorem originates, our proof is diagrammatic and intended to be non-technical.

7.3.1 *Premises of Arrow's Theorem*

Although Arrow originally intended his theorem to apply to democratic voting systems, the EVA-supporter will find that the premises are plausible when applied to theory choice. As stated above, Okasha [18] has also argued that the premises are plausible in this context. Two of the premises, however, are not implied by the EVA account. We present these first. For each other premise we indicate the aspect of the EVA account (EVA-1 to EVA-3) that implies it.

The **Informational Assumption** The only information that can be used in compiling the overall ranking is the order in which the virtues place the theories.
Unrestricted Domain (UD) The rule must provide a ranking for theories with *any combination* of virtues.[3]
Pareto Indifference (PI) For any two theories, T_1 and T_2, if they are ordered as equal with respect to every virtue (i.e. neither has strictly more of any virtue) then the rule must rank them indifferent to one another.
WeakPareto (WP) For any two theories, if one has more of every virtue than another then the rule must rank it higher.

If either PI or WP fail then, despite two theories being ordered in the same way with respect to each virtue, the rule may rank them in the opposite way. In this case, we can hardly be said to be choosing theories on the basis of virtues alone, thus violating EVA-2.

Independence of Irrelevant Alternatives (IIA) For any two distinct theories, T_1 and T_2, the rule must determine their relative output ranking independently of all information other than the amount of virtue of T_1 and T_2.

[3]Michael Morreau [17] explores whether UD or something weaker is true in the theory choice case.

In particular, the ranking does not depend on the virtues of any third theory, T_3. If IIA fails, then how we rank T_1 and T_2 is unaffected by how much virtue other theories have. This would violate EVA-2.

Non-dictatorship (ND) A rule is impermissible if there is some virtue such that if T_1 has more of that virtue than T_2, then the rule always ranks T_1 above T_2.

If there were a dictating virtue, then EVA-3 would be violated: a theory with a slight deficit in the dictating virtue compared to another theory could never be ranked higher than that second theory, no matter how much of the other virtues it had.

7.3.2 Proof of Arrow's Theorem for Two Epistemic Virtues

In this section we sketch a proof of Arrow's theorem for two epistemic virtues and at least three theories; no rules exist that aggregate the two virtues and satisfy all the constraints of Sect. 7.3.1 The sketched proof is adapted to theory choice from the proofs of Blackorby [3] and the reader can find full technical details there. Geanakoplos [8] offers further simple proofs. The primary aim of this section is to give a non-technical demonstration of Arrow's theorem that introduces principles we will use in later arguments.

We represent how much of a particular epistemic virtue theories have by plotting the theories on a smooth line. Together, UD IIA and PI entail that we can tell merely from where we plot theories on these lines how they are ranked relative to one another. In other words, the aggregation rule determines overall ranking as a function of position in the real space constructed by using these smooth lines as axes. This closely resembles Sen's approach [22, 23].[4] Since the informational assumption asserts that the only information that can be used by the aggregation rule is the order of theories' virtue, the aggregation rule must give the same result for any two ways of plotting theories that put them in the same order along each virtue axis. In other words, transforming one way of plotting the theories' virtues to another way of plotting which orders them the same must not yield a different aggregation rule, and we will say that the rule is 'blind' to such information-preserving transformations. So far as the aggregation rule is concerned, *order-isomorphic ways of plotting are equivalently aggregated.*

Given two epistemic virtues we can plot a theory T_1 in the two-dimensional virtue space. Divide theories around T_1 into four quadrants: I, II, III and IV (see Diagram 7.1).

A: The rule must rank each theory in quadrant I above T_1.

By weak Pareto, if one theory has more of every virtue than another, then the rule must rank it higher. Therefore, every theory in I is ranked above T_1, and likewise, every theory in II is ranked below T_1 (Diagram 7.2. The hollow arrows in quadrants

[4]If we take the plotting of the theories on each smooth line to represent a utility function, then our aggregation rule is simply Sen's 'social welfare functional'.

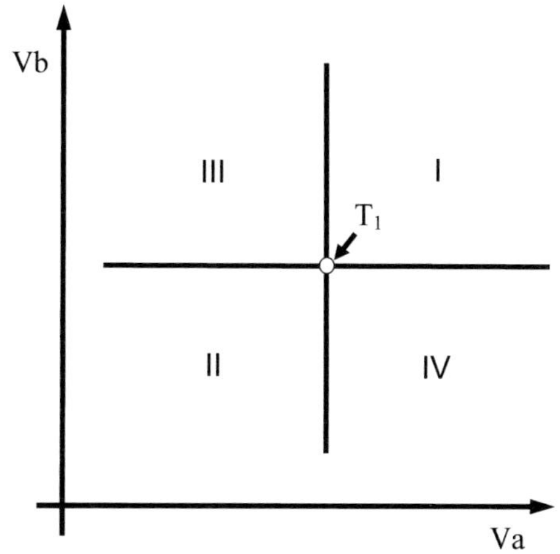

Diagram 7.1 An epistemic virtue space for two virtues V_a and V_b, with some theory T_1

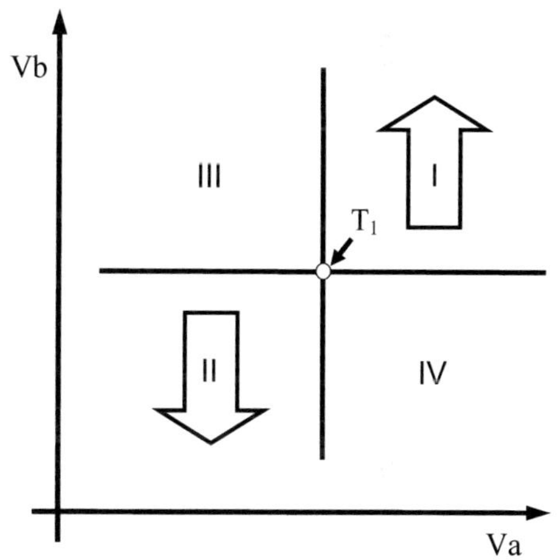

Diagram 7.2 By application of weak Pareto, all theories in quadrants I and II must be ranked above and below T_1 respectively, as indicated by the hollow arrows

I and II indicate that all theories in those quadrants are ranked above and below T_1 respectively).

B: The rule must rank each theory in quadrant III equivalently with respect to T_1. Consider the transformation from T_1 to itself, and $T_2 \rightarrow T_2{}^*$ where $T_2{}^*$ is any theory in quadrant III (see Diagram 7.3). This transformation is order-preserving along both the V_a and the V_b axes. Talk of mapping a theory T to a theory T′ does not commit us to claiming that there actually exists a theory T′ distinct from T with the corresponding virtue scores: we are just using the way of speaking to describe rescaling of the virtue axes. By the informational assumption, the rule must therefore rank T_2 and $T_2{}^*$ identically with respect to T_1. A parallel argument establishes that the rule must rank all theories in IV identically with respect to T_1.

C: The rule cannot rank all theories in quadrant III indifferent to T_1.

Diagram 7.3 Theories in quadrant III cannot be ranked indifferent to T_1

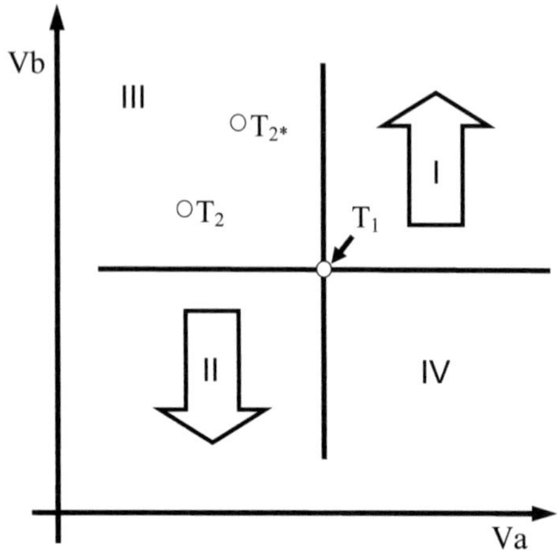

Diagram 7.4 Theories in quadrant III must be ranked opposite to those in quadrant IV with respect to T_1

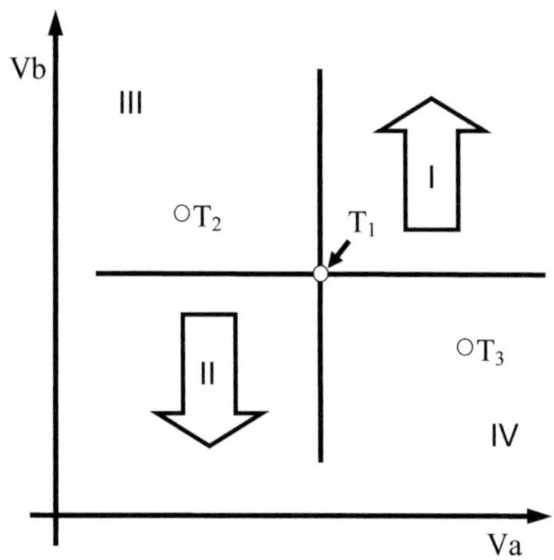

Suppose to the contrary that all theories in III are ranked indifferent to T_1. By transitivity of ranking, all theories in III must be ranked indifferent to one another. However, by UD, there are pairs of theories in III such that one has more of every virtue than the other, and so by WP must be ranked above it (e.g. T_2 and $T_2{*}$ in Diagram 7.3). By contradiction then, theories in III cannot all be indifferent to T_1.

D: The rule must rank theories in quadrants III and IV opposite with respect to T_1.

For any theories T_2 in III and T_3 in IV, the transformation that takes $T_2 \rightarrow T_1$, and $T_1 \rightarrow T_3$ (Diagram 7.4) preserves the order between theories. By the informational assumption, applying this transformation does not change the theories' ranking. Thus, if T_2 is ranked above T_1, then T_1 must be ranked above T_3, and vice versa, if T_2 is ranked below T_1, then T_1 must be ranked below T_3. But, by B, T_2 is ranked above T_1 if and only if all theories in III are ranked above T_1, and T_3 is ranked below T_1 if

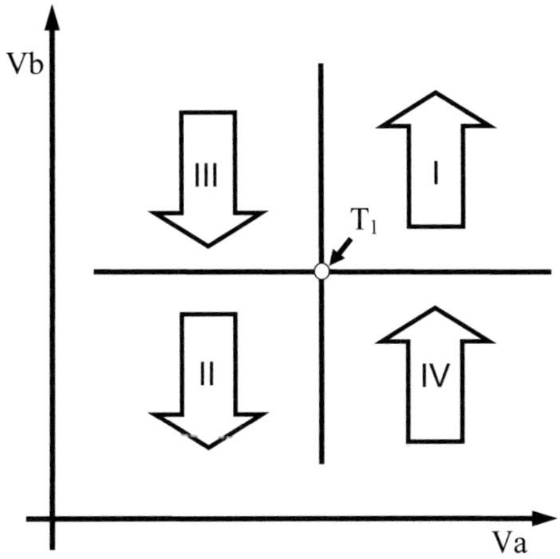

Diagram 7.5 On the first possibility, there is a dictatorship of V_a

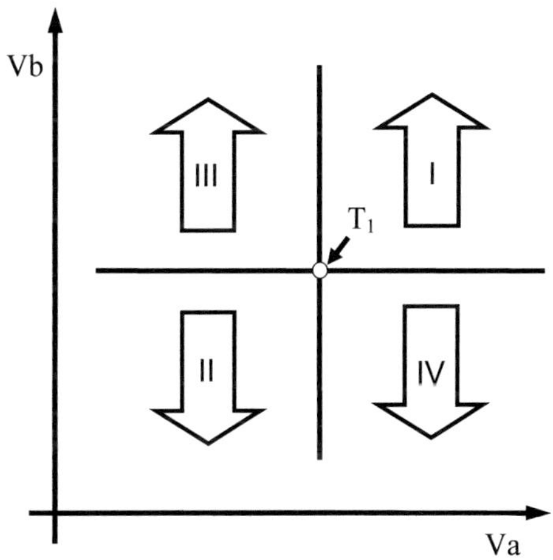

Diagram 7.6 On the second possibility, there is a dictatorship of V_b

and only if all theories in IV are ranked below T_1. Thus, all theories in III are ranked above T_1 if and only if all theories in IV are ranked below it.

Conclusion: the rule must make one virtue a dictator

Two options remain for the ranking of theories in these quadrants, described in Diagrams 7.5 and 7.6. In Diagram 7.5, if any theory has more V_a than T_1 then it is ranked higher, whereas in Diagram 7.6 the same holds for V_b. Thus, the only available rules are dictatorships and these are impermissible. We leave it to the reader to show that any virtue that is dictator over T_1 must also be dictator over all theories in the space.

Our proof above showed that no rule which aggregates theories on the basis of two epistemic virtues satisfies the premises of Arrow's theorem. For conciseness, we briefly indicate how to extend the proof to the case that there are any number n of epistemic virtues. We may generally represent a theory as a point in n-dimensional space. The two-dimensional proof can easily be adapted to show that for any two virtues, any plane of these two virtues in the n-dimensional space, one of those two virtues must be plane dictator (i.e. for any theories on that plane, if one has more of the plane dictating virtue, then it will be ranked higher).[5] Moreover, it's easy to show that for any two virtues, if one virtue is plane dictator over another on one such plane, it must be plane dictator over all such planes. So we can establish that there is a transitive chain of plain dictators. The topmost of these is dictator: if one theory has more of this virtue than another, then that first theory is ranked higher.

7.3.3 The Impossibility Result Is Fatal to EVA

As noted in Sect. 7.3.2, the arguments above establish the impossibility result only if we can consider three or more theories at a time (this was required for step D). This leads Okasha [18, pp. 12–13] to suggest that scientists can avoid the impossibility result by only choosing between two theories at a time. However, this strategy cannot save EVA. There are two circumstances in which we decide between only two theories at a time: (i) where there are many rival theories but we restrict choice to just two at a time, e.g. scientists chose between Aristotelian physics and Newtonian physics and only later between Newtonian physics and Relativity; and (ii) where there are only two rival theories possible, e.g. a choice between geocentrism and non-geocentrism.

In the 'many rival' case, no rule is permissible: whatever rule we use will run into trouble. The arguments of Sects. 7.3.1 and 7.3.2 show that when our rule considers all rival theories, it either contradicts one of the criteria UD, PI, WP, IIA, ND, or it ranks theories cyclically. If our rule contradicts one of the criteria when it considers all rival theories then it will still contradict that criterion when it considers only two theories. For example, if a rule ranks higher any theory with more virtue V_i when there are three theories, then it must also rank higher any theory with more V_i when there are two theories. Whether a theory considers many rival theories or just two has no bearing on whether or not that rule is a dictatorship.

On the other hand, if the rule ranks some theories cyclically when it considers all rivals at once then it won't rank those theories cyclically when it considers them two at a time, since cyclicity requires more than two theories. But, in this case the rule will be impermissible for another reason: it will make which theory we ought to endorse depend on the order in which we consider theories. For example, consider the cyclical ranking: Relativity>Newton>Aristotle>Relativity. Suppose we start by

[5] A $V_i - V_j$ plane in n-dimensional space, for virtues V_i, V_j, is a subspace comprising all the points of the form $(p_1, p_2, \ldots, p_i, \ldots, p_j, \ldots p_n)$ for p_1 to p_n all fixed except p_i, and p_j, which range over all virtue scores in V_i and V_j respectively. Thus it is parallel to the V_i and V_j axes in the sense that each point on the plane is plotted on the same place as each other point on the plane for each virtue other than V_i and V_j. To visualize this for three virtues, consider V_i and V_j as the ground. A $V_i - V_j$ plane is any area parallel to the ground, at any elevation. This is a more specific notion of plane than is standardly used in geometry.

considering Aristotle and Newton; then the rule says we ought to endorse Newton. Next we consider Newton and Relativity theory, and the rule says we ought to endorse Relativity theory. However, if we start by considering Newton vs. Relativity, then the rule says we ought to endorse Relativity. Next we decide between Relativity and Aristotle, the rule says we ought to endorse Aristotle. And surely any rule which makes which theory we endorse depend on the order in which we consider theories is impermissible. This argument undermines Baumann's claim in [2] that we can live with cyclical ranking because discovering more evidence usually resolves the cyclical ranking.

In the case where there are only two rival theories available, some rules will be permissible. But this won't save EVA, because EVA isn't reliable when there are just two rivals. It will often lead us to endorse theories even when those theories are badly wrong. For example, non-geocentrism is less virtuous than geocentrism because non-geocentrism is the disjunction of every theory that contradicts geocentrism. It is not clear that this disjunction should be considered a theory at all, but if it is, then it will certainly not be simple and will make very few empirical claims. So if simplicity and empirical fit are important epistemic virtues, then EVA will wrongly recommend geocentrism. EVA is only a plausible account of theory choice in cases when there are many rival theories.

7.4 Weakening the Informational Assumption

The EVA-supporter may try to avoid the impossibility result by weakening the informational assumption. Indeed, Okasha suggests that this is arguably the most attractive 'escape route' " from the impossibility result [18, p. 97]. For the original informational assumption does seem too strong. It asserts that the *only* information available to the aggregation rule is theories' ordering for each virtue. But for some commonly suggested epistemic virtues, we seem to have *more* information that we can use in the aggregation process to decide the best theory. For example, we seem to know more than that Copernican theory is simpler than Ptolemaic theory; we also know that Copernican theory is *much* simpler than Ptolemaic theory. And we don't simply know that Ptolemaic theory fits the empirical evidence better than Copernican theory; we know it fits the empirical evidence a *little* better. This extra information leads us to believe that EVA-supporters who take simplicity to be an epistemic virtue should weaken the informational assumption.

If the EVA-supporter weakens the information assumption enough then they can avoid the impossibility result, as originally shown by Sen [22, 23] and adapted to the theory choice case by Okasha [18]. Here's one way. Suppose we could have absolute numerical scores for each theory's simplicity, empirical fit, unity, etc.; Copernican theory might have 100 'simplons' and 45 'empiricons' compared to Ptolemaic theory's 40 simplons and 50 empiricons. Given such rich information, it's easy to find permissible, non-cyclic rules. One permissible rule is to multiply the absolute numbers of each virtue a theory possesses. The theory with the highest total score wins.

However, we don't know absolute numerical values for each theory's simplicity, empirical fit, unity, etc. It's not clear whether such absolute values would even be meaningful. Thus, although weakening the informational assumption this much would allow the EVA supporter to avoid the impossibility result, such a weakening is implausible because it presupposes that we have access to information about theories' virtues that is just not attainable.

How much must we weaken the informational assumption, and is such a weakening plausible? Amartya Sen's famous work [22, 23] addressed the former question by giving a taxonomy of the richness of information which might be available, and an extended discussion of how much information is required to avoid the impossibility result. Okasha [18] addressed the latter question by placing Sen's ideas in the theory choice context, and suggesting that the best hope for avoiding Arrow's theorem is to find some way of weakening the informational assumption that is both plausible, and is shown by Sen to avoid the impossibility result. However, Sen and Okasha only consider cases where we have the same amount of information for each voter or virtue. They don't establish how much information is required to avoid the impossibility result when we have different amounts of information for different virtues or voters, and so do not show that it is plausible that we have such rich information.

In Sects. 7.5, 7.6 and 7.7 we argue for our main result, that there is no such desirable 'Goldilocks' level of information: information is either too impoverished to avoid Arrow's impossibility result or too rich to obtain. In order to show this result, we consider a range of weakenings that neither Okasha, nor Sen, nor anyone else does, but which very plausibly obtain in the theory choice case: those where we have different information scales for different virtues. In the current section we lay the groundwork for this argument by describing four information scales.

We now present the new information scales alongside the equivalence classes of plottings that any rule constrained by the corresponding informational assumption must treat in the same way.[6] If we have richer information about theories' virtues, we can make finer-grained distinctions between different ways of plotting theories. Transforming all theories equivalently is just a kind of rescaling of virtue scores, like moving from degrees Fahrenheit to degrees Celsius.

[6] A note on the relation between our taxonomy and Sen's: the original work by Sen [22] distinguishes two ways of changing the informational assumption: by varying restrictions on measurability (e.g. ordinal-scale, cardinal-scale, ratio-scale, unit-scale, absolute-scale), and by varying the restrictions on comparability. However, as Okasha [18, 103–105] points out, in the theory choice setting it is prima facie implausible that there is comparability between virtues. Okasha does observe that there is comparability that arises naturally when we have sufficiently strong measurability. For example, if we have absolute-scale information about two virtues then this entails that they are fully comparable. But his only examples of plausible comparability are of the sort entailed by virtues' measurability. Thus it suffices to talk only about theories' measurability. Furthermore, we are considering cases where virtues have different amounts of measurability. This would make the discussion of the resulting comparability of virtues excessively complex. And, if Okasha is right, this excessive complexity has no benefit since the only plausible comparability between virtues is entailed by the measurability of those virtues. Therefore, we have chosen to not explicitly include comparability in our discussion.

Cardinal-Scale Measurability *A virtue is cardinal-scale measurable if and only if the information available to the aggregation rule about that virtue is the order in which theories fall with respect to that virtue, and the ratios of their differences in values for each virtue.*

A virtue is cardinal scale measurable if we have information about the order of theories with respect to that virtue and about facts such as 'the difference between T_1's virtue and T_2's virtue is three times the difference between T_1's virtue and T_3's virtue.' If a virtue is cardinal-scale measurable then any two ways of plotting a profile of theories are equivalent if and only if they are related by a transformation from v to av+b. These are just those ordinal-scale transformations defined by linear equations.

Unit-Scale Measurability *A virtue is unit-scale measurable if and only if the information available to the aggregation rule about that virtue is the order in which theories fall and absolute differences between virtue scores.*

A virtue is unit-scale, if in addition to all the information available for cardinal-scale virtues, we have access to facts such as 'the difference between T_1's virtue and T_2's virtue is 4.5'. For example, converting temperature in Celsius to Kelvin preserves the absolute differences in temperatures. If a virtue is unit-scale measurable then two ways of plotting a profile are equivalent if they are related by a transformation of the form $v \rightarrow v+b$. These are just the cardinal-scale transformations with a gradient of 1.[7]

Ratio-Scale measurability *A virtue is ratio-scale measurable if and only if the information available to the aggregation rule about that virtue is the order in which theories fall, and ratios between their virtue scores.*

A virtue is ratio-scale measurable if, in addition to all the information available for cardinal-scale virtues, we have information of facts of the form 'T_1 has twice as much virtue as T_2'. For example, kilograms and pounds give different representations of mass, and are related by a ratio-preserving transformation: a 10lb box is twice as heavy as a 5lb one, and remains twice as heavy when we weigh them in kilograms. Unlike ordinal-scale, cardinal-scale and unit-scale virtues, ratio-scale virtues have a meaningful 0 point. If a virtue is ratio-scale then any two ways of plotting a profile are equivalent only if they are related by a transformation of the form $v \rightarrow av$ for any $a > 0$. These are just the cardinal-scale transformations for which $b = 0$.

There are important logical relations between the different scales: cardinal-scale transformations are a limiting case of ordinal-scale transformations. Both unit-scale and ratio-scale transformations are limiting cases of ordinal-scale transformations. Finally, absolute-scale transformations are limiting cases of both unit-scale and cardinal-scale. These logical relations can be seen using graphical examples:

[7]Having all virtues unit-scale is a special case of Sen's 'cardinal-scale-unit-comparability'.

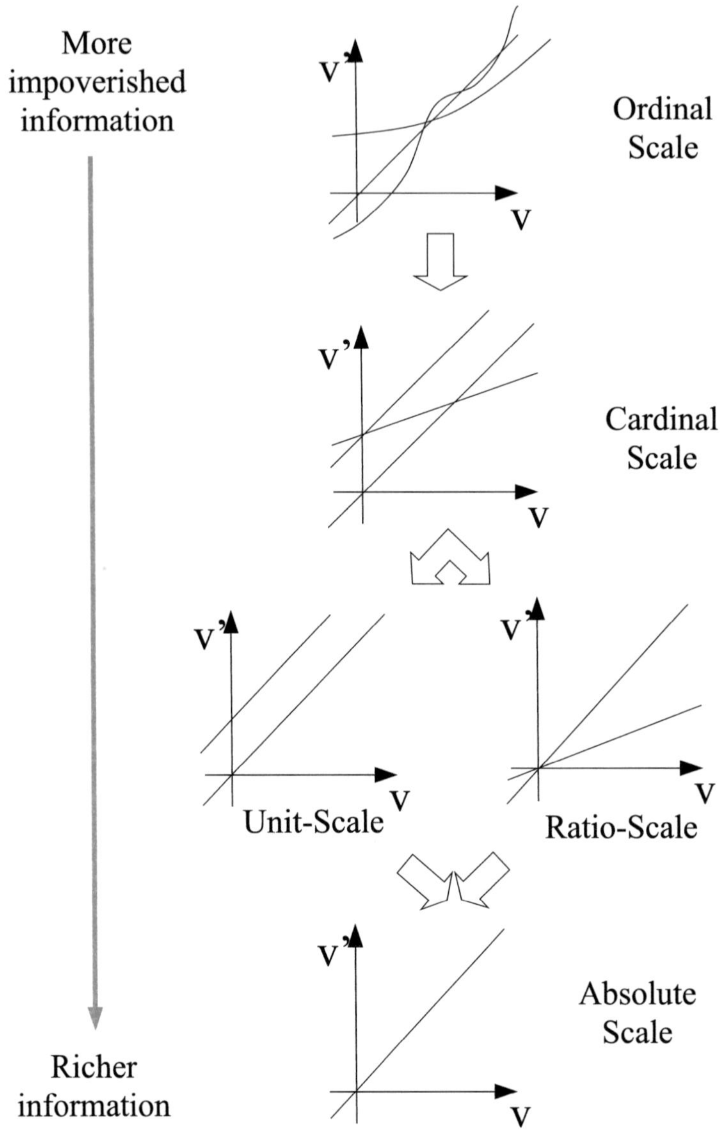

For the rest of this article we assume that we have unit-scale, ratio-scale or weaker information for all epistemic virtues. This is plausible for the commonly suggested epistemic virtues such as simplicity and unity. Indeed, it seems overgenerous to the EVA-supporter. It's difficult to find any epistemic virtues that we plausibly have even ratio-scale or unit-scale information about. Okasha [18] provides an example of how we might get ratio-scale information about empirical fit and simplicity. However, his example concerns line-fitting exercises, and it's not clear how we can find such specific information for choices between more complex theories such as Copernican and Ptolemaic theory.[8] At any rate, the burden of proof lies with the EVA-supporter to show that we have richer information than this.

In the following sections we prove that no aggregation rule is permissible when virtues have different combinations of these information scales. These proofs assume the premises of Sect. 7.3.2. An additional premise is used by these proofs:

Continuity For any point in virtue space, the points in virtue space where a theory would be ranked higher form a closed set, and the points in virtue space where a theory would be ranked lower form a closed set.

Roughly, continuity implies that infinitesimally tiny differences in a virtue cannot make a theory that is worse than our theories suddenly better.

In our argument, continuity is assumed largely for the sake of convenience. The social choice literature often assumes continuity without comment, as if it were merely a technical detail. Continuity has been justified by appeal to the idea that if continuity were false then getting the right ranking may depend on exactly precise measurement of people's preferences, and this is impractical [6]. This justification has been questioned by [1, pp. 323–325]. Parallel justification might be proposed in the theory choice case. However, for the circumstances we consider, the rules which are possible if we allow violations of continuity are every bit as unacceptable as dictatorships, as we will argue. Therefore, for most of the arguments that follow, the assumption of continuity is one of convenience. We sketch in Sect. 7.6 how parallel arguments can be made without the assumption of continuity. The arguments of Sect. 7.7 do rest on substantial assumptions of continuity, and we offer a defense of this assumption there.

7.5 No Rules are Permissible for Cardinal-Cardinal Theory Choice

If all virtues are cardinal-scale, then no aggregation rule is permissible, since all ordinal-scale transformations in the Arrow proof we gave initially (Sect. 7.3) can be cardinal-scale. Thus a parallel argument shows that when there are two or more cardinal-scale virtues, there is no permissible aggregation rule. Another such parallel argument implies that there is no permissible aggregation rule for theory choice on the basis of any combination of cardinal-scale and ordinal-scale virtues.

8Okasha's suggestion appeals to Akaike's theorem. For an in-depth discussion of why Akaike considerations cannot help us choose between theories such as Copernican and Ptolemaic theory see [12]. So, even if Okasha's suggestion were to work out, the arguments of this article will show that the EVA cannot be a good model of theory choice for the central cases supporters of EVA think that it is, but rather only for very simple cases of theory choice. However, Okasha's suggestion may not work out after all. For a discussion that the Akaike considerations cannot tell us about simplicity even in the simple line consideration case, see [19]. For a survey of other problems for this, and similar attempts to characterize simplicity and empirical fit see e.g. [11].

7.6 No Rules are Permissible for Unit-Cardinal Theory Choice or Ratio-Cardinal Theory Choice

In this section, we consider theory choice on the basis of *different* strengths of information about the different virtues. This hasn't been discussed in the extensive literature on Arrow's theorem. As mentioned above, the reason for this is probably that the Arrow's theorem literature has focused on democratic voting, and we naturally assume that we have the same depth of information for all voters. If we find a way to get say ratio-scale information about one voter, then we can also get ratio-scale information about any other voters. Quite the contrary is true for the theory choice case: *in this context, there is no reason to suppose that we have the same depth of information about simplicity, unity, empirical fit, scope etc.* It is therefore worth investigating whether any rules are available when the virtues have different information scales. We present proofs investigating this novel situation of mixed information scales. Our results may be of interest outside the theory choice context, since they will apply to any situation in which an aggregated ordering is decided upon on the basis of factors that are measured on different scales. In the following sections we describe the availability of aggregation rules for each two-virtue combination of the scales defined above. We will use the naming convention 'unit-cardinal measurability' (or simply 'unit-cardinal') to name the scale of information available to the aggregation rule in a two-dimensional virtue space in which one virtue is measured on a unit-scale axis and the other on a cardinal-scale axis. The same convention applies for any pair of information scales.

7.6.1 Unit-Cardinal Theory Choice

We show that when there is one unit-scale virtue V_u, and one cardinal-scale virtue V_c, no rule is permissible. Let T_1 be a theory at any point in the two-dimensional virtue space. Draw quadrants I-IV around T_1 as usual.

A: The rule must rank all theories in quadrant I above T_1 and everything in quadrant II below T_1.

This follows step A in Sect. 7.3 exactly.

B: There is a straight line through III of theories ranking identically with respect to T_1

Let T_2 be a theory in quadrant III. Consider the transformation from $T_1 \rightarrow T_2$ and $T_2 \rightarrow T_3$, where T_3 is chosen such that the difference between T_1 and T_3 is twice the difference between T_1 as T_2 on both virtue axes (see Diagram 7.7). This transformation preserves absolute differences of virtue scores, so is a unit-scale transformation on both virtues V_u and V_c. Since unit-scale transformations are a special case of cardinal-scale transformations, this transformation also is cardinal-scale on V_c. Therefore, the rule must rank T_1 with respect to T_2 the same as it ranks T_2 with respect to T_3. If T_2 is ranked above T_1 then T_3 must be ranked above T_2; thus

Diagram 7.7 Where virtue V_c is cardinal-scale and virtue V_u is ordinal-scale, theories T_2–T_4 must be ranked the same with respect to T_1

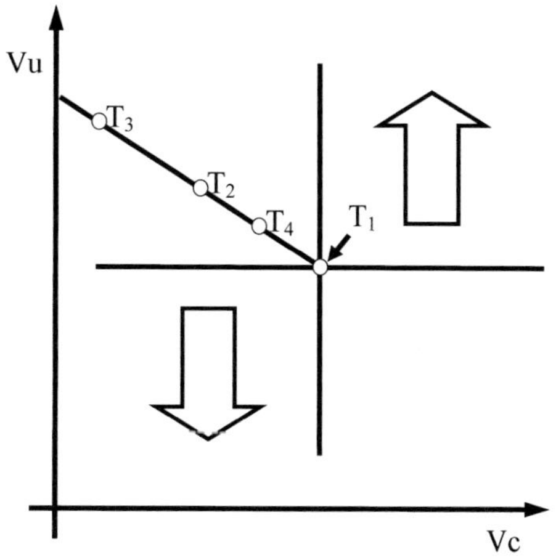

$T_3 > T_2 > T_1$. By transitivity, $T_3 > T_1$. Likewise, if T_2 is ranked below or indifferent to T_1, then T_3 must also be ranked below or indifferent to T_1 respectively.

By a similar procedure, we can find further points that are ranked the same with respect to T_1 as T_2. Consider the transformation from $T_1 \rightarrow T_4$ and $T_4 \rightarrow T_2$, where T_4 is chosen such that the difference between T_1 and T_4 is half the difference between T_1 as T_2 on both virtue axes. This transformation preserves absolute differences of virtue scores, so again the rule is blind to the transformation and must rank T_4 with respect to T_1, the same as it ranks T_2 with respect to T_4. By transitivity, it follows that T_4 and T_2 are ranked the same with respect to T_1. And consider the transformation from $T_2 \rightarrow T_3$ and $T_3 \rightarrow T_5$, where T_5 is chosen such that the difference between T_2 and T_5 is twice the difference between T_2 as T_3 on both virtue axes. This transformation preserves absolute differences of virtue scores, so again the rule is blind to the transformation and, given transitivity, must rank T_5 with respect to T_1 the same as it ranks T_2, T_3 and T_4 with respect to T_1. Iterate this procedure to construct a densely packed set of points along a straight line from T_1 passing through T_2 in quadrant III, each of which are ranked the same with respect to T_1.

The argument above establishes that a densely packed series S of points on a line are all ranked the same with respect to T_1. By the continuity assumption (Sect. 7.4), we establish that the completion of this densely packed set of points, i.e. the full line, is ranked equivalently with respect to T_1. Suppose that all the points in S are ranked above T_1. By continuity, their limit must also be ranked above T_1, therefore the gaps in the densely packed series are also ranked above T_1. Likewise, suppose that all the points in S are ranked below T_1. By continuity, their limit must also be ranked below T_1 too, therefore the gaps in the densely packed series are also ranked below T_1. Finally, suppose that all the points in S are ranked indifferent to T_1. Continuity would also ensure that the intervening points were ranked indifferent to T_1; they couldn't be ranked above or below T_1 since the points in the gap have as their limit a point on S. In any case, weak Pareto rules out the possibility that

Diagram 7.8 T_6 and T_7 must be ranked the same with respect to T_1

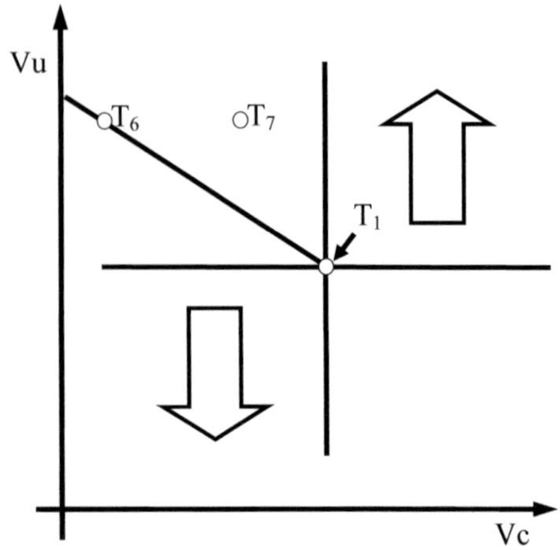

the intervening points are ranked indifferent to T_1.[9] Likewise, a similar line can be constructed passing through quadrant IV.

C: Every point in quadrant III must be ranked the same with respect toT_1.

Let T_6 be any point on the line of points ranked identically with respect to T_1. Consider the transformation that fixes T_1 and sends T_6 to T_7, for T_7 any point in quadrant III with the same amount of V_u as T_6 (see Diagram 7.8). This transformation doesn't change theories' V_u scores and is a cardinal-scale transformation on the V_c axis. Therefore, the rule must rank T_6 and T_7 identically with respect to T_1. A parallel argument shows that any point in quadrant III with the same V_u score as some point on the line must be ranked identically with respect to T_1. That is, the rule must rank all points in quadrant III identically with respect to T_1.

[9] We can drop the continuity assumption here, and in Sect. 7.6.2, by strengthening the non-dictatorship assumption in an inoffensive way. If we drop continuity, the only rules available are still clearly impermissible. Here we briefly sketch how this can be proved. We have the densely packed points in quadrant III, each of which is ranked the same with respect to T_1. By a procedure parallel to that in C we can construct bands throughout quadrant IV such that any theory plotted in these bands must be ranked the same with respect to T_1. (Note that such bands could not have been constructed if neither of the virtues had been cardinal-scale or weaker, as is the case in Sect. 7.7; for that reason we cannot drop the continuity assumption from the proofs of Sect. 7.7, and give a defense of it there.) By applying weak Pareto, we can then 'fill in the gaps'. Suppose that all the points in the bands are ranked above T_1, then we can show by weak Pareto that all points in quadrant III with more of both virtues than any point in these bands must also be ranked above T_1. Suppose that all the points in the bands are ranked below T_1, then we can show by weak Pareto that all points in quadrant III with less of both virtues than any point in these bands must also be ranked above T_1. Thus the only permissible rules are 'near-dictatorship' rules: those where there is a near-dictating virtue such that if one theory has even a fraction more of that virtue than another, then that theory is ranked higher. Such near-dictatorships are clearly impermissible for the same reasons that dictatorships are.

Diagram 7.9 Where virtue V_c is cardinal-scale and virtue V_r is ratio-scale, theories T_2–T_4 must be ranked the same with respect to T_1

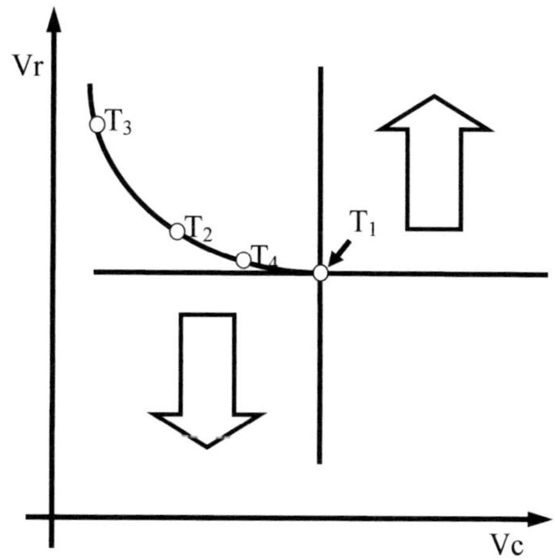

From here the proof proceeds exactly as from stage C in Sect. 7.3. Since cardinal-scale transformations are a special case of ordinal-scale transformations, the same proof will also show that there is no permissible rule for unit-ordinal theory choice.

7.6.2 *Ratio-Cardinal Theory Choice*

When there is one ratio-scale virtue, V_r, and one cardinal-scale virtue, V_c, no rule is permissible. Let T_1 be a theory at any point in the virtue score space. Draw quadrants I—IV about T_1 as usual. Let T_2 be a theory in quadrant III. For brevity, we introduce a labeling convention: for theory T_1 and virtue V_a, label T_1's virtue V_a score 'T_{1a}'.

A: The rule must rank all theories in quadrant I above T_1 and everything in quadrant II below T_1.

This follows A in Sect. 7.3.

B: There is a line through III of theories ranking identically with respect to T_1.

Consider any point T_2 in quadrant III relative to T_1. The ratio-scale co-ordinate for T_1 is T_{1r} and the ratio-scale co-ordinate for T_2 is T_{2r}. We can express T_{2r} as a multiple of T_{1r} to obtain the ratio between them: trivially, $T_{2r} = \frac{T_{2r}}{T_{1r}} T_{1r}$. Since T_2 has less of V_r than T_1, $\frac{T_{2r}}{T_{1r}}$ is less than 1.

Without loss of generality, let T_2 be ranked above T_1. We will show that there is a line starting from T_1 and passing through T_2 such that every point on it is ranked above T_1. We can plot the point T_3 that is such that the ratio between T_1 and T_2 is the same as the ratio between T_2 and T_3 along the ratio-scale axes. There is thus a transformation mapping T_2 to T_3 and T_1 to T_2 that is ratio-preserving along all axes. Since $T_{2r} = \frac{T_{2r}}{T_{1r}} T_{1r}$, in order to preserve the ratio of $\frac{T_{2r}}{T_{1r}}$ between points on the line we have $T_{3r} = \left(\frac{T_{2r}}{T_{1r}}\right)^2 T_{1r}$. But this means that $T_{3r} = \frac{T_{2r}}{T_{1r}} T_{2r}$. By the same process of

Diagram 7.10 T_5 and T_6 must be ranked the same with respect to T_1

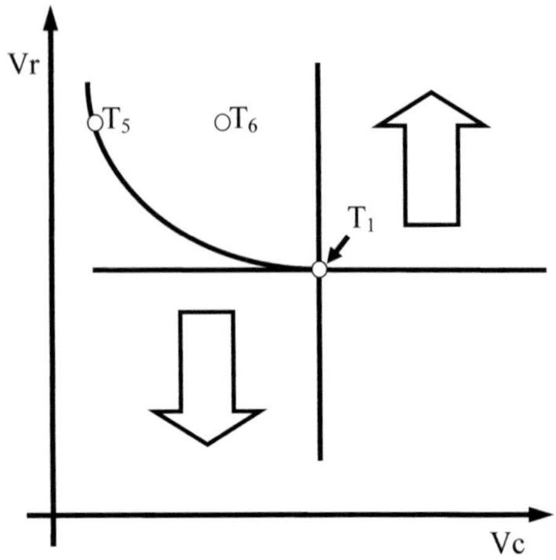

preserving ratios we can find the co-ordinate for T_3's virtue score for the cardinal-scale virtue. Since ratio-scale transformations are a special case of cardinal-scale transformations, the transformation $T_1 \rightarrow T_2, T_2 \rightarrow T_3$ is also cardinal-scale on V_c. For this reason, the rule must rank T_1 with respect to T_2 the same as it ranks T_2 with respect to T_3. By transitivity, T_2 and T_3 are ranked identically with respect to T_1, i.e., both above it.

To show that there is a line of points all of which are ranked above T_1 we need to show that there are theories in between T_1 and T_2 that are densely packed along the line and are ranked above T_1. This is achieved iteratively by taking midpoints of points already known to be on the line. So, we need to find a point $T_4 = (T_{4r}, T_{4c})$ in between T_1 and T_2 which is such that the ratio between T_1 and T_2 and between T_2 and T_3 is the same along the ratio-scale axis. Given this requirement, T_{4r} must therefore be:

$$\left(\frac{T_{2r}}{T_{1r}}\right)^{\frac{1}{2}} T_{2r}.$$

We can find the virtue score of T_4 for the other axis by likewise finding the point in between T_{1c} and T_{2c} that is related to these in the same ratio-preserving way. All these points trace out a line, c=(T_{1r} T_{1c})/r. By continuity, all the points on the line are ranked equivalently with respect to T_1 (Diagram 7.9).

C: Every point in quadrant III must be ranked the same with respect to T_1.

Consider the transformation fixing $T_1 \rightarrow T_1$ and taking $T_5 \rightarrow T_6$ where T_5 is any theory on the line, and T_6 is any theory in quadrant III with the same V_r score as T_5 (see Diagram 7.10). This transformation yields no change in V_r, and is a cardinal-scale transformation on V_c. Therefore, the rule must rank T_5 and T_6 identically with respect to T_1; in this case, both are ranked above it.

A parallel argument shows that any point in quadrant III with the same V_r score as some point on the line must be ranked identically with respect to T_1. So the rule must rank all points in quadrant III with positive scores for ratio-scale virtue V_r identically with respect to T_1. From here the proof proceeds exactly as from stage C in Sect. 7.3. Since cardinal-scale transformations are a special case of ordinal-scale transformations, the same proof will show that there is no permissible rule for unit-ordinal theory choice.

7.6.3 Intermediate Conclusions

We've already established enough to entail conclusions that will be unacceptable to the EVA-supporter. For brevity, distinguish between 'weak' virtues—those about which our information is at most as rich as cardinal-scale; and 'strong' virtues—those about which we have richer than cardinal-scale information, up to and including unit- and ratio-scale information. If all virtues are weak there is no permissible rule, as established in Sect. 7.5. If there is only one strong virtue and all other virtues are weak then there can be no permissible rule either. For in this case, there will only be two sorts of planes: those involving two weak virtues and those involving one strong and one weak virtue. We can use the arguments above to show that on both sorts of planes one of the virtues is a 'dictator', and then use the n-virtue generalization of Arrow's theorem (Sect. 7.3.2) to show that there is no permissible aggregation rule.

We have not yet proved anything about virtue spaces with more than one strong virtue. We now show that if there is more than one strong virtue and all other virtues are weak, then a theory's weak virtues can only make a difference to that theory's ranking in very narrowly defined circumstances. Weak virtues are 'marginalized': that is, they are only used to decide between two theories that are judged indifferent on the basis of all their strong virtues. This is to say that a lexicographic ordering results, which the strong virtue dominates. To prove this, it's sufficient to consider the simple three virtue case with one weak virtue 'V_w' and two strong virtues 'V_{s1}' and 'V_{s2}'. On the V_w-V_{s1} or V_w-V_{s2} planes there are three possible combinations of plane-dictatorships:

(i) The weak virtue plane-dictates over both strong virtues.
(ii) The weak virtue plane-dictates over one strong virtue, and is plane-dictated over by the other.
(iii) The weak virtue is plane-dictated over by both strong virtues.

Any rule that satisfies (i) is a dictatorship. Any rule that satisfies (ii) has cyclical output rankings. But any rule that satisfies (iii) marginalises the weak virtue. We prove each claim in turn.

We know from the generalized proof of Arrow's theorem that amongst weak virtues there is always a transitive chain of plane-dictators, with one as topmost virtue. The dictator on the V_{s1}-V_{s2} plane is either the dictator over all weak virtues or it is plane-dictated over by such a virtue. Either way, we have a dictating virtue for each plane.

Claim (i): Any rule where the weak virtue plane-dictates over both strong virtues is a dictatorship of V_w. To show this, we'll prove that for two theories T_1 and T_2 such that T_2 has more of the weak virtue V_w than T_1 does, T_2 is ranked higher than T_1. For, consider the theory T′ that has the following properties: T′ is on the same V_{s1}-V_{s2} plane as T_1, and T′ is on the same V_w-V_{s2} plane as T_2. T′ has more of virtue V_{s2} than T_1 does, but less of V_w than T_2 does.

Then, since T′ is identical to T_1 except for having greater V_{s2}, it is ranked above or indifferent to T_1, by weak Pareto. But since T′ is on the same V_w-V_{s2} plane as T_2, and has less V_w than it, T′ is ranked below T_2. And so, we have $T_2 > T' \geq T_1$, establishing, by transitivity, the result we wanted.

Claim (ii) Whenever the weak virtue plane-dictates over one strong virtue, and is plane-dictated over by the other, the only aggregation rules give cyclical rankings, and so are impermissible.

To generate such cyclical rankings, consider a single V_{s1}-V_{s2} plane. Find two theories, T_1 and T_2 that are not ranked indifferent to one another, such that one has more V_{s1} and the other more V_{s2}. Such pair of theories exist by the original Arrow proof (see D of Sect. 7.4). Without loss of generality, let T_2 have less V_{s1} and more V_{s2} than T_1, and be ranked above T_1. We now show that there is some theory T′ that must be ranked above T_2 and below T_1. Choose T′ so that it has greater V_w than T_2, and the same V_{s2}. Since V_w dictates over V_{s2}, T_3 must be ranked above T_2. Since V_{s1} dictates over V_w, T_1 must be ranked above T_2. In summary, $T_1 < T_2 < T' < T_1$; a cyclical output ranking. The same argument applies *mutatis mutandis* to show that when T_2 is ranked below T_1, there must be a cyclical output ranking.

Claim (iii): If the weak virtue is plane-dictated over by both strong virtues, then this means that the weak virtue is marginalized: although permissible aggregation rules may exist in this case, they are such that we only appeal to the weak virtue to decide the ranking of theories which have identical scores for both the strong virtues.[10]

In summary, there are three options available to the EVA-supporter. All are unattractive. First, they could bite the bullet and accept that virtues such as simplicity and unifying power which were previously thought important in theory choice are marginalized. But EVA is attractive largely because it seems plausible that simplicity and unifying power *do* typically influence theory choice. Second, EVA-supporters could argue that there is an absolute scale virtue. We rejected this claim as implausible in Sect. 7.4. Third, EVA-supporters could claim that simplicity, unifying power and all other virtues commonly used to determine theory choice are strong virtues; we have richer than cardinal-scale information about them. This seems implausible:

---

[10]To see this, consider any two theories T_1 and T_2, such that T_1 has more V_w. Let T_{1*} be a theory that is the same as T_1except that it has the same amount of V_w as T_2 (i.e. it is the projection of T_1 onto the $V_{s1} - V_{s2}$ plane). The rule must rank T_1 and T_{1*} identically with respect to T_2; this is to say that $T_1 > T_2$ if and only if $T_{1*} > T_2$.

We show this by considering a theory $T_{1'}$ which has the same amount of each virtue as T_{1*} except for slightly less of the strong virtue. We will assume that $T_{1'}$ and T_{1*} are ranked identically with respect to T_2—either both above or both below it. Since the strong virtue dictates, T_1 must be ranked above $T_{1'}$. And so we have $T_1{*} > T_{1'} > T_2$.

claims such as 'T_1 has 10 simplons more than T_2' or 'T_1 is 1.5 times as simple as T_2' seem unknowable or meaningless, as argued above (see Sect. 7.4). Furthermore, the following section shows that even if some virtues are strong, given plausible assumptions there are still no rules available.

7.7 No Rules for Unit-Unit Theory Choice, Ratio-Ratio Theory Choice and Unit-Ratio Theory Choice

In this section, we argue that no rules are permissible for the EVA-supporter for unit-unit, ratio-ratio and unit-ratio theory choice. There are rules compatible with the original Arrow premises (Sect. 7.3.2) in these cases, but all are in serious tension with two assumptions that the EVA-supporter will find hard to deny. These assumptions are described and motivated in Sect. 7.7.1. Then we describe all rules compatible with one assumption (Sect. 7.7.2) and show that they are all incompatible, with the other assumption (Sect. 7.7.3).

7.7.1 Two Assumptions Plausible to the EVA-Supporter

The EVA-supporter will find it hard to deny:

The **trade-off assumption** Two theories with different virtue scores can be ranked indifferent to one another.

Given the EVA-supporter's other commitments, denying trade-off is implausible: the EVA-supporter is already committed to saying that for any two theories, the one with slightly less of some virtue can still be ranked higher if it has sufficiently more of other virtues, else the rule is a dictatorship. If a slight deficit in one virtue can be overcompensated for by a big surplus in other virtues, it seems likely that a deficit in one virtue can be compensated for by a slightly smaller surplus in other virtues. The EVA-supporter will also find it hard to deny:

Strong Pareto (SP) For any two theories, if one has more of at least one virtue and the same amount of all others, then the rule must rank it above the other theory.

Strong Pareto is independently plausible. If, say, simplicity is an epistemic virtue then it seems that a theory's simplicity should also be a reason to prefer it to another theory with the same amount of other virtues. Strong Pareto will be especially plausible to those who appeal to EVA to avoid the underdetermination of theory by evidence. According to all versions of the underdetermination of theory by evidence, there are sometimes two theories that make importantly different claims about the world but have equal empirical fit. This is supposed to challenge realism: if we decide between theories solely on the basis of empirical fit then we cannot know which theory to endorse. A standard realist response has been to appeal to extra-empirical virtues such as simplicity and unity. Thus, such EVA-supporters are committed to something much like strong Pareto. In fact, even weakening strong Pareto significantly won't save EVA, as shown in Sect. 7.7.3

7.7.2 *All Rules Compatible with Trade-Off*

Given the premises in Sect. 7.3.2 and the trade-off assumption for unit-unit, ratio-ratio and unit-ratio theory choice respectively, we now show which rules are permissible.

7.7.2.1 Rules for Unit-Unit Theory Choice

The only permissible rules where both virtues are unit-scale are *weighted utilitarian rules*. A weighted utilitarian rule works by adding up the scores of a theory for each of the virtues. One theory is ranked above another in the output just in case that theory has a higher weighted utility. The virtues may be differently weighted so that an increase in, say, simplicity might be less significant than the same increase in empirical fit. Dictatorship is a limiting case of the weighted utilitarian view, where one virtue has all the weight and the other virtues have none.

The trade-off assumption asserts that some theories with different virtue score are ranked indifferent to one another. Choose T_1 and T_2 to be two such theories. Construct a straight line passing through T_1 and T_2 in the way described in Sect. 7.6.1 (see Diagram 7.11). This is an 'indifference line': theories on this line can be expressed as having the same weighted utility score $W=Ax+By$, where A and B are constants determined by the gradient of the line. Since they have the same weighted utility score, every theory on this line must be ranked the same with respect to T_1, namely, indifferent to it. And since indifference is transitive, every theory on the line must also be ranked indifferent to every other theory on the line.

Given this one indifference line, we can construct another parallel to it. Consider the transformation from $T_1 \rightarrow T_1{}^*$ and $T_2 \rightarrow T_2{}^*$, where $T_2{}^*$ is chosen so that the absolute difference between T_1 and $T_1{}^*$ is the same as that between T_2 and $T_2{}^*$. Because the rule is blind to these transformations it follows from the fact that T_1 is ranked indifferent to T_2 that $T_1{}^*$ is ranked indifferent to $T_2{}^*$. We can now construct a new indifference line passing through $T_1{}^*$ and $T_2{}^*$ by a method parallel to that outlined in the paragraph above, replacing T_1 with $T_1{}^*$ and $T_2$with $T_2{}^*$ mutatis mutandis. Any theory on this line must be ranked indifferent to every other. This line will be parallel to the first line, but will have a higher weighted utility score. By repeating this step, we can establish indifference lines parallel to the first throughout the virtue score space. Since each indifference line corresponds to a weighted utility score, weak Pareto entails that theories on indifference lines with a higher weighted utility score must be ranked higher than those theories with a lower weighted utility score. This completes the proof that in the unit-unit case, where the trade-off assumption is true, the only permissible rules are weighted utilitarian rules.

Diagram 7.11 Where virtues Vu_1 and Vu_2 are unit-scale, T_1 and T_2 share an indifference line, and T_1* and T_2* share an indifference line

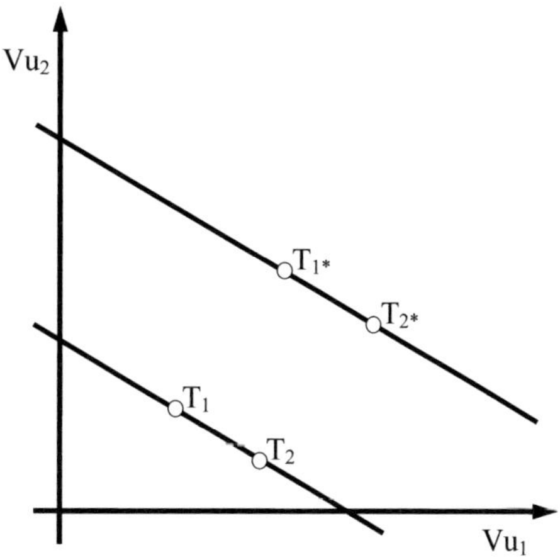

On the other hand, in the unit-unit case where the trade-off assumption is false, the only permissible rules are dominated weighted utilitarian rules. These rules are similar to weighted utilitarian rules. Indeed, for theories that do not share a weighted utility score, a dominated weighted utilitarian rule is just a weighted utilitarian rule: any theory with a higher score (on a higher line) is ranked above a theory with a lower score. For theories that do share a weighted utility score, matters are different. On (undominated) weighted utilitarian rules such theories are ranked indifferent to one another. But plainly that is not possible if the trade-off assumption is false. Instead, on a dominated weighted utilitarian rule, there is some virtue such that for any two theories with the same weighted utility score whichever theory has more of that virtue is ranked above the theory with less of that virtue. We leave it to the reader to adapt the proof above to show this.

7.7.2.2 Rules for Ratio-Ratio Theory Choice

The ratio-ratio case follows almost exactly the same as the unit-unit case. The only permissible rules are Cobb-Douglas rules according to which theories with a higher score $W = x^a y^b$ (for constants a and b) are ranked higher than those with a lower W score. Given the trade-off assumption, theories with exactly the same W score are ranked indifferent to one another. For proofs that these are the only rules compatible with Arrow's assumptions see ([2]; Tsui and Weymark, [24]).

7.7.2.3 Rules for Unit-Ratio Scale Theory Choice

The case where one virtue is unit-scale and the other ratio-scale follows the same pattern as that of ratio-ratio theory choice and unit-unit theory choice. In each of these

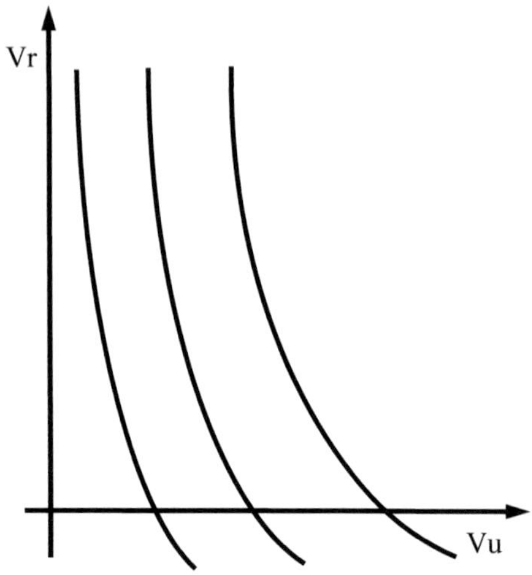

Diagram 7.12 Indifference lines where virtue V_u is unit-scale and virtue V_r is ratio-scale

cases, theory ranking is preserved under transformations that preserve, respectively, the ratios or unit differences between all theories with respect to each axis. In ratio-unit theory choice theory ranking is preserved under those transformations that are ratio-preserving along the V_r-axis, and unit-preserving in the V_u-axis. Given two theories ranked indifferent by ratio-unit aggregation, we construct an indifference line by transforming the indifferent points in a ratio-preserving way along the ratio axis, and a unit-preserving way along the unit axis. This gives us indifference lines of the form $x = \frac{1}{x-n}$ where y is the unit-scale variable and x the ratio-scale variable (Diagram 7.12). Note that this line is invalid when $x = n$, so the unit-ratio rule is going to violate unrestricted domain.

7.7.3 These Rules are Impermissible to the EVA-Supporter

All these rules are either incompatible with strong Pareto, or incompatible with the motivation for strong Pareto. The rules described in Sect. 7.7.1 entail:

Continuity* For any theory T_1, if we draw a segment through virtue space such that theories at one end of the line are ranked above T_1, and theories at the other end are ranked below T_1 then there is some point along this segment at which theories are ranked indifferent to T_1.

To see that continuity* as defined above follows from the rules described in Sect. 7.7.2, notice that any such segment must pass through T_1's indifference line. Thus, there must be some point on it indifferent to T_1.

It's no surprise that the rules just considered entail continuity*, for it is entailed by the continuity assumption. Indeed, several discussions of Arrow's theorem treat

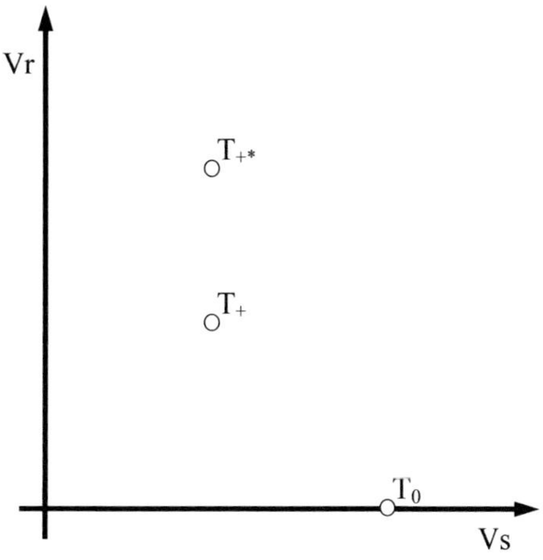

Diagram 7.13 By application of continuity* and strong Pareto, T_0 cannot be ranked indifferent to T_+

continuity as equivalent to continuity* (e.g. [21]). It will be difficult for the EVA theorist to escape the impossibility result by denying continuity*, because that would require endorsing rules according to which infinitesimal changes in theories' virtue can determine which theory is ranked higher. It's prima facie implausible to suggest that such a tiny difference in virtue can make such a significant difference in theory ranking. Suppose the EVA supporter can show that we have unit-scale information about empirical fit. Suppose further that two scientists agree on how much empirical fit Copernican theory has, but only up to the nearest rational number of empirical fit, and agree on the other virtues of Copernican and Ptolemaic theory. If continuity were false, this would be compatible with one scientist being forced to rank Copernican theory above Ptolemaic theory, and the other being forced to rank Ptolemaic theory above Copernican theory, despite only differing in their opinion to an infinitesimally small degree. At any rate, the burden of proof is with the EVA supporter to show that we have such rich information, and that scientists determine which theory is ranked higher on such very fine distinctions of virtue.

Continuity* is incompatible with Strong Pareto. Our proof for this is adapted from Tsui and Weymark ([24]). They discuss only the case where all virtues are ratio-scale. We extend this proof beyond ratio-ratio theory choice, and weaken the assumptions required for the proof to work. Suppose there are two virtues: V_r which is ratio-scale or weaker, and V_s which is either unit-scale, ratio-scale or weaker. We obtain a contradiction by showing that any theory T_0 with none of virtue V_r can be neither above, indifferent to, nor below any theory T_+ with positive amounts of both V_r and V_s.

Claim : T_0 cannot be ranked indifferent to T_+

Suppose the contrary: T_0 is ranked indifferent to T_+. Let T_{+*} be any theory that has the same amount of V_s and more V_r than T_+ (see Diagram 7.13). The transformation from $T_+ \rightarrow T_{+*}$ and $T_0 \rightarrow T_0$ preserves the ratios of theories on V_r and involves no transformation on V_s. Therefore the rule must rank T_0 versus T_+ the same as it

Diagram 7.14 By application of continuity* and weak Pareto, T_0 cannot be ranked above T_+

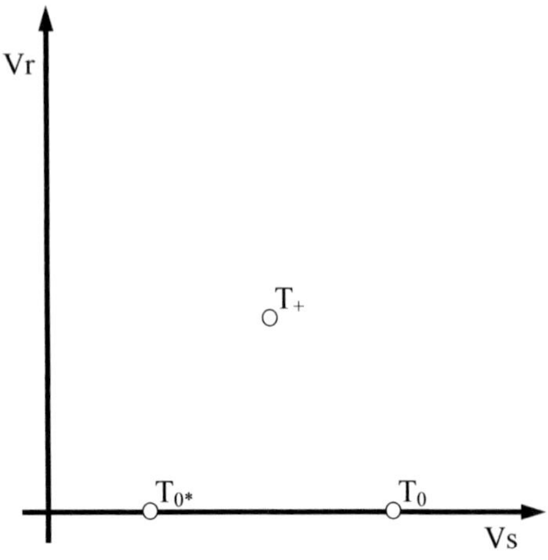

ranks T_0 versus T_{+*}. Since by supposition, T_0 is ranked indifferent to T_+, T_0 must also be ranked indifferent to T_{+*}. Because indifference is transitive, T_+ must be ranked indifferent to T_{+*}. On the other hand, Strong Pareto entails that T_{+*} is ranked above T_+, since it has more of one virtue and less of none. In summary, if T_+ is ranked indifferent to T_0, then T_{+*} is ranked both indifferent to and above T_+. This is impermissible. Therefore, the aggregation rule cannot rank T_+ indifferent to T_0 after all.

Claim : T_0 cannot be ranked indifferent to T_+

Suppose the contrary: T_0 is ranked above T_+. Consider any theory T_{0*} that has less V_s than T_0 and T_+, and V_r (see Diagram 7.14). By weak Pareto, T_0* will be ranked below T_+ since it has less of every virtue. Therefore, there is a line from T_0 to T_0*, at one end of which theories are ranked above T_+ and at the other end of which theories are ranked below T+. The continuity* assumption asserts that there must be a point along this line indifferent to T_+. But, by the argument above, there is no such point. Therefore, T_0 cannot be ranked higher than T_+.

Claim : T_0 cannot be ranked indifferent to T_+

Suppose the contrary: T_+ is ranked above T_0. Strong Pareto entails that T_0* is ranked below T_0, since it has less of one virtue and more of none. Therefore, there is a line from T_0 to T_+, at one end of which theories are ranked above $T_+$and at the other end of which theories are ranked below T_0. The continuity* assumption asserts that there must be a point along this line indifferent to T_0. But this cannot be: we already know by the argument above that there is no theory with positive 'V_r' that is indifferent to T_0. And the only point on this line that does not have positive 'V_r' T_0*. But we know by strong Pareto that T_0* is not ranked indifferent to T_0.

Thus, theory T_0 cannot be ranked above, indifferent, or below T_+. This contradicts Unrestricted Domain, which requires any rule to rank all the theories against all other theories. Therefore, given the trade-off assumption and Strong Pareto, there is no permissible rule for ratio-ratio and unit-ratio theory choice.

This proof doesn't show that there is no permissible rule for unit-unit theory choice. In unit-unit theory choice, the weighted utilitarian rue is the only rule compatible with the trade-off assumption. And, although this rule entails continuity*, continuity* is only inconsistent with strong Pareto when at least one virtue is ratio-scale or weaker. In unit-unit theory choice, no virtues are ratio-scale or weaker. Nevertheless, the unit-unit rule undermines many EVA-supporters' reason for accepting strong Pareto. For, as we argued above, many EVA-supporters accept EVA to avoid underdetermination of theory by evidence. If the weighted utilitarian rule is accepted then an even worse form of underdetermination arises, for each theory sits on an indifference line dense with mutually indifferent theories. This is considerably stronger and more threatening to realism than the relatively harmless trade-off assumption, which claimed that there are at least two theories that ought to be ranked indifferent to one another: for if the weighted utilitarian rule is correct then every theory ought to be ranked indifferent to numerous others. Thus a weighted utilitarian account of theory choice replaces underdetermination of theory by evidence with the even more threatening underdetermination of theory by empirical virtues. This will be unacceptable to many EVA supporters. And we suggest that they should declare the weighted utilitarian rule impermissible.

In Sect. 7.7 we have argued that for unit-unit, ratio-ratio and unit-ratio theory choice all rules are implausible. This has broader consequences. Suppose that there are more than two virtues, all of which are unit-scale or ratio-scale or weaker. Arguments parallel to those showing that there is no permissible rule with unit-unit, ratio-ratio and unit-ratio can be used to show that any plane in this multi-virtue theory choice must have a 'dictator'. So, no matter how many virtues there are or how much information we have about them, there are no aggregation rules.

7.8 Conclusion

We have shown that the EVA-supporter can only avoid the impossibility result either by denying the informational assumption or by denying unrestricted domain. We have also shown that the EVA-supporter cannot plausibly avoid the impossibility result by denying the information assumption. For information is either too impoverished to avoid the impossibility result, or too strong to plausibly obtain. Therefore, it seems that the best options are either to deny unrestricted domain or abandon EVA altogether. Our argument might be extended to apply to other instances of theory choice using virtues, such as truthlikeness ([26]), inference to the best explanation ([15]) and even choices between philosophical theories [14, p. 17].

References

1. Arrow, K., Sen, A., & Suzumura, K. (2011). *Handbook of social choice and welfare* (Vol. 2). Oxford: North-Holand.
2. Baumann, P. (2005). Theory choice and the intransitivity of 'Is a Better Theory Than'. *Philosophy of Science*, *72*, 231–240.
3. Blackorby, C., Donaldson, D., & Weymark, J. A. (1984). Social choice with interpersonal utility comparisons: A diagrammatic introduction. *International Economic Review*, *25*, 327–356.
4. Bonilla, Z. J. (2012). The economics of scientific knowledge. In U. Mäki (Ed.), *Handbook of the philosophy of science: Philosophy of economics*. Elsevier.
5. Brock, W. A., & Durlauf, S. N. (1999). A formal model of theory choice in science. *Economic Theory*, *14*, 113–130.
6. Chichilnisky, G. (1982). Social aggregation rules and continuity. *The Quarterly Journal of Economics*, *25*, 337–352.
7. Duhem, P. (1954). *The aim and structure of physical theory*. Princeton: Princeton University Press.
8. Geanakopolos, J. (1996). Three brief proofs of arrow's impossibility theorem. *Economic Theory*, *26*, 211–215.
9. Howard, D. (1975). Better theories. *Philosophy of Science*, *42*, 20–27.
10. Ivanova, M. (2009). Pierre duhem's good sense as a guide to theory choice. *Studies in History and Philosophy of Science Part A*, *41*, 58–64.
11. Kelly, K. (2007). Ockham's razor, empirical complexity, and truth-finding efficiency. *Theoretical Computer Science*, *383*, 270–289.
12. Kieseppa, I. (1997). Akaike information criterion, curve-fitting, and the philosophical problem of simplicity. *The British Journal for the Philosophy of Science*, *48*, 21–48.
13. Kuhn, T. (1969). *The structure of scientific revolutions*. Chicago: University of Chicago.
14. Ladyman, J., & Ross, D. (2007). *Every thing must go: Metaphysics naturalized*. Oxford University Press.
15. Lipton, P. (1993). *Inference to the best explanation* (2 ed.). Routledge.
16. McMullin, E. (2008). The virtues of a good theory. In S. Psillos & M. Curd (Eds.), *The Routledge companion to the philosophy of science* (pp. 499–508). Routledge.
17. Morreau, M. (2014). Mr. Fit, Mr. Simplicity and Mr. Scope: From social choice to theory choice. *Erkenntnis*, *79*, 1253–1268.
18. Okasha, S. (2011). Theory choice and social choice: Kuhn versus Arrow. *Mind*, *120*, 83–115.
19. Patrick, K. (forthcoming). *Why the Akaike and BIC measures have nothing to do with simplicity*.
20. Psillos, S. (1999). *Scientific realism: How science tracks truth*. London: Routledge.
21. Roemer, J. E. (2004). Eclectic distributional ethics'. *Politics, Philosophy and Economics*, *3*, 267–281.
22. Sen, A. (1970). *Collective choice and social welfare*. San Francisco: Holden-Day.
23. Sen, A. (1977). On weights and measures: Informational constraints in social welfare analysis. *Econometrica*, *45*.
24. Tsui, K.-Y., & Weymark, J. A. (1997). Social welfare orderings for ratio-scale measurable utilities. *Economic Theory*, *10*, 241–56.
25. van Fraassen, B. (1989). *Laws and symmetry*. Oxford: Clarendon Press.
26. Zwart, S., & Franssen, M. (2009). An impossibility theorem for Verisimilitude. *Synthese*, *158*, 75–92.

Chapter 8
Abduction Logics: Illustrating Pitfalls of Defeasible Methods

Diderik Batens

Abstract On the one hand this paper offers an introduction to adaptive logics, focussing on properties that are imposed upon adaptive logics by the fact that they explicate defeasible reasoning. On the other hand new adaptive logics of abduction are presented and employed to illustrate those properties. These logics were developed in view of the criticism to existing adaptive logics of abduction.

Keywords Abduction · Adaptive logics · Defeasible reasoning

8.1 Aim of This Article

This paper has a Janus-faced aim. On the one hand it offers an introduction to the large and diverse set of adaptive logics in terms of their common structure, the *standard format*. The focus is on properties that are imposed on adaptive logics by the fact that they explicate defeasible reasoning. On the other hand new adaptive logics of abduction are presented and employed to illustrate those properties. After some historical comments in Sect. 8.2 and a brief introduction to the logical problem of abduction in Sect. 8.3, adaptive logics are presented as a means to characterize methods in Sect. 8.4. The standard format is introduced stepwise in Sect. 8.5 and each step is illustrated in terms of abductive logics. The need for varying abduction logics and the ease with which this is done in adaptive terms is spelled out in Sect. 8.6. The promised properties are presented in Sect. 8.7 and some closing comments are gathered in Sect. 8.8.

The reason for taking adaptive logics for abduction as an illustration was a note by Frederik Van de Putte. The note argues that all publications on adaptive logics for abduction are affected by certain defects. As the note is not published, I shall neither repeat the contained criticism, nor summarize it, nor comment upon it. In as far as I shall criticize former work on adaptive logics of abduction, the criticism will not be Frederik's. However, and for the record, Frederik's note convinced me to spend time on the matter.

D. Batens (✉)
Centre for Logic and Philosophy of Science, Ghent University, Ghent, Belgium
e-mail: Diderik.Batens@UGent.be

R. Urbaniak and G. Payette (eds.), *Applications of Formal Philosophy*,
Logic, Argumentation & Reasoning 14, DOI 10.1007/978-3-319-58507-9_8

Before starting, some warnings are in place. Adaptive logics are formal logics, but are *not* deductive logics. By a logic, I mean a couple consisting of a language $\mathcal{L}$ and a mapping $\wp(\mathcal{W}) \rightarrow \wp(\mathcal{W})$, in which $\mathcal{W}$ is the set of closed formulas of $\mathcal{L}$—for obvious reasons, the language is often handled implicitly. A logic is formal iff its consequence relation is defined in terms of *logical forms*.[1] Rather than being deductive logics, which guarantee that the consequences of true premises are true, adaptive logics explicate defeasible reasoning processes. The importance of this function should not be underestimated. Most reasoning is defeasible and all knowledge ultimately relies on defeasible reasoning.

All logics considered in this paper will have the standard predicative language schema $\mathcal{L}_s$ or an extension of it as their language schema. The set of open formulas of $\mathcal{L}_s$ will be $\mathcal{F}_s$ and the set of closed formulas $\mathcal{W}_s$.

A warning of a very different sort is that the metatheoretic language will always be classical—metalinguistic logical symbols will have their classical meaning. Moreover, "false" will have its classical meaning in that "A is false" will be identified with the classical negation of "A is true".

The following metavariables, possibly with subscripts, will be used in the *metalanguage*: $A, B, C, \ldots$for (open and closed) formulas, α, β, γ for individual variables and constants, and π for predicates. Where α names a variable and β a constant, $A(\alpha)$ is a formula in which at most α is free and $A(\beta)$ is the result obtained by replacing every free occurrence of α in $A(\alpha)$ by β.

8.2 Historical Note

It seems useful to summarize the origin of inconsistency-adaptive logics because they were the starting point. Devising them in the late 1970s, I relied on previous work in paraconsistency, including an attempt to organise a large set of paraconsistent logics as extensions of a basic system [2]. My tactics was to think about concrete derivations from premise sets, the vague idea being that inconsistencies may occur, but cause a problem for the reasoning process and require a change to this process. The paradigmatic example was a theory that was intended as consistent, but turned out inconsistent, a situation familiar to historians of science.

Work in paraconsistency had revealed that some classically valid inferences do not hold in models that verify one or more inconsistencies. Thus models that verify A as well as $\neg A$ verify $A \vee B$ as well as $\neg A$, even if they falsify B. So Disjunctive Syllogism is invalid in logics that have such models. While reasoning from one's knowledge requires paraconsistency, is does not require one to give up the consistency

[1] Formal logics need not validate Uniform Substitution. Let $\mathcal{S}$, $\mathcal{P}^r$, and $\mathcal{C}$ be respectively the sets of sentential letters, predicates of rank r, and individual constants of $\mathcal{L}$. Let f be a one-one mapping such that $f: \mathcal{S} \longrightarrow \mathcal{S}$, $f: \mathcal{P}^r \longrightarrow \mathcal{P}^r$, and $f: \mathcal{C} \longrightarrow \mathcal{C}$. Extend f first to formulas, $f(A)$ being the result of replacing in A every $\xi \in \mathcal{S} \cup \mathcal{P}^r \cup \mathcal{C}$ by $f(\xi)$; next to sets of formulas: $f(\Gamma) =_{df} \{f(A) \mid A \in \Gamma\}$. That $\Gamma \vdash_{\mathbf{L}} A$ iff $f(\Gamma) \vdash_{\mathbf{L}} f(A)$ is sufficient for **L** to be formal. Incidentally, the Uniform Substitution rule is tiresome at the predicative level [39].

presupposition altogether. One may still consider inconsistencies as abnormalities, which occur only in exceptional situations. If both $A \vee B$ and $\neg A$ are given, then, in the normal situation in which $A \wedge \neg A$ is false, one is justified in deriving B. If, however, it turns out that $A \wedge \neg A$ has to be considered as true in view of the premises, then the derivation of B is unjustified.[2] Some inferences may be considered as unavoidable, for example the inference of A from $A \wedge B$ as well as, given the proper understanding of classical implication, the inference of B from A and $A \supset B$. Meanwhile, other inferences may be seen as contingent on the falsehood of certain inconsistencies, for example the inference of B from $A \vee B$ and $\neg A$, which is contingent on the falsehood of $A \wedge \neg A$, and the inference of $\neg A$ from $\neg B$ and $A \supset B$, which is contingent on the falsehood of $B \wedge \neg B$.[3]

All this led to the idea to consider formulas of the form $A \wedge \neg A$ ($A \in \mathcal{W}_s$) as *abnormalities*, as formulas that are taken to be false unless and until it is shown that the premises require them to be true, or require them to belong to a minimal set of formulas that has at least one true member. The second disjunct refers to the fact that a premise set may entail a disjunction of abnormalities—this will be called a Dab-formula below. For the predicative version, the set of abnormalities is often defined by the form $\exists(A \wedge \neg A)$ ($A \in \mathcal{F}_s$), which is the existential closure of $A \wedge \neg A$, viz. $(A \wedge \neg A)$ preceded by an existential quantifier over every variable free in it. In dynamic proofs of most inconsistency adaptive logics, B is derivable from $\neg A$ and $A \vee B$ on the condition $\{A \wedge \neg A\}$, viz. provided the premises allow one to justifiedly consider $A \wedge \neg A$ as false.

Some hard work was required to avoid circularity and to make the dynamics formally precise. However, clarifying the idea was a pleasure because it only made the idea more convincing, even if the outcome was not the final and most elegant formulation. In 1980 I wrote up my "Dynamic dialectical logics" [3], the pleonastic title meant as a teaser to Richard Routley and Bob Meyer who mention "static dialectical logic" in one of their papers [46]. Unfortunately, the publication of the book for which my paper had been written was badly delayed.

After some twelve years of failing attempts, I managed to devise other corrective adaptive logics around 1995.[4] Around the same time ampliative adaptive logics were formulated, mainly through efforts of Joke Meheus [29], as well as ambiguity adaptive logics, mainly through efforts by Guido Vanackere [48, 49, 51]. In the late 1990s, the *standard format* [6] was formulated—see below. Its major effect was that devising new adaptive logics became an almost obvious task and that integration—characterizing reasoning forms from the literature by an adaptive logic [5]—became

[2]For readers not familiar with paraconsistency, if $A \vee B$ and $\neg A$ are true, then so is $(A \wedge \neg A) \vee (B \wedge \neg A)$. Only the second disjunct entails B. So, in the paraconsistent case, $\{A \vee B, \neg A\}$ does not entail B but only $B \vee (A \wedge \neg A)$. This has been pointed out a long time ago by Newton da Costa, by Alan Anderson and Nuel D. Belnap, and by many others, and it gave rise to inconsistency-adaptive logics.

[3]If both B and $\neg B$ are true, then so is $A \supset B$, even if A is true and $\neg A$ is false.

[4]For present purposes, call an adaptive logic **AL** *corrective* iff all **AL**-consequences of Γ are **CL**-consequences (classical logic consequences) of Γ; *ampliative* iff the converse holds.

an easier and more precise task. From then on, the program seemed to unfold itself in an ever more natural way.

Nothing in the present paper suggests that the adaptive approach to defeasible reasoning would in any way be superior to other approaches. This is a matter for the future to decide, but right now adaptive logics seem to offer an attractive unifying approach. In this paper I shall try to spell out the present state of the art.

8.3 Abduction: Introductory Comments

There is a huge literature on abduction, consisting of some monographs and of many articles and edited works [1, 25, 26, 38, 41]. Actually, several reasoning forms are classified under the name. One of the ways to order them is by distinguishing between what is sometimes called [35] theoretical abductions and practical abductions.

A *theoretical* abduction has the aim to locate a potential explanation for a given fact, either in terms of a specific theory or in terms of any theory that belongs to one's knowledge. If the lamp does not light after the switch was turned on, the lamp may be broken, or it was not plugged in correctly, or a fuse has blown, or there is no electricity on the net, and so on. In general, if one wants an explanation for a given fact, one tries to locate *potential* explanations, facts that, together with one's beliefs about the mechanisms that govern the world, warrant that the *explanandum* obtains. In order to decide that a potential explanation A is a true explanation of the given *explanandum*, you need to find out whether A is true and possibly whether A, rather than different potential explanations that happen to be true, describes the actual 'cause' of the *explanandum*.[5]

Practical abductions serve a different purpose. When you wake up in the morning and the lawn, the trees, and everything else you see around the house is wet, you infer that it has rained during the night. That it has rained is not merely a potential explanation of what you see. It is a statement that you accept as true—some people actually think to *see* that it has rained. So while, in the case of a theoretical abduction, the truth of the potential explanation may be unsettled, the conclusion of a practical abduction extends one's knowledge. Needless to say, it does so in a defeasible way; new information may force one to give up the conclusion.

It is sometimes said that there is a relation between theoretical and practical abductions. If, according to your knowledge, $A_1, \ldots, A_n$ are potential explanations of E and all other potential explanations of E are known to be false, then $A_1 \vee \ldots \vee A_n$ would be a practical abduction that may be inferred from E.

A while ago, I joined those who do not believe in the existence of practical abductions. To be more precise, the inferences classified under that label are often correct inferences, but not abductions. Reconsider the argument from the previous

[5] Both readers familiar with the literature on explanation and readers familiar with detective stories will remember cases of a person who was poisoned at 4 PM and shot at 6 PM, but actually died of a heart attack at 5 PM.

paragraph, let K be the considered knowledge system, and let K ensure that E is true and that $A_1, \ldots, A_n$ are potential explanations of E which are not known to be false. If $E \supset (A_1 \vee \ldots \vee A_n)$ is a deductive consequence of K, then, since K also entails E, the disjunction $A_1 \vee \ldots \vee A_n$ is a *deductive* consequence of K. However, the mere fact that $A_1 \vee \ldots \vee A_n$ is the disjunction of the *known* non-falsified explanations of E is definitely insufficient to infer, by abduction or by anything else, $A_1 \vee \ldots \vee A_n$ from K.

A further problem with practical abductions is that potential explanations are very often ruled out on probabilistic (or similar) grounds. Where that is the case, the justification of the so-called abduction actually depends on the relative probability $\mathcal{P}(A_1 \vee \ldots \vee A_n, E)$. In the example where one infers that it has rained during the night, other potential explanations are physically possible but extremely improbable.[6] So in this case too the inference is not abductive. It does not lead from E to $A_1 \vee \ldots \vee A_n$ in view of the fact that each of $A_1, \ldots, A_n$ explains E, but in view of the fact that the probability of $A_1 \vee \ldots \vee A_n$ on E is sufficiently high.

If practical abductions would exist, they would be defeasible inferences. Where E is shown true and A is shown false, a correct practical abduction should not lead from E to A. So in attempts to devise an adaptive logic for practical abduction [33–35], one of the conditions for deriving A from E is that $E \wedge \neg A$ is not shown to be true. Unfortunately this condition was retained in many papers on adaptive logics for theoretical abduction [17, 19, 24] and that is a mistake. The aim of theoretical abductions is to offer potential explanations and potential explanations need not be true ones. Moreover, one of the conditions for turning a potential explanation into a true one is that the explanation is *shown to be true*. That it was not shown to be false is clearly insufficient. Finally, introducing the requirement that $E \wedge \neg A$ is not shown to be true makes sense in an adaptive logic leading to potential explanations that have not been rejected as true explanations. This is a sensible notion, but not a very central one.

The requirements for being a potential explanation involve a host of logical features that came up in Hempel's relevant work [18] and in its wake. Among the central features are that the explanation should not be 'strengthened': if A is a sufficient condition of E, then so is $A \wedge B$, even if B is not a sensible part of any explanation of E. A ready example: that it rained last night is a sensible potential explanation for the lawn being wet, but that it rained last night and that I was reading a novel is not. Similarly, the *explanandum* should not be 'weakened'. Suppose that A is a sufficient condition for E and that one wants an explanation of B. It would be silly to argue that, although one has no potential explanation of B, one has a potential explanation of the weaker $E \vee B$, viz. A. In more crisp terms: if you want to understand why John is dead, you will not feel helped by the statement that last night's rain explains that John is dead or the lawn is wet. Actually, the statement itself is silly.

Some paragraphs ago, I brought up the role of mechanisms in explanations. There is a variety of such mechanisms and the connected requirements are diverse. A requirement that seems to be shared by all is generality: in order for $A(a)$ to be

[6] I phrase this in terms of probabilities, but any similar approach may be equally acceptable.

an explanation of $E(a)$, the generalization $\forall x(A(x) \supset E(x))$ should belong to the knowledge system. The variety of mechanisms can be summarized as follows: either the implication is replaced, for example by a causal one, resulting in such expressions as $\forall x(A(x) \Rightarrow E(x))$, or else a modality is added, for example expressing a form of lawlikeness, resulting in such expressions as $\Box\forall x(A(x) \supset E(x))$. In the subsequent pages, I shall consider a generic type of approach, one which holds for a large family of expressions of forms considered in the present paragraph. To keep things simple, I shall phrase my proposal in terms of the **CL**-form $\forall x(A(x) \supset E(x))$, leaving to the reader the obvious adjustment to the other forms.

The final question to be considered here is the form of the formulas that we want to derive in a theoretical abduction. The answer is obvious: a theoretical abduction wants to locate potential explanations of a given *explanandum*. So the kind of statement that one wants to derive has the form "$A(a)$ is a potential explanation of $E(a)$". Needless to say, no such statement can be expressed within the standard predicative language $\mathcal{L}_s$. So a logic of theoretical abduction should be phrased in an extension of $\mathcal{L}_s$. Let $\mathcal{L}_+$ be a language under construction that extends $\mathcal{L}_s$ and let us already agree that the signature of $\mathcal{L}_+$ contains the binary operator $\triangleright$. We shall read $A(a) \triangleright E(a)$ as "$A(a)$ is a potential explanation of $E(a)$".

The decision to proceed in this way is partly conventional. The idea is to model the intended logic—provisionally call it **AAL** (adaptive abduction logic)—as an ampliative adaptive logic, viz. an adaptive logic that extends **CL**. This restricts the possible constructions. As all **CL**-consequences of a premise set Γ should be **AAL**-consequences of Γ, it would be awkward to derive $A(a)$ in order to express that $A(a)$ is a potential explanation of $E(a)$. Indeed, it is very well possible that $A(a)$ had been known to be true (or derivable from the knowledge system) all along, but that clearly does not make $A(a)$ a potential explanation of $E(a)$. It is equally awkward to derive $\Diamond A(a)$ or something similar in order to express that $A(a)$ is a potential explanation of $E(a)$. Often several facts are in need of explanation at the same time. We are not interested in the question whether $A(a)$ is a potential explanation, because like any other singular statement, $A(a)$ is bound to be a potential explanation of *something*. What we are interested in is whether $A(a)$ is a potential explanation of $E(a)$. So we better say so by deriving formulas like $A(a) \triangleright E(a)$. And if different *explananda* $E_1(a)$, $E_2(a)$, and $E_3(b)$ are involved, we better are able to distinguish between such statements as $A(a) \triangleright E_1(a)$, $B(a) \triangleright E_1(a)$, $C(a) \triangleright E_2(a)$, $A(b) \triangleright E_3(b)$, and $D(b) \triangleright E_3(b)$.

8.4 Exact Methods as Logics

Many qualitative methods are characterized in a precise way by a logic in the sense of Sect. 8.1. Since the days of the Vienna Circle, certain methods were characterized by criteria in terms of **CL**. So if the method is sufficiently precise, it determines, for any knowledge base Γ, a set of consequences that follow from Γ by the method. In other words, there is a logic **L** such that the **L**-consequence set of Γ comprises the

CL-consequences of Γ plus the set of statements obtained from Γ by the method. As we have seen, the logic **L** is then called ampliative.

Sometimes such a logic **L** shares its language with **CL**, viz. $\mathcal{L}_s$. A so-called inductive method may characterize a logic of inductive generalization, which determines for any consistent set of background knowledge Γ, including a set of singular data, a maximal set of generalizations that, when jointly added to Γ, results in a consistent consequence set. The ampliative consequences, those that are not **CL**-consequences of Γ, are generalizations as well as **CL**-consequences of Γ together with those generalizations. All of them are formulas of $\mathcal{L}_s$. In other cases, the ampliative logic requires an extension of $\mathcal{L}_s$. This is the case for the logic **AAL**, which requires the presence of a binary operator, $\rhd$, to express that some formula is a potential explanation of another.

Logicians should not be prescriptive in methodological matters—that **CL** might offer the required prescriptive authority was part of the confusion that ruined the Vienna Circle program. So logicians should offer a *variety* of potentially useful precise methods and leave it to scientists, or perhaps to philosophers of science, to decide about the justified choices in specific situations. Or, to put it differently, logicians should offer a framework within which philosophers of science might phrase methods in a precise way. The standard format of adaptive logics is intended as a such a framework for defeasible logics **L**. Actually, it seems hard to imagine a method that would define a non-defeasible logic **L**.

Some adaptive logics characterize a method in a direct way—this will be the case for **AAL**. Sometimes a method is characterized by an adaptive logic under a translation. An example is the characterization of "A is compatible with Γ" in terms of the expression $\Gamma^{\Box} \vdash_{\mathbf{COM}} \Diamond A$ [13]. In still other cases, a method is characterized by a combination of adaptive logics.

The specific feature of defeasible reasoning is that some conclusions can only be drawn provisionally. Although it is sensible to draw a conclusion at a certain point in time, it is possible that the conclusion has to be revoked later. The cause of the resulting dynamics may be either that new information became available or that new insights were gained. In the first case, the dynamics is external, viz. external to the reasoning process itself. The external dynamics is the typical feature of non-monotonic reasoning processes. In the second case, the dynamics is internal to the reasoning process. Obtaining more complete and more accurate insights in the premises causes the dynamics. In other words, by deriving more and more conclusions from the premises, we see that a formerly drawn conclusion has to be revoked.[7]

The feature described in the last sentence is well-known to cause a vicious circularity unless special care is taken. In the case of an adaptive logic **AL** the circularity is avoided in a very simple way: **AL** has a 'deductive' kernel, the lower limit logic **LLL**; **LLL**-conclusions are never withdrawn, but may cause other **AL**-conclusions to be withdrawn.

---

[7]People often define their views as the coherent systematization of their explicit views. The connected reasoning displays the internal dynamics. So everyone who has thought about his or her views is familiar with the internal dynamics.

In the case of **AAL**, the lower limit logic is **CL**. As was suggested in Sect. 8.3, that $A(a) \rhd E(a)$ is derivable from a knowledge base Γ requires, apart from $\Gamma \vdash_{\mathbf{CL}} \forall x(A(x) \supset E(x))$,[8] that A is not unnecessarily 'strengthened' and that E is not unnecessarily 'weakened'. We still have to settle the precise meaning of "strengthen" and "weaken" in this context, but it is quite obvious that A is unnecessarily 'strengthened' in case $A(x)$ is $Px \wedge Qx$ and $\Gamma \vdash_{\mathbf{CL}} \forall x(Px \supset E(x))$ holds. In other words, a requirement for $\Gamma \vdash_{\mathbf{AAL}} A(a) \rhd E(a)$ is, in this case, that $\Gamma \nvdash_{\mathbf{CL}} \forall x(Px \supset E(x))$. This requirement unavoidably causes **AAL** to be defeasible. If it is later found out that $\Gamma \vdash_{\mathbf{CL}} \forall x(Px \supset E(x))$,[9] then $A(a) \rhd E(a)$ should be considered as 'defeated'—if it was derived before, it should be considered as non-derived from then on.

For some premise sets and correct conclusions, there is no point in time at which the insights gained into the premises guarantees that the conclusion will not be revoked. So waiting with drawing the conclusion until such guarantee is present, comes to never drawing it. The upshot is that the dynamics is not merely possible but unavoidable.

Let me illustrate this in terms of **AAL**. Let us introduce a severe restriction, to be retained throughout this paper, on expressions of the form $A(a) \rhd E(a)$, requiring that $A(a)$ and $E(a)$ be truth-functions of formulas of the form πa in which $\pi \in \mathcal{P}^1$. As is well known, **CL** is decidable if it is restricted to premise sets and conclusions in which occur only unary predicates. This, however, does not make it decidable whether A is unnecessarily 'strengthened' or E is unnecessarily 'weakened' in $\forall x(A(x) \supset E(x))$. Indeed, $\Gamma \subseteq \mathcal{W}_s$ is the only restriction imposed upon the knowledge base Γ; even if $A(a)$ and $E(a)$ fulfil the restriction, $\mathrm{Cn}_{\mathbf{CL}}(\Gamma) =_{df} \{A \in \mathcal{W}_s \mid \Gamma \vdash_{\mathbf{CL}} A\}$ is in general at best a semi-recursive set. So it is at best semi-decidable whether $\forall x(B(x) \supset E(x)) \in \mathrm{Cn}_{\mathbf{CL}}(\Gamma)$ for a B that is a 'weakening' of A.

The internal dynamics is caused by the absence of a positive test for the consequence relation, in other words by the fact that the consequence set is not even semi-recursive. This too is easily illustrated in terms of **AAL**. We have seen that one of the requirements for $A(a) \rhd E(a)$ is that $\Gamma \nvdash_{\mathbf{CL}} \forall x(B(x) \supset E(x))$ for certain B. However, there is no positive test for **CL**-non-derivability; in other words, the **CL**-consequence relation is semi-recursive but not recursive. So as $\Gamma \vdash_{\mathbf{AAL}} A(a) \rhd E(a)$ requires that $\Gamma \nvdash_{\mathbf{CL}} \forall x(B(x) \supset E(x))$ for certain B, the set $\mathrm{Cn}_{\mathbf{AAL}}(\Gamma)$ is not semi-recursive.

To see why this is so, let us focus on two conditions for $\Gamma \vdash_{\mathbf{AAL}} A(a) \rhd E(a)$. As in the special case considered before, take the positive $\Gamma \vdash_{\mathbf{CL}} \forall x((Px \wedge Qx) \supset E(x))$ and the negative $\Gamma \nvdash_{\mathbf{CL}} \forall x(Px \supset E(x))$, neglecting further negative conditions. In order for $\mathrm{Cn}_{\mathbf{AAL}}(\Gamma)$ to be semi-recursive, there need to be a Turing machine that, if given Γ and C, halts after finitely many steps with the answer YES just in case $C \in \mathrm{Cn}_{\mathbf{AAL}}(\Gamma)$. Let two Turing machines do the job, T1 to test whether $\Gamma \vdash_{\mathbf{CL}} \forall x((Px \wedge Qx) \supset E(x))$ and T2 to test whether $\Gamma \nvdash_{\mathbf{CL}} \forall x(Px \supset E(x))$.

[8]The idea behind **AAL** is that some formulas of the form $A(a) \rhd E(a)$ are derivable from Γ on top of the **CL**-consequences of Γ. So the transition from **CL** to **AAL** cannot have the effect that more generalizations become derivable from Γ.

[9]Note that defeasibility is not connected to derivability but to one's insights in derivability.

Moreover, suppose that indeed $(Pa \wedge Qa) \rhd E(a) \in \mathrm{Cn}_{\mathbf{AAL}}(\Gamma)$. It follows that $\Gamma \vdash_{\mathbf{CL}} \forall x((Px \wedge Qx) \supset E(x))$, whence, as **CL** is semi-recursive, T1 is bound to halt with the answer YES after finitely many steps. It also follows that $\Gamma \nvdash_{\mathbf{CL}} \forall x(Px \supset E(x))$. However, as there is no positive test for **CL**-non-derivability, T2 may never halt. So $\mathrm{Cn}_{\mathbf{AAL}}(\Gamma)$ is not semi-recursive. In general, any **CL**-criterion that involves a condition demanding that a certain statement is not **CL**-derivable from a certain set defines a consequence relation for which there is no positive test. Nearly all methods and methodological criteria involve such conditions. A short list of diverse examples is available elsewhere [7].

The internal dynamics is both more general and more central than the external one. Roughly put, if a logic displays the external dynamics, then its proof theory displays the internal dynamics. The converse does not hold. The Weak consequence relation [42, 43, 45] is a nice example of a monotonic consequence relation for which there is no positive test.[10] So the identification of defeasible reasoning with non-monotonic reasoning is mistaken.

8.5 The Standard Format

As suggested above, the origin of adaptive logics lies with the explication of reasoning. Needless to say, defeasible reasoning cannot be explicated in terms of the usual kind of proofs, which I shall from now on call *static proofs*. Most approaches to defeasible reasoning proceed in semantic terms or in terms of infinite sets of formulas, like the 'extensions' from default logics. Adaptive logics are a rather striking exception. They explicate defeasible reasoning forms in terms of *dynamic proofs*—a theoretic account of this generalization of static proofs was published [11] and a more extensive account will soon be available. Adaptive logics also have a semantics, but I shall start by describing their dynamic proofs.

An adaptive logic **AL** in Standard Format is defined as a triple:

- a *lower limit logic* **LLL**: a logic that has static proofs[11] and has classical disjunction $\check{\vee}$ in its language,[12]
- a *set of abnormalities* Ω, a set of formulas that share a (possibly restricted) logical form or a union of such sets,[13]
- a *strategy* (Reliability, Minimal Abnormality, ...).

---

[10] A is a Weak consequence of Γ iff it is a **CL**-consequence of a consistent subset of Γ. Incidentally, Weak is paraconsistent, $p, \neg p \nvdash_{\mathrm{Weak}} q$, and not reflexive, $p \wedge \neg p \nvdash_{\mathrm{Weak}} p \wedge \neg p$.

[11] The reader may safely identify a logic that has static proofs with a compact Tarski logic, viz. a logic that is compact, reflexive, transitive, and monotonic.

[12] The symbol $\check{\vee}$ is a conventional name to refer to a symbol of the language that has the meaning of classical disjunction.

[13] For a restricted logical form, see the form defining $\mathcal{G}$ below in the text.

The abnormalities of inconsistency-adaptive logics, and of corrective adaptive logics in general, are **CL**-falsehoods. This is not the case for the abnormalities of ampliative adaptive logics are **CL**-contingent. So "abnormality" is a technical term referring to formulas that are presumed to be false until and unless the premises are seen to prevent this.

The expression $\mathrm{Dab}(\Delta)$ will always denote a classical disjunction of the members of a finite $\Delta \subset \Omega$. $\mathrm{Dab}(\Delta)$ is a Dab-consequence of Γ iff $\Gamma \vdash_{\mathbf{LLL}} \mathrm{Dab}(\Delta)$. A Dab-consequence $\mathrm{Dab}(\Delta)$ of Γ is minimal iff there is no $\Delta' \subset \Delta$ such that $\mathrm{Dab}(\Delta')$ is a Dab-consequence of Γ. If $\mathrm{Dab}(\Delta)$ is a minimal Dab-consequence of Γ, Γ requires that at least one member of Δ is true. The strategy determines the way in which an adaptive logic reacts to minimal Dab-consequences of the premise set. The simplest strategy, Reliability, considers all disjuncts of minimal Dab-consequences of Γ as unreliable and accordingly defines $U(\Gamma)$, the set of abnormalities that are unreliable with respect to Γ. The idea behind Reliability is that $\Gamma \vdash_{\mathbf{AL}^r} A$ iff there is a $\Delta \in \Omega - U(\Gamma)$ such that $\Gamma \vdash_{\mathbf{LLL}} A \mathbin{\check{\vee}} \mathrm{Dab}(\Delta)$. So Γ **LLL**-entails $A \mathbin{\check{\vee}} \mathrm{Dab}(\Delta)$; but if $A \mathbin{\check{\vee}} \mathrm{Dab}(\Delta)$ is true, then, as all reliable abnormalities are false, A is true. The Minimal Abnormality strategy is slightly more complicated and will be introduced later.

If the lower limit logic **LLL** is extended with a set of rules or axioms with the effect that every formula is derivable from every member of Ω, then one obtains the so-called *upper limit logic* **ULL**, which is a logic that has static proofs. Note that **ULL** 'rules out' abnormalities, whereas $\mathbf{AL}^r$ 'rules out' abnormalities that are reliable with respect to the premise set. The effect is this: **ULL** extends **LLL** by validating some further rules of inference; $\mathbf{AL}^r$ extends **LLL** by validating certain *applications* of **ULL**-rules.[14]

At this point, we are sufficiently equipped to consider the adaptive logic for (theoretical) abduction, more precisely the variants that have Reliability as their strategy—let us call the logic we are after $\mathbf{AAL}^r$.

Recall that we are interested in deriving statements of the form $A(\alpha) \rhd E(\alpha)$ with $\alpha \in \mathcal{C}$. We have restricted the *explananda* and explanations that will be considered. This may be phrased as: $A(\alpha)$ and $E(\alpha)$ are such that $A(x), E(x) \in \mathcal{F}^{\odot}$, where $\mathcal{F}^{\odot}$ is the set of truth-functions of formulas πx with $\pi \in \mathcal{P}^1$.

Some restrictions to be imposed in this connection were mentioned in Sect. 8.4: the *explanandum* $E(\alpha)$ should not be unnecessarily weakened and the explanation $A(\alpha)$ should not be unnecessarily strengthened. For example, $(Pa \wedge Qa) \rhd (Ra \vee Sa)$ should not be derived if the knowledge base entails $\forall x(Px \supset (Rx \vee Sx))$, or if it entails $\forall x((Px \wedge Qx) \supset Rx)$, and so on. But a further restriction need to be imposed in order to avoid partial self-explanations. As the knowledge base entails the **CL**-theorem $\forall x(Qx \supset Qx)$, the restriction eliminates $Pa \wedge Qa \rhd Qa$. But obviously $Pa \wedge Ra \rhd Qa \wedge Ra$ is equally unacceptable. Indeed, whether the knowledge base entails $\forall x(Px \supset (Qx \vee Rx))$ or not, $Pa \wedge Ra \rhd Qa \wedge Ra$ is unavoidably a case of

[14]Which applications are validated is determined by **LLL**-consequences of the premise set. So adaptive logics display a form of *content-guidance*—Dudley Shapere [47] among others stated and defended the function of this feature in scientific methodology.

partial self-explanation. So we want to impose the requirement on $A(\alpha) \rhd E(\alpha)$ that no predicate occurs in the simplified conjunctive normal forms of both $A(\alpha)$ and $E(\alpha)$—see below.[15]

Needless to say, the considered requirements do not in any way depend on the specific constant α that occurs in $A(\alpha) \rhd E(\alpha)$. The features we are after pertain to the antecedent and consequent of the generalization, and not to the specific object that the potential explanation statement is about. So **AAL**^{r} should enable one to derive $A(\alpha) \rhd E(\alpha)$ (for any $\alpha \in \mathcal{C}$) from a formula $\star\forall\beta(A(\beta) \supset E(\beta))$ which states that the generalization $\forall\beta(A(\beta) \supset E(\beta))$ fulfils the requirements under discussion. Of course **AAL**^{r} is itself under construction and it requires a lower limit logic defined over $\mathcal{L}_+$. Let this lower limit under construction be an extension of **CL**, called **CL**$^+$. **CL**$^+$ should specify the unary $\star$ and this is what I now set out to do.

Let CNF(A) [DNF(A)] be the *simplified* conjunctive [disjunctive] normal form of A.[16] Let π be the alphabetically first predicate (letter) in $\mathcal{P}^1$. Let $w(A)$ be the set of formulas obtained by removing one conjunct from CNF(A), or $\{\pi a \vee \neg\pi a\}$ if CNF(A) has only one conjunct. Let $s(A)$ be the set of formulas obtained by removing one disjunct from DNF(A), or $\{\pi a \wedge \neg\pi a\}$ if DNF(A) has only one disjunct. To save some space, let $\ulcorner\forall x(A(x) \supset B(x))\urcorner$ abbreviate $\{\forall x(C(x) \supset B(x)) \mid C \in w(A)\} \cup \{\forall x(A(x) \supset C(x)) \mid C \in s(B)\}$.[17]

Define $\mathcal{G}$, the set of *clean generalizations*, as the set of formulas of the form $\forall x(A(x) \supset B(x)) \in \mathcal{W}_s$ such that (i) $A(x), B(x) \in \mathcal{F}^{\odot}$, (ii) $A(x)$ and $B(x)$ are **CL**-contingent, and (iii) no [schematic letter for] predicate[s] occurs in both A and B. Note that $\mathcal{G}$ is recursive and that $\ulcorner\forall x(A(x) \supset B(x))\urcorner \subseteq \mathcal{W}_s$ is finite and recursive if $\forall x(A(x) \supset B(x)) \in \mathcal{G}$.

Finally introduce in **CL**$^+$ the axiom schema:

A$\star$ For all $\forall x(A(x) \supset B(x)) \in \mathcal{G}$ and $C \in \ulcorner\forall x(A(x) \supset B(x))\urcorner$,
$C \supset \neg\star\forall x(A(x) \supset B(x))$

and in its semantics the clause:

C$\star$ Where $\forall x(A(x) \supset B(x)) \in \mathcal{G}$ and $C \in \ulcorner\forall x(A(x) \supset B(x))\urcorner$,
if $v_M(C) = 1$, then $v_M(\star\forall x(A(x) \supset B(x))) = 0$.

The **CL**$^+$-semantics is indeterministic, but there is a know method to turn indeterministic semantic systems into strongly equivalent deterministic systems. An alternative is to strengthen the clause to "Where $\forall x(A(x) \supset B(x)) \in \mathcal{G}$, $v_M(\star\forall x(A(x) \supset B(x))) = 1$ iff $v_M(\forall x(A(x) \supset B(x))) = 1$ and $v_M(C) = 0$ for all $C \in \ulcorner\forall x(A(x) \supset B(x))\urcorner$." and to strengthen **CL**$^+$ accordingly in order to safeguard its completeness.

[15]The reasoning in this paragraph applies to language schemas. It applies to languages of which the predicates are conceptually independent.

[16]The simplified normal forms of contingent formulas do not contain irrelevant literals. Example: the simplified disjunctive normal form of $(Px \wedge Qx) \vee (Px \wedge \neg Qx) \vee (Rx \wedge Px)$ is Px.

[17]In the present context, $\ulcorner\forall x(A(x) \supset B(x))\urcorner$ may also be defined as $\{\forall x(C(x)) \mid C \in s(A(x) \supset B(x))\}$. This however would not suit the generic approach, for example in case causal generalizations are considered.

Both approaches provably define a different $\mathbf{CL}^+$, but nevertheless the same adaptive logic $\mathbf{AAL}^r$ in view of the choice of Ω below.

What about the the binary $\rhd$? Clearly, we have pushed all criteria considered so far into the star $\star$. So we can safely add the following definition:

D$\rhd$ $A(\alpha) \rhd B(\alpha) =_{df} \star\forall x(A(x) \supset B(x))$

Let us turn to the dynamic proofs. The lines of annotated dynamic proofs have four elements: a line number, a formula, a justification, and a condition. The first three elements are like in static annotated proofs. The fourth element, the condition, is a set of abnormalities and is obviously specific for (adaptive) dynamic proofs. In order to describe the proofs, we need rules and a marking definition. The rules, which depend on the lower limit logic and on the set of abnormalities, allow one to add lines to a proof. The rules can be reduced to three generic ones. Let

$$A \quad \Delta$$

abbreviate that A occurs in the proof on the condition Δ.

Prem If$A \in \Gamma$:
$$\frac{\ldots\ldots}{A \quad \emptyset}$$

RU If$A_1, \ldots, A_n \vdash_{\mathbf{LLL}} B$:
$$\begin{array}{ll} A_1 & \Delta_1 \\ \ldots\ldots & \\ A_n & \Delta_n \\ \hline B & \Delta_1 \cup \ldots \cup \Delta_n \end{array}$$

RC If$A_1, \ldots, A_n \vdash_{\mathbf{LLL}} B \mathbin{\check{\vee}} \mathrm{Dab}(\Theta)$:
$$\begin{array}{ll} A_1 & \Delta_1 \\ \ldots\ldots & \\ A_n & \Delta_n \\ \hline B & \Delta_1 \cup \ldots \cup \Delta_n \cup \Theta \end{array}$$

In Prem, Γ refers to the premise set. The generic rule RC states that, if $A_1, \ldots, A_n \vdash_{\mathbf{LLL}} B \mathbin{\check{\vee}} \mathrm{Dab}(\Theta)$ and $A_1, \ldots, A_n$ occurs in the proof on conditions $\Delta_1, \ldots, \Delta_n$ respectively, then one may add a line to the proof at which B is derived on the condition $\Delta_1 \cup \ldots \cup \Delta_n \cup \Theta$—the union of the conditions of $A_1, \ldots, A_n$ and of Θ. Note that, under the same circumstances, RU would allow one to derive $B \mathbin{\check{\vee}} \mathrm{Dab}(\Theta)$ on the condition $\Delta_1 \cup \ldots \cup \Delta_n$. RC allows one to push $\mathrm{Dab}(\Theta)$ to the condition—its disjuncts, rather than the disjunction—and to derive the stronger formula B.

Every abnormality is presumed to be false, until and unless it turns out that the premises do not allow one to do so. The conditions of the proof lines comprise the abnormalities that have to be false in order for the formula of the line to be derivable.

Apart from rules to add lines to a dynamic proof, we also need a marking definition. Marked lines are considered as OUT. Not all members of their condition can be considered as false and hence the formula of the line has to be considered as non-derived. Unmarked lines are IN or ALL RIGHT. Whenever a line is added to the proof, a next proof stage is reached, and the effect of the marking definition will depend on this stage[18]: lines that were unmarked at the previous stage may be marked and vice versa. Marking for reliability is straightforward. The Dab-formulas that are derived on the condition $\emptyset$ at stage s determine $U_s(\Gamma)$ in the same way as the Dab-consequences of Γ determine $U(\Gamma)$. So the marking definition proceeds on present insights, viz. on the present estimated set of minimal Dab-consequences of Γ.

Definition 1 Marking for Reliability: where Δ is the condition of line i, line i is marked at stage s iff $\Delta \cap U_s(\Gamma) \neq \emptyset$.

Once more we turn to abduction for illustrations. We already settled that **CL**$^+$ will be the lower limit logic of **AAL**^{r} and Reliability its strategy. What will be the set of abnormalities? An attractive idea seems that every generalization $\forall x(A(x) \supset B(x)) \in \mathcal{G}$ is considered to result in a potential explanation $A(\alpha) \triangleright B(\alpha)$ (for all $\alpha \in \mathcal{C}$) until and unless it is shown that the generalization fails to fulfil the conditions. This means: until it is shown that $\neg\star\forall x(A(x) \supset B(x))$. So we define $\Omega = \{\forall x(A(x) \supset B(x)) \wedge \neg\star\forall x(A(x) \supset B(x)) \mid \forall x(A(x) \supset B(x)) \in \mathcal{G}\}$. Note that the axiom A$\star$ enables one to derive $\neg\star\forall x(A(x) \supset B(x))$ whenever it is shown that $\forall x(A(x) \supset B(x)) \in \mathcal{G}$ does not fulfil the conditions. Let us call the so obtained logic **AAL**^{r}.

Consider a dynamic (toy) proof from the premise set

$$\Gamma_1 = \{\forall x((Px \wedge \neg Qx) \supset Rx), \forall x(Px \supset (Sx \vee Qx \vee Tx)), \forall x(Qx \supset \neg Px)\}.$$

To save some room, let ${}^\star(A(x) \supset E(x))$ abbreviate $\forall x(A(x) \supset E(x)) \wedge \neg\star\forall x(A(x) \supset E(x))$.

1	$\forall x((Px \wedge \neg Qx) \supset Rx)$	premise	$\emptyset$
2	$\forall x(Px \supset (Sx \vee Qx \vee Tx))$	premise	$\emptyset$
3	$\forall x(Qx \supset \neg Px)$	premise	$\emptyset$
4	$\star\forall x((Px \wedge \neg Qx) \supset Rx)$	1; RC	$\{{}^\star((Px \wedge \neg Qx) \supset Rx)\}$
5	$\star\forall x(Px \supset (Sx \vee Qx \vee Tx))$	2; RC	$\{{}^\star(Px \supset (Sx \vee Qx \vee Tx))\}$
6	$\star\forall x(Qx \supset \neg Px)$	3; RC	$\{{}^\star(Qx \supset \neg Px)\}$

We have introduced three star-formulas and no line is marked at stage 6 of the proof. I repeat lines 4–6 and derive some more **CL**$^+$-consequences from Γ_1. Some lines are marked. The superscript of the mark refers to the proof stage at which the mark was introduced.[19]

4	$\star\forall x((Px \wedge \neg Qx) \supset Rx)$	1; RC	$\{{}^\star((Px \wedge \neg Qx) \supset Rx)\}$	$\checkmark^{9}$
5	$\star\forall x(Px \supset (Sx \vee Qx \vee Tx))$	2; RC	$\{{}^\star(Px \supset (Sx \vee Qx \vee Tx))\}$	$\checkmark^{12}$
6	$\star\forall x(Qx \supset \neg Px)$	3; RC	$\{{}^\star(Qx \supset \neg Px)\}$	
7	$\forall x(Px \supset Rx)$	1, 3; RU	$\emptyset$	

[18] A proof stage is a sequence of lines. A proof is a chain of stages, every stage containing the lines of the previous stage in the same order.

[19] The logic under consideration has the unusual property that lines that are marked at a stage remain marked at all subsequent stages.

8	$\neg\bigstar\forall x((Px \wedge \neg Qx) \supset Rx)$	7; RU	$\emptyset$
9	${}^{\star}((Px \wedge \neg Qx) \supset Rx)$	1, 8; RU	$\emptyset$
10	$\forall x(Px \supset (Sx \vee Tx))$	2, 3; RU	$\emptyset$
11	$\neg\bigstar\forall x(Px \supset (Sx \vee Qx \vee Tx))$	10; RU	$\emptyset$
12	${}^{\star}(Px \supset (Sx \vee Qx \vee Tx))$	2, 11; RU	$\emptyset$
13	$\bigstar\forall x(Px \supset Rx)$	7; RC	$\{{}^{\star}(Px \supset Rx)\}$
14	$\bigstar\forall x(Px \supset (Sx \vee Tx))$	10; RC	$\{{}^{\star}(Px \supset (Sx \vee Tx))\}$
15	$Pa \rhd (Sa \vee Ta)$	14; RU	$\{{}^{\star}(Px \supset (Sx \vee Tx))\}$

From 7 one may obviously derive $\forall x((Px \wedge Ux) \supset Rx)$ as well as $\forall x(Px \supset (Rx \vee Ux))$, but the very presence of 7 prevents the derivability of $(Pa \wedge Ua) \rhd Ra$ and of $Pa \rhd (Ra \vee Ua)$. Line 15 will not be marked in any extension of the proof—neither $\forall x(Px \supset Sx)$ nor $\forall x(Px \supset Tx)$ is derivable from Γ_1. Note that, for example, $Qa \rhd \neg Pa$ and $Pa \rhd Ra$ are also derivable on the conditions $\{{}^{\star}(Qx \supset \neg Px)\}$ and $\{{}^{\star}(Px \supset Rx)\}$ and that the lines at which they are so derived will not be marked in any extension of the $\mathbf{AAL}^r$-proof from Γ_1. Also, marked lines of $\mathbf{AAL}^r$-proofs are never unmarked. All this shows that, in comparison to other adaptive logics, $\mathbf{AAL}^r$ is extremely simple.

As suggested in the previous paragraph, dynamic proofs may become stable with respect to certain lines—with respect to specific ways to derive a certain consequence. Indeed, dynamic proofs allow one to distinguish derivability at a stage, which may be very unstable, and final derivability, which is stable in that every set of premises has a unique set of final consequences.

Definition 2 A is *finally derived* from Γ at line i of a finite proof stage s iff (i) A is the second element of line i, (ii) line i is not marked at stage s, and (iii) every extension of the proof in which line i is marked may be further extended in such a way that line i is unmarked.

Definition 3 $\Gamma \vdash_{\mathbf{AL}} A$ (A is *finally* **AL**-*derivable* from Γ) iff A is finally derived at a line of a proof stage from Γ.

Supposing that a characteristic **LLL**-semantics is available, $U(\Gamma)$ can be semantically defined. For each **LLL**-model M, let $\text{Ab}(M) =_{df} \{A \in \Omega \mid M \Vdash A\}$ (the abnormalities verified by M).

Definition 4 M is a *reliable* model of Γ iff M is a **LLL**-model of Γ and $\text{Ab}(M) \subseteq U(\Gamma)$.

Definition 5 $\Gamma \vDash_{\mathbf{AL}^r} A$ (A is an $\mathbf{AL}^r$-consequence of Γ) iff $M \Vdash A$ for all reliable models M of Γ.

So the semantics of adaptive logics is a selection semantics. It selects a subset from the set of **LLL**-models of the premise set. Note that to be a $\mathbf{AL}^r$-model, viz. a reliable model, is a relational property: certain M are reliable models *of* certain premise sets. It is instructive to have a look at Fig. 8.1. If the premise set Γ is normal, the adaptive

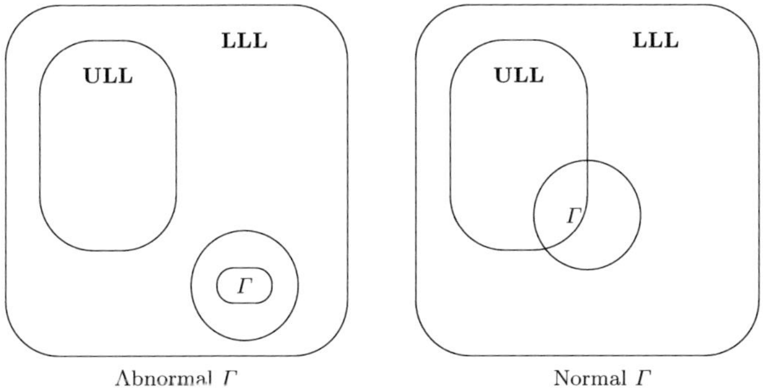

Fig. 8.1 Comparison of Models

models of Γ coincide with the **ULL**-models of Γ[20]; if Γ is abnormal and hence has no **ULL**-models, the adaptive logic still selects a subset of the **LLL**-models of Γ.[21]

The logic $\mathbf{AAL}^r$ is untypical in that $\text{Ab}(M) \neq \emptyset$ for all $\mathbf{CL}^+$-models.

Before closing this long section, I add a few words on the Minimal Abnormality strategy.

Definition 6 M is a *minimally abnormal* model of Γ iff M is a **LLL**-model of Γ and no **LLL**-model M' of Γ is such that $\text{Ab}(M') \subset \text{Ab}(M)$.

Definition 7 $\Gamma \vDash_{\mathbf{AL}^m} A$ (A is an $\mathbf{AL}^m$-consequence of Γ) iff $M \Vdash A$ for all minimally abnormal models M of Γ.

The corresponding marking definition is somewhat tiresome. Consider a proof stage s from Γ. Let $\text{Dab}(\Delta_1)$, $\text{Dab}(\Delta_2)$, ... be the minimal Dab-formulas that were derived on the condition $\emptyset$ at stage s—the formulas $\text{Dab}(\Delta)$ that, at stage s, are derived on the condition $\emptyset$ and are such that, at s, $\text{Dab}(\Delta')$ was not derived on the condition $\emptyset$ for any $\Delta' \subset \Delta$. Define $\Phi_s(\Gamma)$ as the set of minimal choice sets of $\langle \Delta_1, \Delta_2, \ldots \rangle$.[22]

Definition 8 Marking for Minimal Abnormality: where A is the formula and Δ is the condition of line i, line i is marked at stage s iff (i) there is no $\varphi \in \Phi_s(\Gamma)$ such that $\varphi \cap \Delta = \emptyset$, *or* (ii) for some $\varphi \in \Phi_s(\Gamma)$, there is no line at which A is derived on a condition Θ for which $\varphi \cap \Theta = \emptyset$.

In general the Minimal Abnormality strategy yields a richer consequence set than the Reliability strategy. Where $\mathbf{AAL}^m$ is like $\mathbf{AAL}^r$ except in that Reliability is replaced by Minimal Abnormality, it holds that $\text{Cn}_{\mathbf{AAL}^m}(\Gamma) = \text{Cn}_{\mathbf{AAL}^r}(\Gamma)$ for all

[20] The **ULL**-models are the **LLL**-models that verify no abnormality.

[21] The subset is proper, except for some exceptional adaptive logics called flip-flops.

[22] A *choice set* of $\langle \Delta_1, \Delta_2, \ldots \rangle$ comprises one member of every Δ_i ($i \in \{1, 2, \ldots\}$). A choice set of $\langle \Delta_1, \Delta_2, \ldots \rangle$ is *minimal* iff none of its proper subsets is a choice set of $\langle \Delta_1, \Delta_2, \ldots \rangle$.

$\Gamma \subseteq \mathcal{W}_s$. This is due to the fact that for such Γ holds: $\text{Dab}(\Delta) \in \text{Cn}_{\mathbf{CL}^+}(\Gamma)$ iff there is an $A \in \Delta$ such that $A \in \text{Cn}_{\mathbf{CL}^+}(\Gamma)$.[23]

That $\mathbf{AAL}^m$ and $\mathbf{AAL}^r$ are less complex than most adaptive logics,[24] does not reduce them to logics that have static proofs. For example, their consequence sets are not semi-recursive.

8.6 Variant Elements for Abduction

The logics $\mathbf{AAL}^m$ and $\mathbf{AAL}^r$ warrant that potential explanations entail their *explanandum* and moreover avoid certain logical weirdos—untidy strengthening or weakening and sundry forms of self-explanation. Needless to say a host of further requirements may be taken from the literature on explanation. I shall consider just a couple, for fear that this paper expands to a book.

A potential explanation is often sought in terms of a specific theory. The explanation is then confined by the *explanandum* as well as by the theory. Such an outlook is taken by Hintikka and Halonen in their work on the process of explanation [20]. Among the conditions they introduce are that the *explanandum* is not implied by the theory alone and that the explanation is compatible with the theory, i.e. does not falsify the theory. The connected type of abduction, just like the one considered before, does not involve the derivation of any factual statements and is not influenced by the truth of such statements. Where the available theories are T_1, T_2, and so on, the approach under consideration should result in such statements as $A \rhd_1 E$, stating that A explains E in terms of T_1, and similarly for $A \rhd_2 E$ and so on.

In order to incorporate such requirements within an adaptive logic, we need to express that some formulas are not entailed by a certain theory. The simplest way to obtain this effect is to introduce modalities, such as $\Box_i A$ to express that A is a theorem of T_1. Let $\mathcal{L}_m$ extend $\mathcal{L}_s$ with the unary operators $\star$ and $\Box_i$ and the binary operators $\rhd_i$, in both cases for all $i \in \{1, 2, \ldots\}$. Let $\mathbf{CL}^{\mathrm{M}}$ extend $\mathbf{CL}$ with the following axiom schemas and definition—obviously $\alpha \in \mathcal{C}$ in D$\rhd_i$:

A□1 for all **CL**-axiom schemas A, $\Box_i A$
A□2 $\Box_i(A \supset B) \supset (\Box_i A \supset \Box_i B)$
A□3 $\Box_i A \supset A$
A⋆□ for all $\forall x(A(x) \supset B(x)) \in \mathcal{G}$ and $C \in \ulcorner \forall x(A(x) \supset B(x)) \urcorner$,
$\Box_i C \supset \neg \star \Box_i \forall x(A(x) \supset B(x))$
D$\rhd_i$ $A(\alpha) \rhd_i B(\alpha) =_{df} \star\Box_i \forall x(A(x) \supset B(x)) \wedge \neg\Box_i B(\alpha) \wedge \neg\Box_i \neg A(\alpha)$

The $\Box_i$ have the same properties as the $\Box$ in **T** except that I did not care about the meaning of modalities that have modal statements within their scope.

---

[23]The so-called Simple strategy will do for such Γ. Marking for the Simple strategy is defined by: line l is marked at stage s iff, at stage s, a member of the condition of line l is derived on the condition $\emptyset$.

[24]The statement should be taken literally in several senses. It pertains to the proofs, to the computational complexity of the consequence sets, to the semantic selection criterion, ...

Fact 1 *If* $A_1, \ldots, A_n \vdash_{\mathbf{CL}} B$, *then* $\Box_i A_1, \ldots, \Box_i A_n \vdash_{\mathbf{CL}^{\mathrm{M}}} \Box_i B$.

The theories under consideration are obviously **CL**-theories. Where $T_i = \langle \Gamma, \mathbf{CL} \rangle$, let $T_i^{\Box} = \langle \{\Box_i A \mid A \in \Gamma\}, \mathbf{CL}^{\mathrm{M}} \rangle$.

Fact 2 $\Box_i A$ *is a theorem of* $T_i^{\Box}$ *iff* A *is a theorem of* T_i.

In order for $A(a) \rhd_1 E(a)$ to be derivable from theory $T_1^{\Box}$, T_1 should entail $\forall x(A(x) \supset E(x))$, it should not entail any $C \in \lceil \forall x(A(x) \supset B(x)) \rceil$, and it should entail neither $E(a)$ nor $\neg A(a)$. This last requirement, which is not fulfilled by $\mathbf{AAL}^r$, demands an adaptive approach because there is no positive test for non-**CL**-derivability.

Let the adaptive theory-abductive logic $\mathbf{ATAL}^r$ have Reliability as its strategy and $\mathbf{CL}^{\mathrm{M}}$ as its lower limit logic and let its set of abnormalities be defined as the union of three sets[25]:

$$\begin{aligned}\Omega = &\{\Box_i \forall x(A(x) \supset B(x)) \wedge \neg\star\Box_i \forall x(A(x) \supset B(x)) | \forall x(A(x) \supset B(x)) \in \mathcal{G}\}\\ &\cup\\ &\{\Box_i \forall x(A(x) \supset B(x)) \wedge \Box_i B(\alpha) | \forall x(A(x) \supset B(x)) \in \mathcal{G}; \alpha \in \mathcal{C}\}\\ &\cup\\ &\{\Box_i \forall x(A(x) \supset B(x)) \wedge \Box_i \neg A(\alpha) | \forall x(A(x) \supset B(x)) \in \mathcal{G}; \alpha \in \mathcal{C}\}\end{aligned}$$

By transforming and extending Γ_1 we obtain an artificial but instructive toy example of a knowledge system containing two theories. Consider $\Gamma_2 = \{\Box_1 \forall x((Px \wedge \neg Qx) \supset Rx), \Box_2 \forall x(Px \supset (Sx \vee Qx \vee Tx)), \Box_2 \forall x(Qx \supset \neg Px), \Box_2 Sa, Tb\}$. Let ${}^{\star}_{i}(A(x) \supset E(x))$ abbreviate $\Box_i \forall x(A(x) \supset E(x)) \wedge \neg \star \Box_i \forall x(A(x) \supset E(x))$, let ${}^{\alpha}_{i}(A(x) \supset E(x))$ abbreviate $\Box_i \forall x(A(x) \supset E(x)) \wedge \Box_i \neg A(\alpha)$; let $(A(x) \supset E(x))^{\alpha}_{i}$ abbreviate $\Box_i \forall x(A(x) \supset E(x)) \wedge \Box_i E(\alpha)$.

1	$\Box_1 \forall x((Px \wedge \neg Qx) \supset Rx)$	premise	$\emptyset$
2	$\Box_2 \forall x(Px \supset (Sx \vee Qx \vee Tx))$	premise	$\emptyset$
3	$\Box_2 \forall x(Qx \supset \neg Px)$	premise	$\emptyset$
4	$\Box_2 Sa$	premise	$\emptyset$
5	Tb	premise	$\emptyset$
6	$\star\Box_1 \forall x((Px \wedge \neg Qx) \supset Rx)$	1; RC	$\{{}^{\star}_{1}((Px \wedge \neg Qx) \supset Rx)\}$
7	$\star\Box_2 \forall x(Px \supset (Sx \vee Qx \vee Tx))$	2; RC	$\{{}^{\star}_{2}(Px \supset (Sx \vee Qx \vee Tx))\}$
8	$\star\Box_2 \forall x(Qx \supset \neg Px)$	3; RC	$\{{}^{\star}_{2}(Qx \supset \neg Px)\}$
9	$\neg\Box_1 \neg(Pa \wedge \neg Qa)$	1; RC	$\{{}^{a}_{1}((Px \wedge \neg Qx) \supset Rx)\}$
10	$\neg\Box_1 Ra$	1; RC	$\{((Px \wedge \neg Qx) \supset Rx)^{a}_{1}\}$
11	$(Pa \wedge \neg Qa) \rhd_1 Ra$	6, 9, 10; RU	X

The X in 11 abbreviates $\{{}^{\star}_{1}((Px \wedge \neg Qx) \supset Rx), {}^{a}_{1}((Px \wedge \neg Qx) \supset Rx), ((Px \wedge \neg Qx) \supset Rx)^{a}_{1}\}$, the union of the conditions of lines 6, 9, and 10.

---

[25]Note that $\neg\Box_i\neg A$ is adaptively derivable whenever A is compatible with T_i.

It is easily seen that (i) $\Gamma_2 \nvdash_{\mathbf{CL}^{\mathrm{M}}} \Box_i C$ for any $C \in \ulcorner \forall x((Px \wedge \neg Qx) \supset Rx)\urcorner$, (ii) $\Gamma_2 \nvdash_{\mathbf{CL}^{\mathrm{M}}} \Box_1 \neg(Pa \wedge \neg Qa)$, and (iii) $\Gamma_2 \nvdash_{\mathbf{CL}^{\mathrm{M}}} \Box_1 Ra$. So lines 6, 9, 10, and 11 will not be marked in an extension of the proof, whence the formulas derived at these lines are final consequences of Γ_2.

By a reasoning similar to the one resulting in $(Pa \wedge \neg Qa) \rhd_1 Ra$, it is possible to show that $Pa \rhd_2 \neg Qa$. I leave this to the reader as an easy exercise. Let us continue the proof to see what becomes of 2. As the reader will expect, line 7 (and no other of the above lines) will be marked.

7	$\star\Box_2\forall x(Px \supset (Sx \vee Qx \vee Tx))$	2; RC	$\{{}^{\star}_{2}(Px \supset (Sx \vee Qx \vee Tx))\}$ ✓[14]
⋮			
12	$\Box_2\forall x(Px \supset (Sx \vee Tx))$	2, 3; RU	$\emptyset$
13	$\neg\star\Box_2\forall x(Px \supset (Sx \vee Qx \vee Tx))$	12; RU	$\emptyset$
14	${}^{\star}_{2}(Px \supset (Sx \vee Qx \vee Tx))$	2, 13; RU	$\emptyset$
15	$\star\Box_2\forall x(Px \supset (Sx \vee Tx))$	12; RC	$\{{}^{\star}_{2}(Px \supset (Sx \vee Tx))\}$
16	$\Box_2(Sa \vee Ta)$	4; RU	$\emptyset$
17	$(Px \supset (Sx \vee Tx))^{a}_{2}$	12, 16; RU	$\emptyset$
18	$\neg\Box_2\neg Pb$	12; RU	$\{{}^{b}_{2}(Px \supset (Sx \vee Tx))\}$
19	$\neg\Box_2(Sb \vee Tb)$	12; RU	$\{(Px \supset (Sx \vee Tx))^{b}_{2}\}$
20	$Pb \rhd_2 (Sb \vee Tb)$	15, 18, 19; RU	Y

in which Y abbreviates $\{{}^{\star}_{2}(Px \supset (Sx \vee Tx)), {}^{b}_{2}(Px \supset (Sx \vee Tx)), (Px \supset (Sx \vee Tx))^{b}_{2}\}$, the union of the conditions of lines 15, 18, and 19. Line 17 is derived to show that the premise $\Box_2 Sa$ prevents the derivation of $Pa \rhd (Sa \vee Ta)$, as it obviously should.[26] However, neither that premise nor the premise Tb prevents the derivation of $Pb \rhd_2 (Sb \vee Tb)$, and correctly so. Note that this abduction statement would even be derivable if the premise set would contain any consistent subset of $\{Sb \vee Tb, \neg(Sb \vee Tb), Pb, \neg Pb\}$. True as well as false statements have potential explanations; true as well as false statements may serve as explanations for other statements. That an *explanandum* is true and one of its potential explanations is false, does not prevent the latter from being a potential explanation—but it certainly is not the true explanation. In fact, that an *explanandum* is true and all its known potential explanations are false is no objections against their being potential explanations. In the only case not considered so far both Pb and $\neg(Sb \vee Tb)$ are in the premise set. Even then $Pb \rhd_2 (Sb \vee Tb)$ is derivable, but for the pathological reason that the premise set is inconsistent.

The logic $\mathbf{ATAL}^m$ is just like $\mathbf{ATAL}^r$ except that it has Minimal Abnormality as its strategy. As was the case for $\mathbf{AAL}^m$ and $\mathbf{AAL}^r$ and for the same reasons, both logics describe the same consequence relation for $\Gamma \subseteq \mathcal{W}_s$.

Further variant logics are obtained if one allows for explanation in terms of sets of theories. This requires such modalities as $\Box_{\{1,3,4\}}$, governed by $\Box_{S_1} A, \Box_{S_2} B \vdash$

[26]While the premise $\Box_2 Sa$ allows me to easily make the point, its presence is obviously artificial. Note, however, that $\Box_2(\neg Pa \vee Sa \vee Ta)$ is $\mathbf{CL}^{\mathrm{M}}$-derivable from Γ_2 and hence does not have any potential explanations.

$\Box_{S_1 \cup S_2}(A \wedge B)$. Different variants conflate all theories by having $\Box$ without subscript. Such logics ensure that potential explanations are backed by theories rather than by accidental generalizations.

A bit of thinking reveals that the four adaptive logics considered so far do not warrant that the potential explanations are causal. I shall not deal with this problem here, but just point to the way in which it may be solved in an adaptive approach. Causality is clearly related to counterfactuals and similar conditionals. Nicholas Rescher [42, 44, 45] approached counterfactuals in terms of consequence relations defined in terms of maximal consistent subsets of a knowledge system extended with a counterfactual statement. Such consequence relations have been characterized by adaptive logics [5]. Approaches as the one of David Lewis [23, 40] proceed in terms of causal ontology rather than causal criteriology, but as soon as criteria for the distance between possible worlds are provided, they seem to come within reach of adaptive logics. Adaptive approaches to some other views on causality were devised by Maarten Van Dyck and Bert Leuridan [22, 50] among others.

Causal requirements may be build into the adaptive abduction logic. An alternative is that causally directed generalizations are derived independently; they may have the form $\forall x(A(x) \Rightarrow B(x))$, may be established by adaptive means, and may function in an adaptive abduction logic in the same way as $\forall x(A(x) \supset B(x))$ functions in $\mathbf{AAL}^r$ and $\Box_i \forall x(A(x) \supset B(x))$ functions in $\mathbf{ATAL}^r$.

Just for the record, it may be very sensible to study the relation between abduction and why-questions.[27] This may help to understand the circumstances under which a statement may function as an *explanandum* or may function as a potential explanation of a given *explanandum*. Specific forms of abduction may require that certain involved statements are true. It should be kept in mind, however, that most forms of explanation may also be applied to counterfactual statements.

8.7 More Properties of Adaptive Logics

We have seen that defeasible reasoning forms have some awkward properties. The strength of the adaptive logic approach lies in (i) its capacity to incorporate realistic reasoning forms and (ii) the fact that the standard format warrants an impressive set of nice properties. Some of these properties are recalled in this section.

Adaptive logics, even very simple one's like the abduction logics introduced in previous sections, do not have static proofs. This is why one should carefully study the properties of those logics and of dynamic proofs. It is also a good reason to phrase adaptive logics in such a way that Tarski logics and their static proofs are special cases of adaptive logics and dynamic proofs respectively. Below, I shall list a few properties of adaptive logics, all proven as theorems in the literature [10, 16], and comment upon them. Let $\mathcal{M}_\Gamma^{\mathbf{LLL}}$ be the set of **LLL**-models of Γ, $\mathcal{M}_\Gamma^{\mathbf{ULL}}$ the set of

[27] Actually this too involves adaptive features, just like the study of yes/no questions [30, 31].

ULL-models of Γ, $\mathcal{M}^m_\Gamma$ the set of minimally abnormal models of Γ, and $\mathcal{M}^r_\Gamma$ the set of reliable models of Γ.

The following two theorems show the adequacy of the logics. The lemmas provide the central clue.

Lemma 1 $\Gamma \vDash_{\mathbf{AL}^r} A$ *iff* $\Gamma \vDash_{\mathbf{LLL}} A \,\check{\vee}\, \text{Dab}(\Delta)$ *and* $\Delta \cap U(\Gamma) = \emptyset$ *for a finite* $\Delta \subset \Omega$.

Theorem 3 $\Gamma \vdash_{\mathbf{AL}^r} A$ *iff* $\Gamma \vDash_{\mathbf{AL}^r} A$. *(Soundness and Completeness for Reliability)*

Lemma 2 $M \in \mathcal{M}^m_\Gamma$ *iff* $M \in \mathcal{M}^{\mathbf{LLL}}_\Gamma$ *and* $\text{Ab}(M) \in \Phi(\Gamma)$.

Theorem 4 $\Gamma \vdash_{\mathbf{AL}^m} A$ *iff* $\Gamma \vDash_{\mathbf{AL}^m} A$. *(Soundness and Completeness for Minimal Abnormality)*

If a model of the premiss set is not selected, then this is justified by the fact that a selected model of the premiss set is less abnormal. This property is a corollary to Strong Reassurance, also known as Stopperedness or Smoothness. It basically means that there are no infinite sequences of ever less abnormal models.

Theorem 5 *If* $M \in \mathcal{M}^{\mathbf{LLL}}_\Gamma - \mathcal{M}^m_\Gamma$, *then there is a* $M' \in \mathcal{M}^m_\Gamma$ *such that* $\text{Ab}(M') \subset \text{Ab}(M)$. *(Strong Reassurance for Minimal Abnormality)*

Theorem 6 *If* $M \in \mathcal{M}^{\mathbf{LLL}}_\Gamma - \mathcal{M}^r_\Gamma$, *then there is a* $M' \in \mathcal{M}^r_\Gamma$ *such that* $\text{Ab}(M') \subset \text{Ab}(M)$. *(Strong Reassurance for Reliability)*

Some comments on Theorem 7. 1: Minimal Abnormality defines an adaptive logic that is at least as strong as Reliability (and often stronger). 3, 4: a well known and important property. 5: no new abnormalities are derivable by going adaptive; so adaptive logics minimize abnormalities. 6: truly defeasible consequence relations cannot be transitive because some of the consequences may prevent other formulas from being derivable; yet, extending a premise set with some of its consequences never results in new consequences. 7: truly defeasible consequence relations cannot be monotonic because a consequence may be eliminated when new premises are added; yet what is derivable remains derivable when consequences are added to the premise set.

Theorem 7 *Each of the following obtains:*

1. $\mathcal{M}^m_\Gamma \subseteq \mathcal{M}^r_\Gamma$ *and* $\text{Cn}_{\mathbf{AL}^r}(\Gamma) \subseteq \text{Cn}_{\mathbf{AL}^m}(\Gamma)$.
2. *If* $\text{Dab}(\Delta)$ *is a minimal* Dab-*consequence of* Γ *and* $A \in \Delta$, *then some* $M \in \mathcal{M}^m_\Gamma$ *verifies A and falsifies all members (if any) of* $\Delta - \{A\}$.
3. $\mathcal{M}^m_\Gamma = \mathcal{M}^m_{\text{Cn}_{\mathbf{AL}^m}(\Gamma)}$ *and* $\text{Cn}_{\mathbf{AL}^m}(\Gamma) = \text{Cn}_{\mathbf{AL}^m}(\text{Cn}_{\mathbf{AL}^m}(\Gamma))$. *(Fixed Point for Minimal Abnormality)*
4. $\mathcal{M}^r_\Gamma = \mathcal{M}^r_{\text{Cn}_{\mathbf{AL}^r}(\Gamma)}$ *whence* $\text{Cn}_{\mathbf{AL}^r}(\Gamma) = \text{Cn}_{\mathbf{AL}^r}(\text{Cn}_{\mathbf{AL}^r}(\Gamma))$. *(Fixed Point for Reliability)*
5. *For all* $\Delta \subseteq \Omega$, $\text{Dab}(\Delta) \in \text{Cn}_{\mathbf{AL}}(\Gamma)$ *iff* $\text{Dab}(\Delta) \in \text{Cn}_{\mathbf{LLL}}(\Gamma)$. *(Immunity)*

6. *If* $\Gamma' \subseteq \mathrm{Cn}_{\mathbf{AL}}(\Gamma)$, *then* $\mathrm{Cn}_{\mathbf{AL}}(\Gamma \cup \Gamma') \subseteq \mathrm{Cn}_{\mathbf{AL}}(\Gamma)$. *(Cautious Cut, Cumulative Transitivity)*
7. *If* $\Gamma' \subseteq \mathrm{Cn}_{\mathbf{AL}}(\Gamma)$, *then* $\mathrm{Cn}_{\mathbf{AL}}(\Gamma) \subseteq \mathrm{Cn}_{\mathbf{AL}}(\Gamma \cup \Gamma')$. *(Cautious Monotonicity)*

Comments on Theorem 8. 1: for premise sets that entail no Dab-formula, adaptive consequence sets coincide with upper limit consequence sets.[28] 2: if a premise set is abnormal but has lower limit models, its upper limit consequence set is trivial but its adaptive consequence set is not. 3: the general relation between the four kinds of logics. 4–7: cases where the inclusions are proper. 8: going adaptive never results in triviality.

Theorem 8 *Each of the following obtains:*

1. *If* Γ *is normal, then* $\mathcal{M}^{\mathbf{ULL}}_{\Gamma} = \mathcal{M}^{m}_{\Gamma} = \mathcal{M}^{r}_{\Gamma}$ *and* $\mathrm{Cn}_{\mathbf{AL}^r}(\Gamma) = \mathrm{Cn}_{\mathbf{AL}^m}(\Gamma) = \mathrm{Cn}_{\mathbf{ULL}}(\Gamma)$.
2. *If* Γ *is abnormal and* $\mathcal{M}^{\mathbf{LLL}}_{\Gamma} \neq \emptyset$, *then* $\mathcal{M}^{\mathbf{ULL}}_{\Gamma} \subset \mathcal{M}^{m}_{\Gamma}$ *and* $\mathrm{Cn}_{\mathbf{AL}^r}(\Gamma) \subseteq \mathrm{Cn}_{\mathbf{AL}^m}(\Gamma) \subset \mathrm{Cn}_{\mathbf{ULL}}(\Gamma)$.
3. $\mathcal{M}^{\mathbf{ULL}}_{\Gamma} \subseteq \mathcal{M}^{m}_{\Gamma} \subseteq \mathcal{M}^{r}_{\Gamma} \subseteq \mathcal{M}^{\mathbf{LLL}}_{\Gamma}$ *and* $\mathrm{Cn}_{\mathbf{LLL}}(\Gamma) \subseteq \mathrm{Cn}_{\mathbf{AL}^r}(\Gamma) \subseteq \mathrm{Cn}_{\mathbf{AL}^m}(\Gamma) \subseteq \mathrm{Cn}_{\mathbf{ULL}}(\Gamma)$.
4. $\mathcal{M}^{r}_{\Gamma} \subset \mathcal{M}^{\mathbf{LLL}}_{\Gamma}$ *iff* $\Gamma \cup \{A\}$ *is* **LLL**-*satisfiable for some* $A \in \Omega - U(\Gamma)$.
5. $\mathrm{Cn}_{\mathbf{LLL}}(\Gamma) \subset \mathrm{Cn}_{\mathbf{AL}^r}(\Gamma)$ *iff* $\mathcal{M}^{r}_{\Gamma} \subset \mathcal{M}^{\mathbf{LLL}}_{\Gamma}$.
6. $\mathcal{M}^{m}_{\Gamma} \subset \mathcal{M}^{\mathbf{LLL}}_{\Gamma}$ *iff there is a (possibly infinite)* $\Delta \subseteq \Omega$ *such that* $\Gamma \cup \Delta$ *is* **LLL**-*satisfiable and there is no* $\varphi \in \Phi_{\Gamma}$ *for which* $\Delta \subseteq \varphi$.
7. *If there are* $A_1, \ldots, A_n \in \Omega$ $(n \geq 1)$ *such that* $\Gamma \cup \{A_1, \ldots, A_n\}$ *is* **LLL**-*satisfiable and, for every* $\varphi \in \Phi_{\Gamma}$, $\{A_1, \ldots, A_n\} \nsubseteq \varphi$, *then* $\mathrm{Cn}_{\mathbf{LLL}}(\Gamma) \subset \mathrm{Cn}_{\mathbf{AL}^m}(\Gamma)$.
8. $\mathrm{Cn}_{\mathbf{AL}^m}(\Gamma) \neq \mathcal{W}$ *and* $\mathrm{Cn}_{\mathbf{AL}^r}(\Gamma) \neq \mathcal{W}$ *iff* $\mathrm{Cn}_{\mathbf{LLL}}(\Gamma) \neq \mathcal{W}$. *(Reassurance)*

Compare Cumulative Indifference with Cumulative Transitivity, Cautious Monotonicity, and Fixed Point.

Theorem 9 *If* $\Gamma' \subseteq \mathrm{Cn}_{\mathbf{AL}}(\Gamma)$, *then* $\mathrm{Cn}_{\mathbf{AL}}(\Gamma \cup \Gamma') = \mathrm{Cn}_{\mathbf{AL}}(\Gamma)$. *(Cumulative Indifference)*

An important feature of dynamic proofs is that, in whichever way a proof starts off, all final consequences are derivable in it and nothing else is finally derivable in it.

Theorem 10 *If* $\Gamma \vdash_{\mathbf{AL}} A$, *then every* **AL**-*proof from* Γ *can be extended in such a way that* A *is finally derived in it. (Proof Invariance)*

Equivalence between premise sets agrees with the following (and several other) standard and intuitive criteria.

Theorem 11 *If* $\Gamma' \in \mathrm{Cn}_{\mathbf{AL}}(\Gamma)$ *and* $\Gamma \in \mathrm{Cn}_{\mathbf{AL}}(\Gamma')$, *then* $\mathrm{Cn}_{\mathbf{AL}}(\Gamma) = \mathrm{Cn}_{\mathbf{AL}}(\Gamma')$. *(Equivalent Premise Sets)*

---

[28] So inconsistency-adaptive logics assign all **CL**-consequences to consistent premise sets.

It is remarkable that so rich a set of properties obtains for an general approach to defeasible reasoning, notwithstanding its computational complexity. Many more properties were studied, for example the Deduction Theorem which does not hold in general for Reliability, but holds for Minimal Abnormality. There may be quite some variance between adaptive logics, as was illustrated by the relative simplicity of $\mathbf{AAL}^r$ and $\mathbf{ATAL}^r$.

8.8 What It Is All About

Whenever a premise set belongs to a decidable fragment of the lower limit logic, for example finite premise sets of the propositional fragment of most Tarski logics, its adaptive consequence set is decidable. If one moves to the full predicative level, there even is no positive test for final derivability. It should be stressed that this is a fact of life which is caused by the consequence relations, not by the fact that the consequence relations are explicated by adaptive logics. The study of adaptive logics has provided insights in the domain of defeasible logics. There are no static proofs for defeasible consequence relations. So dynamic proofs are not a complication but a simplification: they are a means to obtain a realistic grasp on defeasible reasoning in its full complexity. Thinking about dynamic proofs, including infinite stages, is still simpler than thinking about models.[29] The latter moreover do not explicate defeasible reasoning.[30]

Even at the predicative level, there are *criteria* for final derivability. Such criteria were phrased in terms of 'block proofs' and the 'block semantics' [4], in terms of tableau methods [14, 15], and in terms of prospective proofs [8, 53], which are a proof *procedure* that provides a criterion for final derivability. Needless to say such criteria hold in some cases but cannot hold generally.

So what if no criterion applies? A first question is whether the dynamics of the proofs goes anywhere. Here the block analysis of proofs offers useful insights. A proof stage provides an insight in the premises. Every step of the proof will be either informative or non-informative. It is informative iff, by the step, more insight in the premises is gained—the matter can be formally defined [4]. Sensible proofs are composed of informative steps only, but no insight is lost if non-informative steps are added to a proof. So as a dynamic proof proceeds, the insights in the premises accumulate. This, however, does not guarantee that, as the dynamic proof proceeds, the set of formulas derived at the stage converges to the final consequence set. This obviously cannot be guaranteed because adaptive consequence sets are up to Π^1_1-complex [36, 37, 52].

[29]That insights gained in terms of dynamic proofs may be rephrased in semantic terms does not undermine the avail of dynamic proofs.

[30]Several points from this paragraph were clarified by a weird discussion in the literature [12, 21, 52] that started off as an attack on adaptive logics, presumably on the dynamic proofs, although that was never very clear.

Have sets of consequences-at-a-stage any use if no criterion applies? Yes in that the full set of final consequences may not required by the application context. Consider the paradigmatic example from Sect. 8.2, a theory intended as consistent that turns out to be inconsistent. As one starts reasoning about the theory, certain abnormalities will soon be located. These abnormalities will be narrowed down, for example in view of personal constraints—the operation proceeds smoothly by adaptive means [9]. Locating abnormalities and narrowing them down will go on until a clear idea for replacement is available. Such an idea may suffice to launch a hypothesis for a replacement. A nice example is that several apparently consistent set theories were formulated as alternatives for Frege's set theory in view of the Russell paradox but before the Curry paradox was discovered. Another nice but slightly more complex example is Clausius' removal of an inconsistency from thermodynamics [27, 28, 32].

The aim of applications of adaptive logics in creative processes is to arrive at sensible hypothetical proposals. The function of the logics is to provide an analysis that suggests such hypotheses. In that respect $\mathrm{Cn}_{\mathbf{AL}}(\Gamma)$ is an ideal, which is often beyond reach in real life situations. That it is only an ideal should not prevent us from studying it in order to show that the applied mechanism is coherent and conceptually sound.

The situation is similar for defeasible consequence relations that characterize methodological phenomena like abduction. The human condition demands that we decide on the basis of information that is incomplete and unsettled in empirical respects and that is incomplete and unsettled in logical respects as well. This is not a reason to stop doing epistemology or philosophy of science. And it is not a reason to give up trying to phrase methods in a precise way and to study those precise formulations by means of the best available metatheoretic means. Only morons believe that all decisions are equally good if no decision can be justified in an absolute way. Sensible people better keep an eye on them.

Acknowledgements I am indebted to Joke Meheus and especially to Frederik Van De Putte for comments on a draft of this paper.

References

1. Aliseda, A. (2006). *Abductive Reasoning.*, Logical investigations into discovery and explanation Dordrecht: Springer.
2. Batens, D. (1980). Paraconsistent extensional propositional logics. *Logique et Analyse*, *90–91*, 195–234.
3. Batens, D. (1989). Dynamic dialectical logics. In G. Priest, R. Routley, & J. Norman (Eds.), *Paraconsistent logic* (pp. 187–217)., Essays on the inconsistent München: Philosophia Verlag.
4. Batens, D. (1995). The clue to dynamic aspects of logic. *Logique et Analyse*, *150–152*, 285–328. Appeared 1997.
5. Batens, D. (2000). Towards the unification of inconsistency handling mechanisms. *Logic and Logical Philosophy*, *8*, 5–31. Appeared 2002.

6. Batens, D. (2001). A general characterization of adaptive logics. *Logique et Analyse, 173–175*, 45–68. Appeared 2003.
7. Batens, D. (2004). The need for adaptive logics in epistemology. In D. Gabbay, S. Rahman, J. Symons, & J. P. V. Bendegem (Eds.), *Logic* (pp. 459–485)., Epistemology and the unity of science Dordrecht: Kluwer Academic Publishers.
8. Batens, D. (2005). A procedural criterion for final derivability in inconsistency-adaptive logics. *Journal of Applied Logic*, *3*, 221–250.
9. Batens, D. (2006). Narrowing down suspicion in inconsistent premise sets. In J. Malinowski & A. Pietruszczak (Eds.), *Essays in Logic and Ontology* (Vol. 91, pp. 185–209)., Poznań studies in the philosophy of the sciences and the humanities Amsterdam/New York: Rodopi.
10. Batens, D. (2007). A universal logic approach to adaptive logics. *Logica Universalis*, *1*, 221–242.
11. Batens, D. (2009). Towards a dialogic interpretation of dynamic proofs. In Cédric Dégremont, Laurent Keiff & Helge Rückert (Eds.), *Dialogues, logics and other strange things. essays in honour of shahid rahman*, (pp. 27–51). London: College Publications, 558 p.
12. Batens, D., De Clercq, K., Verdée, P., & Meheus, J. (2009). Yes fellows, most human reasoning is complex. *Synthese*, *166*, 113–131.
13. Batens, D., & Meheus, J. (2000a). The adaptive logic of compatibility. *Studia Logica*, *66*, 327–348.
14. Batens, D., & Meheus, J. (2000b). A tableau method for inconsistency-adaptive logics. In R. Dyckhoff (Ed.), *Automated Reasoning with Analytic Tableaux and Related Methods* (Vol. 1847, pp. 127–142). Lecture Notes in Artificial Intelligence, Springer.
15. Batens, D., & Meheus, J. (2001). Shortcuts and dynamic marking in the tableau method for adaptive logics. *Studia Logica*, *69*, 221–248.
16. Batens, D., Straßer, C., & Verdée, P. (2009). On the transparency of defeasible logics: Equivalent premise sets, equivalence of their extensions, and maximality of the lower limit. *Logique et Analyse*, *207*, 281–304.
17. Beirlaen, M., & Aliseda, A. (2014). A conditional logic for abduction. *Synthese*, *191*, 3733–3758.
18. Carl G. Hempel. *Aspects of Scientific Explanation and Other Essays in the Philosophy of Science*. The Free Press, New York, 1965.
19. Gauderis, T. (2013). Modelling abduction in science by means of a modal adaptive logic. *Foundations of Science*, *18*, 611–624.
20. Halonen, I., & Hintikka, J. (2005). Toward a theory of the process of explanation. *Synthese*, *143*, 5–61.
21. Horsten, L., & Welch, P. (2007). The undecidability of propositional adaptive logic. *Synthese*, *158*, 41–60.
22. Leuridan, B. (2009). Causal discovery and the problem of ignorance. An adaptive logic approach. *Journal of Applied Logic*, *7*, 188–205.
23. Lewis, D. (1973). *Counterfactuals*. Mass.: Harvard University Press, Cambridge.
24. Lycke, H. (2012). A formal explication of the search for explanations: the adaptive logics approach to abductive reasoning. *Logic Journal of the IGPL*, *20*, 497–516.
25. Magnani, L. (2001). *Abduction, reason, and science processes of discovery and explanation*. New York: Kluwer Academic / Plenum Publishers.
26. Magnani, L., Carnielli, W. & Pizzi, C. Eds. (2010). *Model-based reasoning in science and technology. abduction, logic, and computational discovery* (Vol. 314). Studies in Computational Intelligence, Heidelberg: Springer.
27. Meheus, J. (1993). Adaptive logic in scientific discovery: The case of Clausius. *Logique et Analyse*, *143–144*, 359–389. Appeared 1996.
28. Meheus, J. (1999a). Clausius' discovery of the first two laws of thermodynamics. A paradigm of reasoning from inconsistencies. *Philosophica*, *63*, 89–117. Appeared 2001.
29. Meheus, J. (1999b). Deductive and ampliative adaptive logics as tools in the study of creativity. *Foundations of Science*, *4*, 325–336.

30. Meheus, J. (1999b). Erotetic arguments from inconsistent premises. *Logique et Analyse*, *165–166*, 49–80. Appeared 2002.
31. Meheus, J. (2001). Adaptive logics for question evocation. *Logique et Analyse*, *173–175*, 135–164. Appeared 2003.
32. Meheus, J. (2002). Inconsistencies in scientific discovery. Clausius's remarkable derivation of Carnot's theorem. In H. Krach, G. Vanpaemel, & P. Marage (Eds.), *History of modern physics* (pp. 143–154)., Acta of the XXth International Congress of History of Science Turnhout (Belgium): Brepols.
33. Meheus, J. (2011). A formal logic for the abduction of singular hypotheses. In D. Dieks, W. J. Gonzalez, S. Hartmann, T. Uebel, & M. Weber (Eds.), *Explanation* (pp. 93–108)., Prediction, and confirmation Dordrecht: Springer.
34. Meheus, J., Adaptive logics for abduction and the explication of explanation-seeking processes. In Pombo and Gerner [41], pp. 97–119.
35. Meheus, J., & Batens, D. (2006). A formal logic for abductive reasoning. *Logic Journal of the IGPL*, *14*, 221–236.
36. Odintsov, S. P., & Speranski, S. O. (2012). On algorithmic properties of propositional inconsistency-adaptive logics. *Logic and Logical Philosophy*, *21*, 209–228.
37. Odintsov, S. P., & Speranski, S. O. (2013). Computability issues for adaptive logics in multi-consequence standard format. *Studia Logica*, *101*(6), 1237–1262. doi:10.1007/s11225-013-9531-2.
38. Paul, G. (2000). AI approaches to abduction. In D. M. Gabbay & P. Smets (Eds.), *Handbook of defeasible reasoning and uncertainty management systems* (pp. 35–98)., Abductive reasoning and learning Dordrecht: Kluwer Academic Publishers.
39. Pogorzelski, W. A., & Prucnal, T. (1975). The substitution rule for predicate letters in the first-order predicate calculus. *Reports on Mathematical Logic*, *5*, 77–90.
40. John, L. (1976). *Pollock*. Dordrecht: Subjunctive Reasoning. Reidel.
41. Pombo, O., & Gerner, A. (Eds.). (2007). *Abduction and the Process of Scientific Discovery*. Lisboa: Centro de Filosofia das Ciências da Universidade de Lisboa.
42. Rescher, N. (1964). *Hypothetical Reasoning*. Amsterdam: North-Holland.
43. Rescher, N. (1973). *The Coherence Theory of Truth*. Oxford: Clarendon.
44. Rescher, N. (2005). *What If?*. New Brunswick, New Jersey: Transaction Publishers.
45. Rescher, N., & Manor, R. (1970). On inference from inconsistent premises. *Theory and Decision*, *1*, 179–217.
46. Routley, R., & Meyer, R. K. (1976). Dialectical logic, classical logic, and the consistency of the world. *Studies in Soviet Thought*, *16*, 1–25.
47. Shapere, D. (2004). Logic and the philosophical interpretation of science. In P. Weingartner (Ed.), *Alternative logics* (pp. 41–54)., Do sciences need them? Berlin, Heidelberg: Springer.
48. Vanackere, G. (1997). Ambiguity-adaptive logic. *Logique et Analyse*, *159*, 261–280. Appeared 1999.
49. Vanackere, G. (1999). Minimizing ambiguity and paraconsistency. *Logique et Analyse*, *165–166*, 139–160. Appeared 2002.
50. Van Dyck, M. (2004). Causal discovery using adaptive logics. Towards a more realistic heuristics for human causal learning. *Logique et Analyse*, *185–188*, 5–32. Appeared 2005.
51. Van Kerckhove, B., & Vanackere, G. (2003). Vagueness-adaptive logic: A pragmatical approach to Sorites paradoxes. *Studia Logica*, *75*, 383–411.
52. Verdée, P. (2009). Adaptive logics using the minimal abnormality strategy are Π^1_1-complex. *Synthese*, *167*, 93–104.
53. Verdée, P. (2013). A proof procedure for adaptive logics. *Logic Journal of the IGPL*, *21*, 743–766. doi:10.1093/jigpal/jzs046.

Chapter 9
A Multimodal Pragmatic Analysis of the Knowability Paradox

Massimiliano Carrara, Daniele Chiffi and Davide Sergio

Abstract The *Knowability Paradox* starts from the assumption that every truth is knowable and leads to the paradoxical conclusion that every truth is also actually known. Knowability has been traditionally associated with both contemporary verificationism and intuitionistic logic. We assume that classical modal logic in which the standard paradoxical argument is presented is not sufficient to provide a proper treatment of the verificationist aspects of knowability. The aim of this paper is both to sketch a language $\mathcal{L}^P_{\Box,K}$, where alethic and epistemic classical modalities are combined with the pragmatic language for assertions $\mathcal{L}^P$, and to analyse the result of the application of our framework to the paradox.

Keywords Knowability · Logic for pragmatics · Multimodality

9.1 Introduction

A logical argument, known as the *Knowability Paradox*, starts from the assumption that *every truth is knowable* and leads to the paradoxical conclusion that *every truth is actually known*.

M. Carrara
FISPPA Department - Section of Philosophy, University of Padua, Padua, Italy
e-mail: massimiliano.carrara@unipd.it

D. Chiffi
Ragnar Nurkse Department of Innovation and Governance,
Tallinn University of Technology, Tallinn, Estonia
e-mail: chiffidaniele@gmail.com

D. Sergio (✉)
Freelance Logician, Lecce, Italy
e-mail: sdavide73@gmail.com

R. Urbaniak and G. Payette (eds.), *Applications of Formal Philosophy*,
Logic, Argumentation & Reasoning 14, DOI 10.1007/978-3-319-58507-9_9

The idea that every truth is knowable is traditionally associated with a verificationist perspective, a perspective that assumes intuitionistic logic as the "correct" one.

The paradox is formulated in a classical modal logic. However, the fact that classical and verificationist/intuitionistic notions are involved into its formulation makes classical modal logic not adequate to handle properly the specific features of the notions of *truth*, *proof* and *knowledge*.

Our idea is to provide a treatment of the paradox in an intuitionistic-like system—a multimodal extension $\mathcal{L}^P_{\Box,K}$ of the pragmatic logic for assertions $\mathcal{L}^P$—compatible with classical systems; a system where a verificationist interpretation of classic (modal) propositions together with intuitionistic connections (operations) on them is given.[1]

The paper is divided into 7 sections. Section 9.2 is devoted to briefly outlining the structure of the knowability paradox. In Sects. 9.3, 9.4 and 9.5 we present our pragmatic language $\mathcal{L}^P_{\Box,K}$. In Sect. 9.6 we analyse the result of the application of our framework to the paradox.

9.2 The Knowability Paradox

The *Knowability Paradox* (KPx) is based on two principles: the *Knowability Principle* (KP) and the *Principle of Non-Omniscience* (Non-Om). (KP) is usually expressed in the following way:

(KP) $\forall p(p \to \Diamond Kp)$

while (Non-Om) is formulated as:

(Non-Om) $\exists p(p \wedge \neg Kp)$

The expression "Kp" reads "p is, has been, or will be known by somebody". Specifically, the paradox arises because (KP) $\to \neg$ (Non-Om) is a theorem of classical modal logic, that is:

(KPx) $[\forall p(p \to \Diamond Kp)] \to [\forall p(p \to Kp)]$.

Assume the following two properties of knowledge:

1. the distributive property over conjunction (Dist), i.e., if a conjunction is known, then its conjuncts are also known, and *vice versa*;
2. the factivity (Fact), i.e., if a proposition is known, then it is true.

Moreover, assume the following two standard modal claims, which can be formulated using the usual modal operators $\Diamond$ ("it is possible that") and $\Box$ ("it is necessary that"). The first is the *Rule of Necessitation*:

---

[1] There is a certain number of multi-modal approaches to the paradox in the literature proposed from a variety of perspectives. See for example: [1, 2, 8, 10, 13, 19, 20].

(Nec) If p is a theorem, then $\Box p$

The second rule establishes the interdefinability of the modal concepts of necessity and possibility:

(ER) $\Box\neg p$ is logically equivalent to $\neg\Diamond p$

From (KP) and (Non-Om) a contradiction follows. The first argument:

(1)	$p \wedge \neg Kp$	from (Non-Om)
(2)	$(p \wedge \neg Kp) \rightarrow \Diamond K(p \wedge \neg Kp)$	substitution of "$p \wedge \neg Kp$" for p in (KP)
(3)	$\Diamond K(p \wedge \neg Kp)$	from (1) and (2) by *Modus Ponens*

The second, independent, argument:

(4)	$K(p \wedge \neg Kp)$	assumption
(5)	$Kp \wedge K\neg Kp$	distributivity of K
(6)	$Kp \wedge \neg Kp$	factivity of K
(7)	$\bot$	contradiction
(8)	$\neg K(p \wedge \neg Kp)$	*reductio*, discharging (4)
(9)	$\Box\neg K(p \wedge \neg Kp)$	(Nec)
(10)	$\neg\Diamond K(p \wedge \neg Kp)$	(ER)

From (3) and (10) a contradiction follows.

Fitch [14] and Church (we follow here [17]) proved that

(*) $\forall p \neg\Diamond\ K\ (p \wedge \neg Kp)$

is a theorem. But if (*) and (Non-Om) hold, then (KP) has to be rejected since the substitution of $p \wedge \neg Kp$ for p in (KP) leads to a contradiction.

Hence, if (KP) is accepted, then (Non-Om) must be denied. However, the negation of (Non-Om) is classically equivalent to $\forall p(p \rightarrow Kp)$.

Therefore, (KP) is sufficient to obtain the paradoxical conclusion of (KPx), a conclusion that is to be particularly problematic for the holders of antirealism who accept knowability.

However, (KPx) is a *classical* (modal) theorem, and since (KP) has been traditionally associated with both contemporary verificationism and intuitionistic logic, it seems that (KP) is not the correct formalization of knowability. Indeed, according to us, the paradoxical reading of (KPx) can be avoided as soon as we express knowability in an adequate form; to do this we need to take into account the verificationist and intuitionistic features of an antirelist version of knowability.

In order to explain the verificationist and intuitionistic features in a setting *compatible* with classical systems, we introduce a muldimodal pragmatic language for assertions $\mathcal{L}^P_{\Box,K}$. This language is given by an extension of the expressiveness of the pragmatic language for assertion $\mathcal{L}^P$, developed in (Dalla Pozza & Garola, 1995), from propositional contents to modal (propositional) contents, and, in particular, to assertions on (classical) alethic and epistemic contents. Moreover, $\mathcal{L}^P_{\Box,K}$ preserves the main characteristic of $\mathcal{L}^P$; that is, the *integrated* perspective about truth and proof according to the Justification Principle (JP); a principle which captures the intuition that the notion of proof presupposes the classical notion of truth as a regulative concept, since a proof of a proposition amounts to a proof that its truth value is the value "true":

(JP) The assertion of α is justified iff a proof exists that α is true.

In this way, by means of (JP), a verificationist reading of classical propositions is introduced.

9.3 An Outline of $\mathcal{L}^P$

$\mathcal{L}^P$ is a *language for assertions* mainly inspired by Frege and Dummett and by Austin's theory of illocutionary acts. Roughly speaking, the idea is to distinguish propositions from judgments: A proposition is either true or false, while a *judgment*, that can be expressed through the speech act of an *assertion*,[2] is—according to Dalla Pozza and Garola's view—either *justified* or *unjustified*. A justified assertion is defined in terms of the existence of a proof that the asserted content is true. Although the concept of proof is meant to be *intuitive* and *unspecified*, it must always be understood as *correct*: a proof is a proof of a truth. The key ideas behind the language are both the explication of the notion of assertion of a content in terms of justification-values, and the definition of justification-values in terms of the existence of proof for the truth of the asserted content. Therefore, the existence of a proof *is* the ground for a (justified) assertion.

A pragmatic language is the (disjoint) union of two sets of formulas: radicals and sentences. The set of radicals is the descriptive part of the language, while the set of sentences is its pragmatic part.

Radicals represent *propositional contents* of sentences. Sentences express illocutionary acts. A sentence is either *elementary*, i.e., obtained by prefixing a sign of pragmatic mood to a radical, or *complex*, i.e., obtained from other sentences by means of logico-pragmatic connectives ($\sim$, $\cap$, $\cup$, $\supset$, $\equiv$; signs for *pragmatic negation, conjunction, disjunction, implication*, and *equivalence* respectively).

Radicals and sentences have different syntactic forms and are interpreted in different ways. A pragmatic interpretation of a pragmatic language consists of both a semantic interpretation of its descriptive part and a pragmatic interpretation of its pragmatic part. Radicals are interpreted semantically in terms of truth-values, thus every radical is either *true* or *false*. Sentences are interpreted pragmatically in terms of justification-values, so that every sentence is either *justified* or *unjustified*.

[2]Notice that Frege's analysis is extendable to other speech acts such as asking, questioning, etc. So is $\mathcal{L}^P$. Languages where $\mathcal{L}^P$ is expanded so to give rise of other pragmatics acts have been studied. See, for example, [3].

In the pragmatic language for assertions $\mathcal{L}^P$ the descriptive part $\mathcal{L}$ is identified with the language of classical propositional logic (the set of propositional formulas), and the set of sentences is a set of assertions. Elementary sentences are thus built up using only the sign of pragmatic mood of assertion, $\vdash$. So, for example, if α_1 and α_2 are propositional formulas, then $\vdash \alpha_1$ and $\vdash \alpha_2$ are elementary assertions, while $\vdash \alpha_1 \cap \vdash \alpha_2$ or $\vdash \alpha_1 \cup \vdash \alpha_2$ are complex assertions. Intuitionistic language in $\mathcal{L}^P$, is represented as the fragment built up starting from elementary sentences with atomic radicals by means of pragmatic connectives, and, therefore, no classical connective falls under the scope of an assertion [9].

In $\mathcal{L}^P$, there are no assertions whose contents are *modal* propositions. In order to overcome such a limitation, we introduce $\mathcal{L}^P_{\Box,K}$, a pragmatic language for assertions of *modal* propositional contents. In particular, the descriptive part $\mathcal{L}_{\Box,K}$ of $\mathcal{L}^P_{\Box,K}$ is the *fusion* [5] $\mathcal{L}_\Box \oplus \mathcal{L}_K$ of two modal languages, $\mathcal{L}_\Box$ and $\mathcal{L}_K$, endowed with two independent boxes, $\Box$ and K, interpreted as "it is proved that"[3] and "it is known that" respectively.[4] In this way, we have a language that allows us to combine alethic, epistemic and verificationist features within a classically understood framework.

9.4 $\mathcal{L}^P_{\Box,K}$ and Its Interpretations

The set of radical formulas and the set of assertive formulas of $\mathcal{L}^P_{\Box,K}$ are respectively defined recursively through the following formation rules:

$$\alpha := p \mid \top \mid \bot \mid \neg\alpha \mid \alpha_1 \wedge \alpha_2 \mid \alpha_1 \vee \alpha_2 \mid \alpha_1 \rightarrow \alpha_2 \mid \alpha_1 \leftrightarrow \alpha_2 \mid \Box\alpha \mid K\alpha$$
$$\delta := \vdash \alpha \mid \sim \delta \mid \delta_1 \cap \delta_2 \mid \delta_1 \cup \delta_2 \mid \delta_1 \supset \delta_2 \mid \delta_1 \equiv \delta_2$$

In order to get a pragmatic interpretation of $\mathcal{L}^P_{\Box,K}$, we have to semantically interpret $\mathcal{L}^P_{\Box,K}$. This amounts to an interpretation of its descriptive part $\mathcal{L}_{\Box,K}$.

As a matter of fact, the semantics of the fusion $\mathcal{L}_1 \oplus \mathcal{L}_2$ of two modal languages, $\mathcal{L}_1$ and $\mathcal{L}_2$, endowed with two independent boxes, $\Box_1$ and $\Box_2$, is given within the class of frames of the form $\langle W, R_1, R_2\rangle$ where $\langle W, R_1\rangle$ and $\langle W, R_2\rangle$ are frames for $\Box_1$ and $\Box_2$ respectively. The axiomatic presentation through Hilbert calculus is obtained by merging the axioms and the inference rules of both logics. Moreover, *Bridge Principles* (PBs) can be added, i.e. axioms intended to logically connect the independent boxes, e.g., $\Box_1\alpha \rightarrow \Box_2\alpha$.[5]

We are assuming $\mathcal{L}_{\Box,K}$ to be the *fusion* $\mathcal{L}_\Box \oplus \mathcal{L}_K$ of $\mathcal{L}_\Box$ and $\mathcal{L}_K$ endowed with $\Box$ and K intuitively interpreted as "it is proved that" and "it is known that" respectively.

---

[3]Where 'proof' has to be understood in its intuitive sense.

[4]The fusion $\mathcal{L}_1 \oplus \mathcal{L}_2$ of two modal languages, $\mathcal{L}_1$ and $\mathcal{L}_2$, endowed with two independent boxes, $\Box_1$ and $\Box_2$, is the *smallest* modal language generated by both boxes. Note also that the fusion of modal languages is commutative.

[5]BPs can be equivalent to conditions on the relations between accessibility relations [5].

Hence, it is intuitive to consider relational structures of the form $\langle W, R_{\Box}, R_K \rangle \in \mathcal{C}$ where $\mathcal{C}$ is the class of frames with W a set of possible worlds, and $R_{\Box}, R_K \subseteq W$ x W binary accessibility relations on W such that $R_{\Box}$ is reflexive and transitive, while R_K is reflexive, symmetric and transitive. In this way, $\Box$ is a **S4**-like alethic modality, and K is a **S5**-like epistemic modality.

In addition, we introduce the following *Bridge Principle* (BP):

(BP) $\quad \Box\alpha \rightarrow \neg K \neg\alpha$

which can be intuitively read as "if it is the case that α is proved to be true, then it is not the case that α is known to be false". (BP) gives a logical connection between $\Box$ and K that turns out to be equivalent to the condition that $R_{\Box} \subseteq R_K$ [6]. Namely, (BP) is valid (on an appropriate frame) if and only if $R_{\Box} \subseteq R_K$.

The idea behind (BP) can be made clearer if we consider its equivalent formulation in terms of conjunction:

(BP′) $\quad \neg(\Box\alpha \wedge K\neg\alpha)$

(BP′) identifies the relation expressing a minimal condition holding between proof and knowledge according to our pre-theoretical insights. That is, there must be a logical incompatibility between the proof that α is true and the knowledge that α is false.

The assertable modal contents are given by combining (the *fusion* of) two modal languages, $\mathcal{L}_{\Box}$ and $\mathcal{L}_K$, endowed with two independent boxes, $\Box$ for the alethic modality and K for the epistemic one. These two boxes are intuitively interpreted as "it is proved that" and "it is known that" and formally realized by means of a **S4**-like and a **S5**-like modality respectively.

Given the intuitive interpretation of the $\Box$, it is possible to read classical alethic contents intuitionistically. We interpret it in this way for two reasons. The first one is a technical reason, and it is related to the fact that any pragmatic language for assertions is a intuitionistic-like system. Indeed, since the pragmatic connectives are interpreted intuitionistically, any pragmatic language is essentially an intuitionistic one. What we are given here is a verificationist interpretation, by means of (JP), of classical modal propositions in terms of assertions together with intuitionistic-like connections of them defined via pragmatic connectives. In this way, a set of Pragmatic Bridge Principles (PBPx) explaining the relations between classical connectives, pragmatic arguments and modal operators is obtained.

The second one is related to the paradox. Indeed, if alethic notions have an intuitionistic-like semantics, then "it is possible that", the dual notion of "it is necessary that", can be interpreted as "there is no proof that not". In such a way, the possibility of something to be true is reduced to the (actual) absence of a proof of its falsity, and $\Diamond K\alpha$ becomes "(at this moment in time), there is no proof that $K\alpha$ is false".

The semantic and the pragmatic interpretation of $\mathcal{L}^P_{\Box,K}$ are given through the following definitions.

(Def.1) [Semantic interpretation of $\mathcal{L}^P_{\Box,K}$]

Let $\mathcal{C}$ be the class of frames $F = \langle W, R_{\Box}, R_K \rangle$ such that W is a set of possible worlds, $R_{\Box}, R_K \subseteq W \text{x} W$ are binary accessibility relations on W, $R_{\Box}$ is reflexive and transitive, R_K is reflexive, symmetric and transitive, and $R_{\Box} \subseteq R_K$.

Let $\mathcal{V}_F$ be the class of valuations

$$v : \begin{cases} PROP \to \wp(W) \\ p \mapsto v(p) \subseteq W \end{cases}$$

on a frame $F \in \mathcal{C}$ where $PROP$ is the set of atomic propositional radicals.

Let $\mathcal{M} = \{M = \langle F, v \rangle \mid F \in \mathcal{C} \,\&\, v \in \mathcal{V}_F\}$ be the class of models on a frame F.

Let $M = \langle \langle W, R_{\Box}, R_K \rangle, v \rangle \in \mathcal{M}$.

Then, a *semantic interpretation* σ_v of $\mathcal{L}^P_{\Box,K}$ on M is any function

$$\sigma_v : \begin{cases} (\mathcal{L}_{\Box} \oplus \mathcal{L}_K) \text{x} W \to \{T, F\} \\ (\alpha, w) \mapsto \sigma_v(\alpha, w) \in \{T, F\} \end{cases}$$

which satisfies the following truth-rules:

(TR1) Let $p \in PROP$ and $w \in W$. Then:
(i) $\sigma_v(\top, w) = T$
(ii) $\sigma_v(\bot, w) = F$
(iii) $\sigma_v(p, w) = T \Leftrightarrow p \in v(p)$

(TR2) Let $\alpha, \alpha_1, \alpha_2 \in \mathcal{L}_{\Box,K}$ and $w \in W$. Then:
(i) $\sigma_v(\neg\alpha, w) = T \Leftrightarrow \sigma_v(\alpha, w) = F$
(ii) $\sigma_v(\alpha_1 \wedge \alpha_2, w) = T \Leftrightarrow \sigma_v(\alpha_1, w) = T$ and $\sigma_v(\alpha_2, w) = T$
(iii) $\sigma_v(\alpha_1 \vee \alpha_2, w) = T \Leftrightarrow \sigma_v(\alpha_1, w) = T$ or $\sigma_v(\alpha_2, w) = T$
(iv) $\sigma_v(\alpha_1 \to \alpha_2, w) = T \Leftrightarrow \sigma_v(\alpha_2, w) = T$ whenever $\sigma_v(\alpha_1, w) = T$
(v) $\sigma_v(\alpha_1 \leftrightarrow \alpha_2, w) = T \Leftrightarrow \sigma_v(\alpha_1 \to \alpha_2, w) = T$ and $\sigma_v(\alpha_2 \to \alpha_1, w) = T$

(TR3) Let $\alpha \in \mathcal{L}_{\Box,K}$ and $w \in W$. Then:
(i) $\sigma_v(\Box\alpha, w) = T \Leftrightarrow$ for all v belonging to W, $\sigma_v(\alpha, v) = T$ whenever $w R_{\Box} v$
(ii) $\sigma_v(K\alpha, w) = T \Leftrightarrow$ for all v belonging to W, $\sigma_v(\alpha, v) = T$ whenever $w R_K v$

(Def. 2) [Pragmatic Interpretation of $\mathcal{L}^P_{\Box,K}$]

Let σ_v be a *semantic interpretation* of $\mathcal{L}^P_{\Box,K}$ on a model M. Then a *pragmatic interpretation* π_{σ_v} of $\mathcal{L}^P_{\Box,K}$ on M is any (partial) function

$$\pi_{\sigma_v} : \begin{cases} \mathcal{L}^P_{\Box,K} \text{x} W \to \{J, U\} \\ (\delta, w) \mapsto \pi_{\sigma_v}(\delta, w) \in \{J, U\} \end{cases}$$

which satisfies the following Justification Rules (JRs) and the Correctness Criterion (CC):

(JR1) Let $\alpha \in \mathcal{L}_{\Box,K}$ and $w \in W$. Then:
$\pi_{\sigma_v}(\vdash \alpha, w) = J \Leftrightarrow$ a proof exists that $\sigma_v(\alpha, w) = T$

Hence, $\pi_{\sigma_v}(\vdash \alpha, w) = U \Leftrightarrow$ no proof exists that $\sigma_v(\alpha, w) = T$

(JR2) Let $\delta, \delta_1, \delta_2 \in \mathcal{L}^P_{\Box,K}$ and $w \in W$. Then:
(i) $\pi_{\sigma_v}(\sim \delta, w) = J \Leftrightarrow$ a proof exists that $\pi_{\sigma_v}(\delta, w) = U$
(ii) $\pi_{\sigma_v}(\delta_1 \cap \delta_2, w) = J \Leftrightarrow \pi_{\sigma_v}(\delta_1, w) = J$ and $\pi_{\sigma_v}(\delta_2, w) = J$
(iii) $\pi_{\sigma_v}(\delta_1 \cup \delta_2, w) = J \Leftrightarrow \pi_{\sigma_v}(\delta_1, w) = J$ or $\pi_{\sigma_v}(\delta_2, w) = J$
(iv) $\pi_{\sigma_v}(\delta_1 \supset \delta_2, w) = J \Leftrightarrow$ a proof exists that $\pi_{\sigma_v}(\delta_2, w) = J$ whenever $\pi_{\sigma_v}(\delta_1, w) = J$
(v) $\pi_{\sigma_v}(\delta_1 \equiv \delta_2, w) = J \Leftrightarrow \pi_{\sigma_v}(\delta_1 \supset \delta_2, w) = J$ and $\pi_{\sigma_v}(\delta_2 \supset \delta_1, w) = J$

(CC) Let $\alpha \in \mathcal{L}_{\Box,K}$ and $w \in W$. Then $\pi_{\sigma_v}(\vdash \alpha, w) = J \Rightarrow \sigma_v(\alpha, w) = T$

9.5 Semantic and Pragmatic Validity for $\mathcal{L}^P_{\Box,K}$

We consider now the semantic and pragmatic notions of validity for $\mathcal{L}^P_{\Box,K}$. Intuitively, a radical is semantically valid (s-valid) for $\mathcal{L}^P_{\Box,K}$ if it is a *classical logical law* for its descriptive part.

It worth noting that by taking into account soundness and completeness results about modal logics [4] and their fusions [12, 15]—given the fact that the axiomatic presentation through Hilbert calculus of the fusion of modal logics is obtained by merging the axioms and the inference rules of the fused logics, and considering our (BP)—it is possible to introduce definitions and to obtain results about s-validity into the following proposition.

(Def. 3) [Semantic validity for $\mathcal{L}^P_{\Box,K}$]
Let $\Lambda \subseteq \mathcal{L}_{\Box,K} := \mathcal{L}_\Box \oplus \mathcal{L}_K$ be the normal modal logic given by the following axioms and closed under the following rules:

(TAUT) all the schemas of propositional tautologies belong to Λ;
(AX) the following axiom schemas belong to Λ:
(1) $\Box(\alpha_1 \rightarrow \alpha_2) \rightarrow (\Box\alpha_1 \rightarrow \Box\alpha_2)$ (P-Distributivity)
(2) $\Box\alpha \rightarrow \alpha$ (Correctness)
(3) $\Box\alpha \rightarrow \Box\Box\alpha$
(4) $K(\alpha_1 \rightarrow \alpha_2) \rightarrow (K\alpha_1 \rightarrow K\alpha_2)$ (K-Distributivity)
(5) $K\alpha \rightarrow \alpha$ (Factivity)
(6) $K\alpha \rightarrow KK\alpha$ (Positive Introspection)
(7) $\neg K\alpha \rightarrow K\neg K\alpha$ (Negative Introspection);
(BP) $\Box\alpha \rightarrow \neg K\neg\alpha$ (Proof-Knowledge Compatibility);
(MP) If $\alpha_1 \in \Lambda$ and $\alpha_1 \rightarrow \alpha_2 \in \Lambda$, then $\alpha_2 \in \Lambda$ (Modus Ponens);
(N) If $\alpha \in \Lambda$, then $\Box\alpha,\ K\alpha \in \Lambda$ (Generalization);
(UF) Uniform substitution.

Let $\alpha \in \mathcal{L}_{\Box,K}$. Then

(SV) $\alpha \in \mathcal{L}_{\Box,K}$ is s-valid for $\mathcal{L}^P_{\Box,K} \Leftrightarrow \alpha \in \Lambda$

Similarly for the s-validity for radicals; since pragmatic sentences are interpreted in terms of justification-values, it makes sense to define a sentence as pragmatically valid (p-valid) if it is justified in any case. Therefore:

(Def. 4) An assertion $\delta \in \mathcal{L}^P_{\Box,K}$ is p-valid iff $\pi_{\sigma_v}(\delta, w) = J$ for every π_{σ_v}.

Moreover, it turns out that the two notions of validity are closely related to each other by the *Justification Lemma*. The lemma is based on a *modal translation* of the pragmatic part of $\mathcal{L}^P_{\Box,K}$ into its descriptive part $\mathcal{L}_{\Box,K}$, *viz.* a syntactic translation of any assertion into a *modal* radical formula. There are two intuitive motivations behind the translation. The first one is that an (elementary) assertion is justified just in case there is a proof, intuitively left unspecified, that what is asserted is true. The second reason is that the way proofs, and so justification-values, are combined into the (JRs) is captured by the BHK interpretation of intuitionistic logic [18]. Therefore, the justification of assertions can be formalized by means of the Gödel-McKinsey-Tarski *modal* translation of intuitionistic logic in terms of a **S4**-like modality [16]. Statements such as "a proof exists that α is true" can be translated with the formula $\Box\alpha$, and any complex assertion can be translated into a modal formula according to the modal translation of its BHK reading. This clarifies also the way the pragmatic connectives are introduced. That is, *via* the intended reading of Gödel-McKinsey-Tarski translation which captures the BHK proof interpretation of intuitionistic logic.

The following definition makes the translation precise.

(Def. 5) [Modal Translation of $\mathcal{L}^P_{\Box,K}$ into $\mathcal{L}_{\Box,K}$]
Let $\mathcal{L}^P_{\Box,K}$ and let $(-)^*$ be the function:

$$(-)^* : \begin{cases} \mathcal{L}^P_{\Box,K} \to \mathcal{L}_{\Box,K} \\ \delta \mapsto (\delta)^* \end{cases}$$

such that:

(MT1) Let $\alpha \in \mathcal{L}_{\Box,K}$. Then:
$(\vdash \alpha)^* = \Box\alpha$

(MT2) Let $\delta, \delta_1, \delta_2 \in \mathcal{L}^P_{\Box,K}$. Then:
(i) $(\sim \delta)^* = \Box\neg(\delta)^*$
(ii) $(\delta_1 \cap \delta_2)^* = (\delta_1)^* \wedge (\delta_2)^*$
(iii) $(\delta_1 \cup \delta_2)^* = (\delta_1)^* \vee (\delta_2)^*$
(iv) $(\delta_1 \supset \delta_2)^* = \Box((\delta_1)^* \to (\delta_2)^*)$
(v) $(\delta_1 \equiv \delta_2)^* = \Box((\delta_1)^* \leftrightarrow (\delta_2)^*)$

(Lemma 1) [Justification Lemma for $\mathcal{L}^P_{\Box,K}$]
Let $\delta \in \mathcal{L}^P_{\Box,K}$ and $(\delta)^*$be its modal translation. Then, for every *pragmatic interpretation* π_{σ_v} of $\mathcal{L}^P_{\Box,K}$ we have that:

(JL) $\pi_{\sigma_v}(\delta, w) = J \Leftrightarrow \sigma_v((\delta)^*, w) = T$

On the basis of the justification lemma, we get a criterion for p-validity. The idea is the following: if justification-values can be reduced to truth-values, then p-validity can be reduced to s-validity as well.

(Lemma 2) [Pragmatic validity for $\mathcal{L}^P_{\Box,K}$]
Let $\delta \in \mathcal{L}^P_{\Box,K}$ and $(\delta)^*$be its modal translation. Then:

(PV) δ is p-valid for $\mathcal{L}^P_{\Box,K} \Leftrightarrow (\delta)^*$ is s-valid for $\mathcal{L}^P_{\Box,K}$

[Remark 1] Let us show an application of (PV). Consider (BP). It is not difficult to see that from $\Box\alpha \rightarrow \neg K\neg\alpha$, it is possible to derive $\Box(\Box\alpha \rightarrow \Box\neg K\neg\alpha)$, and that $\Box(\Box\alpha \rightarrow \Box\neg K\neg\alpha) = (\vdash \alpha \supset \vdash \neg K\neg\alpha)^*$. Namely, $\Box(\Box\alpha \rightarrow \Box\neg K\neg\alpha)$ is the modal translation of $\vdash \alpha \supset \vdash \neg K\neg\alpha$. It follows that $\vdash \alpha \supset \vdash \neg K\neg\alpha$ is p-valid, and it could be read as the pragmatic version of (BP).

[Remark 2] In $\mathcal{L}^P_{\Box,K}$ two versions of *Modus Ponens* can be formulated in the following ways:

(PMP1) If $\vdash (\alpha_1 \rightarrow \alpha_2)$ and $\vdash \alpha_1$, then $\vdash \alpha_2$
(PMP2) If $\delta_1 \supset \delta_2$ and δ_1, then δ_2

[Remark 3] Here it is a list of p-valid formulas and *Pragmatic Bridge Principles* that explain the relations between the semantic logical operators and the pragmatic ones.

(PBP1) $(\vdash \neg\alpha) \supset (\sim\vdash \alpha)$
(PBP2) $\vdash (\alpha_1 \wedge \alpha_2) \equiv (\vdash \alpha_1 \cap \vdash \alpha_2)$
(PBP3) $(\vdash \alpha_1 \cup \vdash \alpha_2) \supset \vdash (\alpha_1 \vee \alpha_2)$
(PBP4) $\vdash (\alpha_1 \rightarrow \alpha_2) \supset (\vdash \alpha_1 \supset \vdash \alpha_2)$
(PBP5) $\vdash (\alpha_1 \leftrightarrow \alpha_2) \supset (\vdash \alpha_1 \equiv \vdash \alpha_2)$
(PBP6) $\vdash \Box\alpha \equiv \vdash \alpha$
(PBP7) $\vdash K\alpha \supset \vdash \alpha$
(PBP8) $\vdash K\alpha \supset \vdash \Box\alpha$
(PBP9) $\vdash K\alpha \equiv \vdash KK\alpha$
(PBP10) $\vdash \alpha \supset \vdash \neg K\neg\alpha$
(PBP11) $\vdash \neg K\alpha \supset \vdash K\neg K\alpha$

Notice that $\vdash \alpha$, $\vdash \Box\alpha$, and $\vdash \Box\Box\alpha$ are p-equivalent assertions, that $\vdash K\alpha$ and $\vdash KK\alpha$ are p-equivalent as well, but neither $\vdash K\alpha$ nor $\vdash KK\alpha$ is p-equivalent to $\vdash \alpha$. Indeed, $\Box(\Box K\alpha \leftrightarrow \Box\alpha)$ is not s-valid since $\Box(\Box K\alpha \rightarrow \Box\alpha)$ is s-valid, but $\Box(\Box\alpha \rightarrow \Box K\alpha)$ is not. Therefore, by (PV), (PBP7) $\vdash K\alpha \supset \vdash \alpha$ is p-valid, but $\vdash \alpha \supset \vdash K\alpha$ is *not* p-valid.

In other words, because factivity holds, a proof that α is known to be true is transformed into a proof that α is true, but not vice versa.

[Remark 4] Here it is a list of some p-valid formulas that could be of some interest.

(P1) $\delta_1 \supset (\delta_2 \supset \delta_1)$
(P2) $(\delta_1 \supset (\delta_2 \supset \delta_3)) \supset ((\delta_1 \supset \delta_2) \supset (\delta_1 \supset \delta_3))$

(P3) $(\delta_1 \cap \delta_2) \supset \delta_{1/2}$
(P4) $\delta_1 \supset (\delta_2 \supset (\delta_1 \cap \delta_2))$
(P5) $\delta_{1/2} \supset (\delta_1 \cup \delta_2)$
(P6) $(\delta_1 \supset \delta_3) \supset ((\delta_2 \supset \delta_3) \supset (\delta_1 \cup \delta_2 \supset \delta_3))$
(P7) $(\delta_1 \supset \delta_2) \supset ((\delta_1 \supset \sim \delta_2) \supset \sim \delta_1)$
(P8) $(\delta_1 \equiv \delta_2) \supset ((\delta_1 \supset \delta_2) \cap (\delta_2 \supset \delta_1))$
(P9) $((\delta_1 \supset \delta_2) \supset ((\delta_2 \supset \delta_1) \supset (\delta_1 \equiv \delta_2))$

It is worth noting that it is now possible to formalize the verificationist features of knowledge and proof in our multimodal pragmatic language $\mathcal{L}^P_{\Box,K}$. Furthermore, these verificationist features are compatible with the classical interpretation of intuitionism, as indicated by the modal translation in *(Def. 4)*.

9.6 The Paradox of Knowability in $\mathcal{L}^P_{\Box,K}$

We use $\mathcal{L}^P_{\Box,K}$ to provide a fine-grained analysis of the Knowability Paradox (KPx). Notice that only justified assertions can be expressed in $\mathcal{L}^P_{\Box,K}$ while in $\mathcal{L}_{\Box,K}$ it is also possible to express unjustified assertions by means of their modal translation. Observe also that the validity of a deduction in $\mathcal{L}^P_{\Box,K}$ is guaranteed by making reference to the corresponding multimodal logical steps expressible in $\mathcal{L}_{\Box,K}$ in virtue of the *Justification Lemma*.

The Paradox is based on the Knowability Principle (KP) and the Principle of Non-Omniscience (Non-Om), and it is of particular interest from an antirealist perspective. As we have seen, the view that all truths are knowable is logically incompatible with the reasonable idea that we are non-omniscient. Specifically, the paradox arises because (KP) $\rightarrow \neg$ (Non-Om) is a theorem of a classical modal logic:

(KPx) $[\forall p(p \rightarrow \Diamond Kp)] \rightarrow [\forall p(p \rightarrow Kp)]$

Hence, leaving quantification aside, it turns out that $(p \rightarrow \Diamond Kp) \rightarrow (p \rightarrow Kp)$ is a logical law of the descriptive part of the language $\mathcal{L}_{\Box,K}$, i.e., it is s-valid. And, by Justification Lemma, $\vdash [(p \rightarrow \Diamond Kp) \rightarrow (p \rightarrow Kp)]$ is p-valid.

Consider the following derivation:

(1) $\vdash [(p \rightarrow \Diamond Kp) \rightarrow (p \rightarrow Kp)] \supset$
$(\vdash (p \rightarrow \Diamond Kp) \supset \vdash (p \rightarrow Kp))$ (PBP4)
(2) $\vdash [(p \rightarrow \Diamond Kp) \rightarrow (p \rightarrow Kp)]$ $p-$ valid
(3) $\vdash (p \rightarrow \Diamond Kp) \supset \vdash (p \rightarrow Kp)$ (PMP2) 1,2
(4) $\vdash (p \rightarrow Kp) \supset (\vdash p \supset \vdash Kp)$ (PBP4)
(5) $\vdash (p \rightarrow \Diamond Kp) \supset (\vdash p \supset \vdash Kp)$ transitivity of $\supset$ 3, 4

Thus, 5 $\vdash (p \rightarrow \Diamond Kp) \supset (\vdash p \supset \vdash Kp)$ is the pragmatic reading of the paradox. It is clear that, in $\mathcal{L}^P_{\Box,K}$, in order to obtain $\vdash Kp$, that is, *a proof that* p *is actually known*, both a justification of the asserted version of (KP), (KPPC) $\vdash (p \rightarrow \Diamond Kp)$, *and* of $\vdash p$ are required. In this way, in (5), the paradoxical conclusion that *every truth is actually known* seems to disappear because the conclusion is $\vdash p \supset \vdash Kp$, i.e., *every asserted truth is an assertion of an actually known truth.*

However, even if $\vdash p \supset \vdash Kp$ would be considered a paradoxical conclusion, there would be a problem with the justification of $\vdash (p \rightarrow \Diamond Kp)$. The only justification an antirelist could give for it would be the one depending on the fact that an antirealist has, somehow, to accept it. And, according to us, this does not count as a proof.

Moreover, given that the antirealist conception of truth is paired with intuitionistic logic, (KPPC) does not seem to be an adequate formulation of the antirealist version of Knowability. Indeed, as shown in [9], intuitionistic language is represented as the fragment built up starting from elementary sentences with only atomic radicals by means of pragmatic connectives. Therefore, no classical connective has to fall under the scope of an assertion. Hence, if a justification of (KPPC) would be available, then it would be a justification of something that does seem compatible with the antirealist form of knowability.

A version of (KP) compatible with antirealism is needed. Consider the following formulation:

(KPI) for all p, if p is intuitionistically true, then $\Diamond Kp$ is intuitionistically true.

It is not entirely clear how a modality should be interpreted from the intuitionistic point of view. However, the intuitionistic double negation ($\sim\sim$) seems to be a form of possibility. Indeed, there is a bridge between the intuitionistic double negation and the verificationist reading the (classical) possibility of a proposition p, $\Diamond p$, obtained by (JP). That is, $\sim\sim\vdash p \supset \vdash \Diamond p$ is p-valid. It must also be noticed that the verificationist reading of the (classical) possibility of a proposition p, $\Diamond p$, introduced by (JP) - i.e., the assertion of the truth of $\Diamond p$ - cannot be reduced to intuitionistic logic because $\vdash \Diamond p \equiv \sim\vdash \neg p$ is p-valid. And, the classical negation that falls under the scope of an assertion cannot be eliminated.

Hence, if Kp is considered as an atomic radical proposition, then we have an intuitionistically valid representation of (KP):

(KPPI) for all p ($\vdash p \supset \sim\sim\vdash Kp$).[6]

Moreover, since ($\vdash p \supset \sim\sim\vdash Kp$) implies ($\vdash p \supset \vdash \Diamond Kp$) and $\vdash \Diamond Kp \equiv \sim\vdash \neg Kp$, also

(KPPI') for all p, ($\vdash p \supset \vdash \Diamond Kp$)

and

(KPPI'') for all p, ($\vdash p \supset \sim\vdash \neg Kp$),

[6]On (KPPI) see [11].

would be more adequate versions of an antirealist form of Knowability than (KP).[7]

However, neither of them, (KPPI), (KPPI') and (KPPI"), would be sufficient to obtain the paradoxical conclusion.

To summarize. In classical (modal) logic, (KP) is *sufficient* to obtain the paradoxical conclusion of (KPx), i.e., ¬(Non-Om). Within the framework of $\mathcal{L}^P_{\Box,K}$, the paradoxical conclusion *every truth is actually known* seems to disappear because in 5 the conclusion is $\vdash p \supset \vdash Kp$, where the *proofs* of both the truth of p and of Kp are involved, not only the mere being truth of these formulas.

Nonetheless, assuming that $\vdash p \supset \vdash Kp$ is a paradoxical conclusion since, according to us, there exist no poof of (KP), the conclusion could not be given either. And, even if a proof would be available, then it would turn out to be a justification for (KPPC). But (KPPC) is not an adequate representation of an antirelist version of Knowability. Hence, according to 5, the paradoxical conclusion would follow from an assumption that it is not an antirealist version of knowability. A version of (KP) compatible with intuitionistic logic should be (KPPI). But then it would not be sufficient to get the (supposed) paradoxical conclusion either. And the same holds for the other versions of knowability compatible with antirealism, that is (KPPI') and (KPPI").

9.7 Conclusions

In the paper we have outlined a multimodal pragmatic language $\mathcal{L}^P_{\Box,K}$ for the analysis of the Knowability Paradox. The language is based on the identification, by the Justification Principle (JP), of the verificationist notion of truth with the notion of justified assertion - the proof of a classical (modal) truth. This verificationist interpretation of classical (modal) propositions in terms of assertions is integrated with intuitionistic-like connections of them defined via pragmatic connectives. In this way, a set of Pragmatic Bridge Principles (PBPx), explaining the relations between classical connectives, pragmatic ones and (classical) modal operators, is obtained. We have assumed a knowledge-proof compatibilist Bridge Principle (BP). Justifications do not always warrant for knowledge, but, according to (BPB7), justified knowledge is a sufficient condition for the lack of any proof of the contrary. That is to say that justified knowledge is a sufficient condition for the assertion of the falsity being unjustified: there is not any proof of the contrary of what it is known.

Because of its analytic expressiveness, $\mathcal{L}^P_{\Box,K}$ provides a fine-grained analysis of the notions of truth, proof, knowledge, and their relations. Our approach takes into account the fact that the Knowability Principle is mainly associated with a verificationist perspective in epistemology, and that an important connection between verificationism and intuitionistic logic is usually recognized. Indeed, in our pragmatic language it is possible to develop a compatibilist perspective on classical and intuitionistic systems: a notable aspect concerning the communicability among dif-

[7] On (KPPI') and (KPPI") see [7].

ferent logics that comes to be an advantage for the interpretation of the paradox from the antirelistic perspective.

Within $\mathcal{L}^{P}_{\Box,K}$, the Knowability Paradox seems to disappear because the pragmatic version of the paradoxical conclusion that *every truth is actually known* turns out to be *every asserted truth is an assertion of an actually known truth*, i.e., $\vdash p \supset \vdash Kp$, where the proofs of the truth of p and of the truth of Kp are required, not only their being true. Even assuming that $\vdash p \supset \vdash Kp$ is a paradoxical conclusion since, according to us, there exist no proof of (KP), the conclusion could not be given either. And, if a proof was available, then it would be a justification for (KPPC), which is not an adequate representation of the antirelist version of Knowability. Hence, the paradoxical conclusion would follow from the assumption that it is not an antirealist version of knowability. A version of (KP) compatible with intuitionistic logic should be (KPPI). But then it would not be sufficient to get the (supposed) paradoxical conclusion either. And the same holds for the other versions of knowability compatible with antirealism, that is (KPPI′) and (KPPI″).

Thus, we conclude that the pragmatic reading of the paradoxical conclusion is not really paradoxical because the notion of justification is involved. Nonetheless, if it is assumed that it is paradoxical, then one has to face the problem of the correct formulation of knowability. Within our pragmatic multimodal language, we have listed four pragmatic representations of it: (KPPC), (KPPI), (KPPI′) and (KPPI″). On the one hand, despite the fact that (KPPC) would be sufficient to obtain the supposed paradoxical conclusion, it is not an antirealistically adequate representation of knowability, so it cannot be taken into account. On the other hand, even if (KPPI), (KPPI′) and (KPPI″) are antirealistically adequate representations of knowability, none of them is sufficient to obtain the supposed paradoxical conclusion, and, therefore, the paradox does not arise.

Acknowledgments * We would like to thank the referees of the volume for their helpful comments and suggestions. The research of Daniele Chiffi is supported by the Estonian Research Council, PUT1305 2016-2018, PI: Pietarinen. Massimiliano Carrara's research was conducted while he was in his sabbatical year.

References

1. Artemov, S., & Protopopescu, T.(2013). Discovering knowability: a semantic analysis. *Synthese*, *190*(16), 3349–3376.
2. Beall, J.C. (2005). Knowability and possible epistemic oddities. In J. Salerno, (Ed.), *New essays on the knowability paradox*, (pp. 105–125). Oxford: OUP Press.
3. Bellin, G., & Biasi, C. (2004). Towards a logic for pragmatics. Assertions and conjectures. *Journal of Logic and Computation*, *14*, 473–506.
4. Blackburn, P., de Rijke, M., & Venema, Y. (2001). *Modal Logic*. Cambridge: Cambridge University Press.
5. Carnielli, W., Coniglio, M.E. (2011). Combining logics. In Edward N. Zalta (Ed.), *The Stanford Encyclopedia of Philosophy*. Winter 2011 edition.
6. Carnielli, W., Pizzi, C., & Bueno-Soler, J. (2008). *Modalities and multimodalities* (vol. 12). Springer.

7. Carrara, M., & Chiffi, D. (2014). The knowability paradox in the light of a logic for pragmatics. In R. Ciuni, H. Wansing, & C. Willkommen (Eds.), *Recent Trends in Philosophical Logic*, (pp.33–48). Springer.
8. Costa-Leite, A. (2006). Fusions of modal logics and Fitch's paradox. *Croatian Journal of Philosophy*, *17*, 281–290.
9. Dalla Pozza C., & Garola, C. (1995). A pragmatic interpretation of intuitionistic propositional logic. *Erkenntnis*, *43*, 81–109.
10. De Vidi, D., & Solomon, G. (2001). Knowability and intuitionistic logic. *Philosophia*, *28*(1), 319–334.
11. Dummett, M. (2009). Fitch's paradox of knowability. In J. Salerno, (Ed.), *New essays on the knowability paradox*, (pp. 51–52). Oxford: OUP Press.
12. Fine, K., & Schurz, G. (1991). Transfer theorems for stratified multimodal logics. In B. J. Copeland (Ed.), *Logic and Reality* (pp. 169–213). Oxford: Clarendon Press.
13. Fischer, M. (2013). Some remarks on restricting the knowability principle. *Synthese*, *190*(1), 63–88.
14. Fitch, F. B. (1963). A logical analysis of some value concepts. *The Journal of Symbolic Logic*, *28*(2), 135–142.
15. Gabbay, D.M., & Shehtman, V.B. (1998). Products of modal logics, part 1. *Logic Journal of IGPL*, *6*, 73–146.
16. Gödel, K. (1933). Eine interpretation des intuitionistischen aussagenkalkuls. *Ergebnisse Eines Mathematischen Kolloquiums*, *4*, 39–40.
17. Salerno, J. (2009). *New essays on the knowability paradox*. Oxford: OUP.
18. Troelstra, A.S., & Schwichtenberg, H. (2000). *Basic proof theory* (No. 43). Cambridge: Cambridge University Press.
19. Wansing, H. (2002). Diamonds are a philosopher's best friends. *Journal of Philosophical Logic*, *31*(6), 591–612.
20. Williamson, T. (1992). On intuitionistic modal epistemic logic. *Journal of Philosophical Logic*, pp. 63–89.

Part III
Closer to the Core

Chapter 10
Philosophical Aspects of an Alleged Connection Between the Axiom of Choice and Predicting the Future

Pawel Pawlowski

Abstract In 2008 Christopher Hardin and Alan Taylor published an article titled "Peculiar connection between the axiom of choice and predicting the future" in which they claim that if some system can be described as a function from a set of some instants of time to some set of states, then there is a way to predict the next value of the function based on its previous input. Using their so-called μ-strategy one can randomly choose an instant t and the probability that the strategy is correct at t (i.e. that the output for a strategy for input t is exactly the same as the value of the function) equals 1. Mathematical aspects of this article are sound, but the background story about the correlation between theorems and philosophical aspects of predicting the future faces certain problems. The goal of my paper is to bring them up.

Keywords Axiom of choice · Computability on the real numbers · Probability

10.1 Introduction

The main goal of this paper is to show that the interpretation of Hardin and Taylor's mathematical results (from their *A peculiar connection between the axiom of choice and predicting the future*) in terms of prediction, while perhaps pedagogically useful, is not philosophically adequate.

To be fair, the authors pointed out explicitly that their account does not give a practical means to predict the future. But the philosophical issues that I will discuss reach deeper than just practical limitations. The criticized paper is written in a way as if, if we somehow could overcome practical impossibilities, the whole account would be sound and predicting the future would be possible. My goal is to show that this is not the case—that the problems are not only practical, but also conceptual.

It might be also the case that the authors were aware of the philosophical drawbacks of their interpretation. Nevertheless, it is still interesting to point them out explicitly, especially focusing on unobvious philosophical assumptions which have to be made to make the proposed interpretation seem to work.

P. Pawlowski (✉)
Centre for Logic and Philosophy of Science, Ghent University, Ghent, Belgium
e-mail: haptism89@gmail.com

R. Urbaniak and G. Payette (eds.), *Applications of Formal Philosophy*,
Logic, Argumentation & Reasoning 14, DOI 10.1007/978-3-319-58507-9_10

The paper is divided into 4 parts. The main aim of the first part is to introduce Hardin's and Taylor's theorems and their interpretation as a means to predict the future. I will abbreviate this interpretation as IAF (interpretation about future). The next part discusses some problems with the concept of randomness and the notion of prediction. The third part focuses on the notion of knowledge which is needed to prove that μ-strategy guesses well on a set of full measure. The last part is devoted to some problems with the notion of a probability measure extended to uncountable sets and general remarks about the axiom of choice. At the end of this paper in the Sect. 10.4 I will write a few words about the computational aspect of μ-strategy.

10.1.1 Mathematical Background

Let's first present the mathematical apparatus used by Hardin and Taylor for defining their μ-strategy and obtaining their theorems about it. Fix two sets T and S, where S has at least two elements, and T is linearly ordered by a relation $<$. Let S^T denote the set of all functions from T to S. For each element $t \in T$ we define an equivalence relation $\approx_t$ on S^T by saying that two elements f, g of S^T are in this relation iff they both agree on $< t = \{s \in T : s < t\}$, that is when both functions have the same output up to t.

Hardin and Taylor presented an interpretation of the above theorems in terms of predicting the future. In the interpretation T represents a set of all instants of time and S is supposed to encode all possible states of the world. For them, the world is a system which develops through time and can be described as a function from some set of instants of time to a set of states S. They call each element v of S^T a scenario. A unique scenario $v_{tr} \in S^T$ represents the actual way the world develops.

Each instant of time t can also be seen as an agent who tries to predict the future based on the knowledge of the history of the world up to t. The equivalence relation $\approx_t$ defined by an element t in IAF is understood as the set of scenarios consistent with the knowledge of an agent at t. To make a prediction, each agent has to select an equivalence class of those functions which agree with the true scenario up to t. To do that she has to know the values of all functions up to t.

Now, fix a well-ordering $\preceq$ on S^T. In order to do so we have to assume Zermelo's Theorem which says that for any set there is a relation which well-orders this set. Zermelo's theorem is equivalent to the Axiom of Choice on which I will elaborate later on. Note also that the order is fixed in the sense that it is global for the whole argument.

We define μ-strategy as the function from the set of all equivalence classes

$$STR = \{[v]_t : v \in S^T, t \in T\}$$

to S^T, where $[v]_t$ denotes the abstraction class of v under $\approx_t$, such that $\mu([v]_t)$ is the least element according to the well-ordering $\preceq$ of $[v]_t$. Let $v_{tr} \in S^T$. In terms of IAF, each equivalence class generated by t corresponds to information about what happens in the world up to t and μ-strategy simply takes each such piece of information and spits out a scenario for all moments in time. The way μ -strategy works is it picks the $\preceq$-least element of a given abstraction class as the prediction, where $\preceq$ is just some fixed well-ordering on all scenarios. (Observe that $\preceq$ doesn't depend on the ordering of moments of time $<$.)

Now for their main result. Suppose that $T = \mathbb{R}$ and $<$ is the standard ordering relation on reals. The probability that for a randomly chosen element t of T the value of the function $v_{tr}(t)$ (which tells us what really happens at t) is the same as $\mu([v]_t)(t)$ (that is, what μ-strategy tells us will happen at t) equals 1.[1]

The proof proceeds by showing that the set W_0 of all times for which the strategy guesses incorrectly (that is, $\{t \in T \mid \mu(v_t)(t) \neq v_{tr}(t)\}$) has measure 0.

This theorem can be extended in two ways. The first strengthening says that μ strategy is correct (has the same output as v_{tr}) with probability 1 on some interval $[t; t + \epsilon)$ where $\epsilon > 0$.

The second strengthening is obtained by redefining the equivalence relation on S: f, g are in relation $\approx'_t$ iff there is an $s < t$ such that f and g agree on $(s; t)$. Then the probability that μ-strategy is correct still equals 1.

According to IAF, given the values of v_{tr} on $(-\infty, t)$ an agent tries to predict $v_{tr}(t)$. Such a prediction in Hardin and Taylor's opinion can be naturally understood as predicting the present. If agent t tries to guess the values of v_{tr} on $[t, t + \epsilon)$,[2] this can be understood as predicting the future. The above theorems in such an interpretation mean that using the above strategy we might guess based on what previously happened what will happen on $(t, t + \epsilon)$ with probability 1.

10.2 Problems with Randomness and Predicting

The authors explain: "For a fixed true scenario, if one randomly selects an instant t in the interval [0,1] (or in $\mathbb{R}$ under a suitable probability distribution), then Corollary 3.4 does tell us that the μ-strategy will be correct at t with probability 1." [4]. Hardin and Taylor say that we randomly choose an instant t and by random they mean "under suitable probability distribution".

Yet, it is hard to say what they mean by a suitable distribution. One can give at least two different accounts of what a suitable distribution is. On the first one, by a suitable distribution they might mean a uniform distribution. On the second one a distribution is suitable if every infinite countable set has measure 0. (Maybe Hardin and Taylor used another definition of randomness but they are not explicit about it.)

Both accounts are problematic. The first account, because there is no uniform distribution on $\mathbb{R}$: there is no uniform distribution on $\mathbb{R}$, $\mathbb{Q}$ (rationals), and on $\mathbb{N}$ which would work in this context.

---

[1] All proofs can be found in [4].

[2] Where $\epsilon > 0$ and $[t, t + \epsilon)$ is some small interval of time. For every t such ϵ exists in the sense that given t we can match a suitable ϵ.

For suppose that we divide $\mathbb{R}$ into countably many intervals $[z, z+1)$ where $z \in \mathbb{Q}$. Every such interval has equal length, and $\mathbb{R} = \bigcup_{z \in \mathbb{Q}} [z, z+1)$. The uniform distribution should assign the same probability c to every interval. Suppose that $c = 0$. Then by σ-additivity we obtain $Pr(\bigcup_{z \in \mathbb{Q}} [z, z+1)) = 0$ which implies that $Pr(\mathbb{R}) = 0$ which contradicts the axioms. If, on the other hand, probabilities of intervals $[z, z+1)$ are greater than 0, then by σ-additivity we have that $Pr(\mathbb{R}) = \infty$ which again contradicts with the axioms. For natural numbers and rational numbers the arguments are analogous.

Why is the second account problematic? It is problematic because the existence of non-empty sets which have measure 0 combined with the notion of randomness defined on uncountable spaces is unintuitive—some fundamental results about probability distributions over uncountable sets, when applied to the real world (to finite or potentially infinite sets), seem to be false.

Suppose that set A consists of uncountably many socks and countably many skirts. According to the second account of what a suitable distribution is, the probability that a randomly chosen item from A is a skirt equals 0. But intuitively speaking it should not, because sooner or later we should draw a skirt—after all, there are infinitely many of them!

Another example: suppose that someone wants to randomly draw one natural number from the set $A = \{1, 2, 3, 4, 5, 6, 7, 8, 9, 10\}$. The natural assumption is that I can draw every element from A with non-zero probability, because every element of A is in A. But suppose someone says that the probability of drawing a natural number less than 5 equals 0. In an important sense it is not a random choice of a natural number since it is not the case that all natural numbers have the same chance of being drawn.

The same phenomenon occurs with real numbers. We want to randomly choose one element of $\mathbb{R}$ but the probability of choosing, for example, a natural number, equals 0, while if we rely on our intuitions about finite probability spaces, it should be greater than 0. The problems arise because our intuitions about randomness are limited only to finite sets and when one tries to extended them to uncountable spaces it turns out that they are highly problematic.

One can make an analogy between the theorems under discussion and the Banach-Tarski paradox. Both theorems sound paradoxical but that is because we do not have reliable intuitions about uncountable sets.

What is interesting about Hardin and Taylor's strange results is that sometimes we have theorems about finite or countable sets and intuitive problems arise when we try to extrapolate them to uncountable sets. Here it is in some sense the other way around. We have a strange theorem about uncountably infinite sets (probability spaces) and problems arise when we want to apply that result to more mundane phenomena.

Another important issue about the IAF interpretation is the notion of ϵ, which is a small value greater than 0 for which μ-strategy is correct on $[t, t+\epsilon)$. From the theorems, we know that for every t such an $\epsilon > 0$ exists, but we do not know how far it goes.

Moreover, ϵ depends on t. For each t it might be a different distance, so there is no unique distance from t for which the strategy is guaranteed to work.

We also do not know the algorithm for calculating such an ϵ. We do know, however, that for some instants there is no algorithm for calculating ϵ (there are only countably many algorithms).

It may also happen that ϵ is so small that it does not have any impact on our world. For instance, if there is a minimal period of time needed for something to change then ϵ could have been smaller than it.

The next problematic issue is with the concept of prediction. The process of prediction which takes place according to the authors' procedure is the following: the first step is to randomly choose a t to build the set $(-\infty, t)$, then to check which functions belong to $[v]_t$, and then to pick a function g according to μ-strategy (that is, the $\preceq$-least g in a given abstraction class) such that the probability that $g(t) = \mu([v]_t)(t) = v_{tr}(t)$ equals 1.

In order to apply μ-strategy in practice one has to know the values of the function on $(-\infty, t)$ for each t that can be randomly chosen. Otherwise we might not be able to determine the output for μ-strategy.

As pointed out in [3] there is a difference between saying that (i) for a randomly chosen t the probability that the strategy is correct equals 1, and saying that (ii) for a fixed t the probability that for a randomly chosen scenario the strategy is correct equals 1.

Theorems proved by Hardin and Taylor show that (i) is the case but they do not assure us that (ii) must be true. Generally, (ii) might fail. What is striking is the fact that the first one has nothing to do with what we usually mean by prediction. A prediction of what will happen at a time t fits the second account better.

First, consider an example: a physical experiment of the form "if we use this and that on x then bla bla will happen". Imagine that we want to predict the result of the experiment. Then, the context determines the interesting set of instants of time. Generally, when we make a prediction it is not the case that the set of instants of time is arbitrary. The context usually restricts or fixes the set of instants of time under consideration. But for such an account the theorems do not hold.

Notice that (i) is not what we usually mean by prediction. For example, let the current moment be time $t = a$ and suppose someone wants to use μ-strategy to predict the future. Suppose he knows the values of v_{tr} on $(-\infty, a)$. Now he randomly chooses t'. What if the chosen t' belongs to (a, ∞)? If this happens, the agent cannot determine which functions belong to the equivalence class of $[v_{tr}]_a$ since he has only information about the values up to t and $t < a$. Yet, Hardin and Taylor seem to suggest that it is an intuitive way of understanding the notion of predicting the future.

10.3 Problems with the Concept of Knowledge

Another troublesome issue is the concept of knowledge involved in the application of μ-strategy. Hardin and Taylor distinguished two cases: first, when we know the values of v_{tr} on $(-\infty, t)$ and second, when we know the values of v_{tr} on (s, t) for some s.

At a first glance, the second conception seems to be weaker. Consider two functions $f(x)$ and $g(x)$, where

$$f(x) = \begin{cases} 1 \text{ if I am smiling at } x, \\ 0 \text{ if I am not smiling at } x, \end{cases} \tag{10.1}$$

$$g(x) = \begin{cases} f(x) \text{ if } x \in (s; t), \\ 12 \quad \text{if } x \notin (s; t) \text{ and I'm smiling at } s, \\ 15 \quad \text{if } x \notin (s; t) \text{ and I'm not smiling at } s, \end{cases} \tag{10.2}$$

Now, the first conception when we have information about $(-\infty, t)$ allows us to determine the values for both $f(x)$ and $g(x)$ on $(-\infty, t)$ because I know what happened on (s, t), so I know if I was smiling at x or not. The second conception not only does not allow one to determine the values for $g(x)$, $f(x)$ on $(-\infty; t)$, but also I cannot determine the values on (s, t) for $g(x)$ because I do not know whether I was smiling at s. In this sense the second conception is weaker.

But on the other hand in both conceptions we have to somehow store information about what happened on an uncountable set of instants of time. So, both conceptions assume that someone who is guessing can process uncountably many values of a given function. To predict what will happen at t, someone who is guessing has to know the values of a function on the interval $A = (\infty, t)$ or $B = (s, t)$ for some s. It is not problematic as long as there is a general pattern for determining the values. But if there is no such a pattern, then in order to make a guess we have to separately process and know what happened at uncountably many instants of time. The notion of knowledge used here presupposes that someone or something can store and has operative access to uncountable sets of propositions, where most of them cannot be formulated as sentences in a language due to the countability of symbols of the language used.

Moreover, to determine the output of μ-strategy for each t one has to eliminate functions which are inconsistent with what he knows about values of v_{tr} on (s, t), because in order to apply μ-strategy one has to determine which functions belong to $[v]_t$ in order to pick up the $\preceq$-least element of $[v]_t$. This cannot be effectively done, because for a given t one has to compare functions on values from (s, t), which involves uncountably many comparisons of infinitely many functions. So in order to effectively apply μ-strategy we have to know the values of all functions for all intervals of the form $(-\infty, t)$ for any t.

Note also that the theorems obviously do not give any way of constructing a well-order on S^T. There is no clear algorithm to decide which functions belong to which equivalence class even if we know their values. A similar thing goes for ϵ, the theorems do not give any way of computing it given a t.

One reason for this is that from the beginning we strongly rely on an unrestricted version of the Axiom of Choice which is known to not give any method for the construction to generate well-ordering.

But, even if we restrict ourselves to the case in which we do not need full axiom of choice, it may still happen that things go in a similar fashion. For instance, there is no formula in ZFC which defines a well-ordering on reals.

In general—it is not the case that we need full formulation of the Axiom of Choice—but in order to prove the result for reals we have to have some formulation of the axiom.

10.4 A Few Words About Computability

Usually when we talk about computability we talk about some functions on natural numbers. There is no commonly accepted mathematical notion of computability over real numbers. There are some well developed theories[3] [1] but none of them is widely accepted. Moreover, computer programs always compute in finite time on finite sets of data.

Of course, sometimes we can intuitively say that a function on reals is computable or that it is not. For instance a function which takes two arguments and gives the greater number as the output is intuitively computable. μ-strategy is not intuitively computable in this sense, because it is based on the well-ordering theorem which states that for every non-empty set there is a relation which well-orders this set. Finding such an ordering sometimes is very hard (it is worth noting that the existence of definable well-ordering of real numbers is independent of ZFC see [2], p. 130). For many uncountable sets no one ever defined such orderings, in particular no one ever defined such an ordering on reals. It is not trivial if such orderings can be defined.

Moreover, even if we find a well ordering, it is not a trivial matter to decide if a given function belongs to a given abstraction class. In some cases depending on the set of states we have to use Zermelo theorem on sets which are bigger than the set of all reals, which is also problematic for the intuitive notion of computation.[4] Uncomputability of a strategy of prediction is problematic. Almost always we understand prediction effectively—if a strategy is not computable and we do not have any other way of determining its output, in an important sense it is not a prediction at all.

[3]The most popular one is the Blum-Shub-Smale (BSS) machine.

[4]There is no point to talk about the computability of μ-strategy when it is defined on natural numbers because when time is not dense the strategy does not predict with probability 1.

10.5 Conclusions

The mathematical aspects discussed in the criticized paper are interesting and sound. Alas, IAF not only does not give a *practical* means to predict the future, but also more theoretical links between IAF and prediction face some problems. The theorem works mainly because of the density of $\mathbb{R}$, and probability on uncountable sets is very problematic. The notion of randomness used in the interpretation is troublesome, due to the lack of a uniform distribution on one hand, and because we normally do not randomly choose the time for which we want to make a prediction. Moreover the strategy uses Zermelo's theorem to establish a well-order and does not give a way to construct such an ordering (it can be the case that the required constructive ordering does not exists). What is also crucial, it is impossible to understand prediction without any connection to effective computability. If μ-strategy is not computable (and it rather clearly is not), we do not really deal with prediction.

Acknowledgments I want to thank Rafal Urbaniak for his time and endless comments concerning this paper. I would also thank two anonymous referees, whose comments were very helpful. This research has been funded by Research Foundation Flanders (FWO).

References

1. Braverman, M., & Cook, S. (2006). Computing over the reals: Foundations for scientific computing. *Notices of the American Mathematical Society*, *53*, 318–329.
2. Devlin, K. (1994). *The joy of sets: Fundamentals of contemporary set theory*. Berlin: Springer.
3. George, A. (2007). A proof of induction? *Philosophers' Imprint*, *7*(2), 1–5.
4. Hardin, C. S., & Taylor, A. D. (2008). A peculiar connection between the axiom of choice and predicting the future. *American Mathematical Monthly*, *115*, 91–96.

Chapter 11
Counterpossibles, Impossible Worlds, and the Notion of Similarity

Maciej Sendłak

Abstract The subject of this paper is a world-semantic analysis of counterpossibles, i.e., counterfactuals with impossible antecedents. We focus on the notion of similarity between worlds, which determines truth-value of counterfactuals. There are two commonly accepted assumptions about this notion. According to the first one, every possible world is more similar to the actual world than any impossible world. According to the second one, the trivial world (world where everything is true) is the most dissimilar to the actual world. Considering the notion of similarity we argue for a negative thesis and a positive thesis. The negative thesis is that both of these assumptions are false, and as such should not be taken as a "guide" to our understanding of similarity. The positive thesis is an alternative interpretation of the notion of similarity. The interpretation is based on an analogy of the inference to the best explanation and on the assumption that similarity is a ternary relation satisfied by the actual world, a non-actual world and a given factor of similarity. Similarity understood in this manner is a notion which requires an indication of a rule which supports the truth of the antecedent and explains its connection with the consequent.

Keywords Counterfactuals · Counterpossibles · Impossible worlds · Abduction

11.1 Introduction

Many contemporary philosophers of modality have been arguing that the standard analysis of counterfactuals in the framework of possible world semantics is insufficient when it comes to counterpossibles, i.e., counterfactuals with impossible antecedents.[1] As an alternative to the traditional approach, they proposed an extended account that is based on world-semantics which is committed to both possible and

[1]By standard analysis we mean theories delivered by Robert Stalnaker [24] and David Lewis [11].

M. Sendłak (✉)
Institute of Philosophy, University of Warsaw, Warsaw, Poland
e-mail: maciej.sendlak@gmail.com

R. Urbaniak and G. Payette (eds.), *Applications of Formal Philosophy*,
Logic, Argumentation & Reasoning 14, DOI 10.1007/978-3-319-58507-9_11

*im*possible worlds. One of the main aims of this extension was to satisfy the need of reasoning about what is taken to be impossible.[2] Introducing impossible worlds raises many philosophical questions, and even though one can find various analyses of the logical structure and ontological status of impossible worlds and their application, very few of them discuss the important notion of similarity between worlds.[3] The importance of this notion lies in its role, which is to determine whether a given counterfactual (with possible or impossible antecedent) is true or false.[4]

Although "the discussion developed so far should show that the issue of the structure, closeness and ordering of impossible worlds is quite open" [2], there are two claims, which are in some sense the core of the standard understanding of the notion of similarity. The first one is commonly shared among advocates of impossible worlds; the second one raises some doubts. According to the first one, the trivial world, i.e., the world where everything is true, is the most dissimilar to the actual world (@). In other words, every non-trivial world (possible or impossible) is closer (or more similar) to the actual world than the trivial world. We will call this claim the Dissimilarity of the Trivial World (DTW). The second assumption about similarity and impossible worlds is the Strangeness of Impossible Condition (SIC), according to which every possible world is closer to the actual world than any impossible one. Both of these claims were formulated by Daniel Nolan [15].

In this paper we will argue for a negative thesis and a positive thesis. The negative thesis is that both DTW and SIC are false, and as such should not be taken as a "guide" to our interpretation of similarity. The positive thesis arises as a reaction to the negative one, and it is an alternative interpretation of the notion of similarity. Our aim is not to say which counterpossibles are true and which are false. This question is answered by (CF*), which is presented below. The question that we are going to give an answer for is "*what* makes given worlds more similar than others?"

The result of our investigation should be as general as possible, and because of that we will not discuss any particular account of impossible world. Nevertheless, we will base our analysis on two heuristic assumption, (i) the actual world is ruled by the classical logic, (ii) postulating impossible worlds should not lead to changes in the logic of the actual world. This view is shared by Daniel Nolan [15] and David Vander Laan [25], among others. These assumptions will help us to point out a problem with DTW. Even though the acceptance of DTW has particular consequences for the

---

[2][3, 4, 8, 14, 15, 22, 25–27].

[3]For a comprehensive analysis of ontological status of impossible worlds see [2, 16].

[4]Earlier versions of this material were presented in Bratislava (Slovak Academy of Science) at "Issues on the Impossible Worlds" in May 2014, in Warsaw (University of Warsaw) at "Philosopher's Rally" in July 2014, in Ghent (Centre of Logic and Philosophy of Science) at "Entia et Nomina" in July 2014, and in New York at Graham Priest's graduate student seminar in November 2014 (CUNY Graduate Center). I am grateful to the participants of these meetings for their helpful comments and discussions. I would like to thank to the anonymous reviewers for this volume for their comments concerning the earlier versions of the paper.

This material is based on work supported by the Polish National Center of Science under Grant No. 2012/05/N/HS1/02794. Thanks to the Polish-U.S. Fulbright Commission I had the opportunity to develop the ideas presented here during my stay at CUNY Graduate Center.

advocates of the two above mentioned assumptions, as we shall see, philosophers who choose a different logic are in no better situation. The final result of our investigation, i.e., and interpretation of the notion of similarity should be applicable regardless of a chosen theory of impossible worlds.

11.2 Counterfactuals

Counterpossibles can be represented as sentences of the form: "If A were true, then C would be true" ($A > C$), in which it is stated that the truth of an impossible antecedent (A) leads to a given consequent (C); e.g.,

(1) If the Monadology were true, then Leibniz would be right about metaphysics.
(2) If the Monadology were true, then Leibniz would be wrong about metaphysics.
(3) If intuitionistic logic were true, then the Law of Excluded Middle would be false.
(4) If intuitionistic logic were true, then the Law of Excluded Middle would be true.

Each of the counterfactuals above contains impossible (necessarily false) antecedents. It means that there are no possible worlds where these antecedents are true. This is due to the fact that if one assumes the truth of a given metaphysics or logic in the actual world, then one should also assume that it is necessarily true.[5] According to the standard analysis of counterfactuals:

(CF) "$A > C$" is true in @ iff either (i) there is no world where A is true or (ii) there is a world w where A and C are true and w is more similar to the actual world than any world w', where A is true but C is false.

In virtue of (CF), sentences (1)–(4) are true, since all of them satisfy condition (i). On the contrary, we would rather like to consider only some of them true and others false. For this reason a more sensitive analysis of their truth conditions is required.

To solve this problem, many philosophers have been arguing that one needs to invoke impossible worlds, i.e., worlds where what is impossible (necessarily false) in the actual world is true. They claim that, just as for every possibility, there is a world, which represents it, so for every impossibility there is a world which represents what is impossible from the actual world's point of view (e.g., [27]). As a consequence, the advocates of impossible worlds postulate worlds where, for example, a round square exists, Monadology is true, 10 is a prime number, it is raining and not raining at the same time, etc.

To avoid the trivial consequences of postulating worlds where what is in the actual worlds considered to be necessarily false is true, one should assume that these worlds are elements of different *spaces* than the space of possible worlds. It is worth noting that because of that, modal terms should be taken to be indexical with respect to given logical spaces: what is impossible in our logical space (i.e., in all worlds which are

[5] Although the above example assumes that the metaphysics of Monadology and intuitionistic logic are incorrect, one can easily change examples.

ruled by classical logic) is possible in some other logical spaces (e.g., paraconsistent spaces). In this sense, every impossibility is true in some world, but the world has to be outside the set of possible worlds.[6]

Of course, there is no parity between different impossible worlds. Some of them are closer (more similar) to the actual one than others. As we already mentioned, there are issues with determining how to measure similarity between worlds. It was also not easy in the case of standard analysis, and now, when one introduces a plenitude of impossible worlds it is even more puzzling. Nevertheless, it seems that we can point out a claim, which at least tells us what is the most dissimilar world:

> First, it is intuitive to claim that some impossible worlds are more similar to the actual world @ than others. For instance, the explosion world (call it *e*) at which everything is the case, that is, at which every sentence is true, seems to be as far from @ as one can imagine, provided one can actually imagine or conceive such an extremely absurd situation. Now, pick the impossible world, *t*, at which everything is as in @, except that I wear an impossible t-shirt which is white all over and black all over. Intuitively, *t* is closer to @ than *e* [2].[7]

Regardless of any detailed account of similarity, the existence of a plentitude of possible as well as impossible worlds and their sets allows us to avoid vacuous truth of counterfactuals with necessarily false antecedents. Thanks to such worlds, one can easily extend the standard analysis by claiming that every impossibility is true in some impossible world:

(CF*) "$A > C$" is true in @ iff there is a (possible or impossible) world w, where A and C are true and w is more similar to the actual world than any world w', where A is true but C is false.

This extension should keep the analysis of counterfactuals from being insensitive to the problem of counterpossibles. Sentences such as (1) and (3) are considered to be true because there is an *im*possible world in which the antecedent and the consequent of these counterfactuals are both true, and this world is more similar to the actual one than any world where the antecedent and consequent of (2) and (4), respectively, are true. Thanks to this, one can present non-vacuously correct reasoning from necessarily false claims.

11.3 Troubles with Similarity

The above extension works well for most examples of counterpossibles. Nevertheless, it seems that when it comes to the trivial world, troubles arise. Although it might be bizarre enough, postulating the existence of this world is a simple consequence of the claim that for every impossibility there is a world where it is true. If we

---

[6]It is worth to notice that indexical use of "possible" and 'impossible" allows to avoid the risk of believing that the actual world is one of impossible worlds. After all, only a possible world could be actual.

[7]See also [15].

agree that it is impossible that everything is true, then there is an impossible world where everything is true—the trivial world. Since we assumed that the actual world is ruled by classical logic, considering the trivial world it is worth to square it with the fundamental rule of this logic, the so-called Rule of Explosion also known as *ex contradictione sequitur quodlibet* (ECQ). It is usually expressed as an implication $[A \wedge \neg A] \rightarrow B$ and states that from contradiction everything follows. The reason we mention it here is that there is only one world where B mentioned above is true, and it is the trivial world.

The analysis of a relationship between implication and counterfactuals has a great history in philosophical literature ([1]: 20–44), but beyond many differences in various approaches to this issue, lately one claim seems to be commonly accepted. It can be expressed as $A \rightarrow B \vdash A > B$, and it states that "any logical truth of the form $A \rightarrow B$, gives rise to true conditional $A > B$" ([21]: 331).[8] This connection between implication and conditionals allows us to consider the following sentences:

(5) If there were a true contradiction, then everything would be true.
(6) If there were a true contradiction, then (still) not everything would be true.

Assume that the antecedent and the consequent of (5) are true in w_1, while those of (6) are true in w_2. From classical logic's (i.e., the actual world's) point of view, the antecedents of both of these counterfactuals express an impossibility, so in order to evaluate their truth we should assume that both w_1 and w_2 are impossible worlds. The important difference between them is that w_1 is the trivial world, while w_2 is a non-trivial one. Assuming that the actual world is ruled by classical logic, we would rather like to admit the truth of (5) than of (6). After all, the first one is just a counterfactually expressed ECQ. If it is truly so, then (according to (CF*)) we have to admit that w_1 is more similar to the actual world than w_2. But as it has been stressed above, one of the basic assumptions in theories of impossible worlds is that the trivial world is the most dissimilar from the actual world. If we assume DTW and admit that the trivial world (w_1) is the most dissimilar to the actual world, then w_2 is more similar than w_1. As a result, (6) becomes a true counterfactual and (5) should be taken to be false. If (5) is false then ECQ is such as well.[9] In consequence, this analysis of counterpossibles leads to the rejection of one of the fundamental rules of classical logic, which would mean that classical logic is false in the actual world. In virtue of the above, we will argue that if DTW implies the inadequacy of classical logic in the actual world, we should reject DTW.

One might argue that ECQ is a logical law, and as such it remains valid in every possible world regardless of the truth value of (5) or (6). In this way, the falsehood of (5) would not result in the falsehood of ECQ. Nevertheless, a consequence of it

[8] See also [6, 10]: 87–9. It should be stressed that it does *not* mean that any true conditional results in a true implication.

[9] This result is consistent with those theories of impossible worlds, which are based on paraconsistent logic ([14, 18, 22]). Nevertheless, one can modify the example in such a way that the acceptence of DTW will imply the claim that ECQ is false in the actual world according to classical logic. Regardless to what we believe to be the true logic of the actual world, one should not believe that according to classical logic ECQ is false.

would imply that one could maintain that $A \rightarrow A$ is true in the actual world while $A > A$ is false, which surely should not be the case. Moreover it seems that (5) expresses exactly the same claim, which is expressed in ECQ, so it is difficult to imagine what could be a better way of expressing ECQ in the natural language than (5). As Priest pointed out: "Conditionals may not express laws of logic; but which conditional holds may certainly depend on logical laws. Thus, $[A \wedge B] > A$ since $[A \wedge B]$ entails A." [21]: 330).

One may believe that taking (5) to be true undermines the impossible world analysis of counterpossibles in general. After all, this entire framework was meant to make sense of non-vacuously correct reasoning based on what is impossible, and (5) shows us that from a contradiction everything follows. In this sense, every sentence that is both true and false should imply everything, and this seems to contradict the basic motivations for introducing impossible worlds in the first place. Now the question is: how can one accept (5) and at the same time make non-vacuous inferences relying on a paraconsistent logic?

To answer this question we should notice that there is an important difference between assuming a contradiction in classical logic on the one hand, and assuming a contradiction true in one of the worlds in the space of a paraconsistent logic on the other. When we are thinking about a contradiction which does not lead to the truth of everything, we are considering the last option. In this sense every non-vacuously true counterpossible with a contradiction as the antecedent is (*implicitie* or *explicite*) assumed to be true in a world of paraconsistent logic. Consider two examples:

(7) If it were raining and not raining at the same time, then everything would be true.
(8) If it were raining and not raining at the same time, then not everything would be true.

Both of them contain impossible antecedents, and it seems that we can find two different contexts in which they have different truth-values. If we try to analyze them with an assumption that classical logic is true, then (7) would be true and (8) would be false, just as in the case of (5) and (6). On the other hand, if the counterfactuals above were preceded by a claim such as "Assuming the truth of paraconsistent logic, ..." then obviously we would say that (8) is true and (7) is false. After all, that is what the advocates of paraconsistent logics would like to claim. In other words, one can find a reason to believe that there is a context in which (7) is true and others where it is false. In this sense just because we take (5) to be true does not mean we treat every contradiction in the same way. Especially not those which are assessed in a world ruled by a paraconsistent logic. This shows that the only possible way of taking ECQ to be true in the actual world and (5) to be false (and consequently (6) to be true), is if we consider the antecedent to be true in a world of paraconsistent logic. But as we have seen above, it is clearly not the trivial world.

Although we started with the problem of the dissimilarity of the trivial world, before we sketch our interpretation of similarity, we will analyze the Strangeness of Impossibility Condition (SIC). According to this condition "any possible world is more similar (nearer) to the actual world than any impossible world" ([15]: 550).[10] In this sense, a world where there is no woodpecker (which is a possible world) is more similar to the actual world than a world where a round square exists. Although *prima facie* it is reasonable to believe in SIC, there is also a good reason to reject this condition. Contrary to the claim of the dissimilarity of the trivial world, SIC is not very widely accepted and some philosophers doubt its validity. We will join them and show that SIC should be taken to be false.

Let's start with an analogy. Consider three objects: a ball, a tomato and a ladder. If one asks, "What is more similar to the ball? A tomato or a ladder?" most of us would probably answer "the tomato". When asked why so, we can say that they both have the same shape. This is the correct answer, but only if we understood our question as "What is more similar to the ball *when it comes to having the same shape*?"

But if one presents the question in a different way, say, "What is more similar to the ball *when it comes to having the same nature*?" the answer would be different. In this case we should say that the ladder is more similar. After all, a ladder and a ball are artifacts, while the tomato is not. This shows that it is very difficult to think about similarity *per se*. Usually our understanding of similarity between objects depends on a chosen feature that we take to be the most important. In this sense, each time we compare objects we (either explicitly or implicitly) focus on a given feature. Without this restriction, the result of such a comparison might be misleading. Similarity understood in this way is in a fact a ternary relation: $S(a, b, F)$—*a is similar to b because of factor (property) F*. In this sense two objects are similar if they share the same property. A ladder is similar to a ball because they are both artifacts, and a tomato is similar to a ball because they are both round. By analogy, being more similar (MS) is a quaternary relation $MS(a, b, c, F)$—*because of factor F, a is more similar to b than c is*.

Consider the possible world mentioned above where there are no woodpeckers (but where no circle is a square) and impossible worlds where a round square exists (but where woodpeckers also exist). When it comes to the lack of round squares (and presumably, being possible), we can say that the former is more similar to the actual world than the latter. Nevertheless, we can also say that considering the number of woodpeckers the last one is more similar to the actual world than the first one is. In this sense, similarity between worlds depends on the choice of an aspect. If the most important feature of a world is to have an adequate number of woodpeckers, and one does not care about geometrical impossibilities, then one can say that there is an impossible world that is more similar to the actual world than one of the possible worlds.

Someone who would like to save the validity of SIC might argue that the most important feature of a world is whether it is possible or impossible. Surely this is an important feature of a world, especially when we are dealing with the analysis of modality. Accepting this assumption, SIC might easily be taken to be true. Nevertheless it would change the meaning of SIC. After all, it would then state that *considering the feature of being possible*, every possible world is more similar to the

[10] See also [9].

actual world than any impossible world is. Although it is hard to argue against this claim, it presupposes that the most important feature of a world is being possible or impossible and, as we have seen, we do not have to compare worlds only because of this feature. This allows us to believe that SIC in its original formulation is false, and as such should not be used as a guide for a proper understanding of the notion of similarity.

One might argue that the above considerations do not help in an analysis of counterpossibles. After all, we were comparing @ with two worlds (the one with woodpeckers and the one with a round square), which are very different from each other. Both (CF) and (CF*) state that in order to determine the truth-value of a given counterfactual, we should compare worlds, where what is expressed by the antecedent of this counterfactual is true. Because of that, to give a more complete view we will focus on examples of counterpossibles with the same antecedents. In the next few pages, we will present a philosophical interpretation of similarity, which should (a) help to explain why we consider some counterpossibles to be non-vacuously true, and (b) allow us to keep the validity of ECQ in classical logic at the same time.

11.4 Abduction and the Notion of Similarity

It is not an easy task to give any interpretation of the notion of similarity without any pre-theoretical intuitions. That's why we will base our investigations on some counterpossibles, which we take to be true, and we will try to point out their common features. We will start with two pairs of counterfactuals. The first pair shares a mathematically impossible antecedent; the second one will share a metaphysically impossible one.

(9) If $2 + 2 = 8$, $2 + 3 = 10$, and $2 + 4 = 12$, then $2 + 5 = 14$.
(10) If $2 + 2 = 8$, $2 + 3 = 10$, and $2 + 4 = 12$, then $2 + 5 = 7$.

To analyze them we should assume that there are two impossible worlds such that:

At w_3: 2+2=8, 2+3=10, 2+4=12, 2+5=14.
At w_4: 2+2=8, 2+3=10, 2+4=12, 2+5=7.

Both of these worlds are impossible since what is true in them is (from the mathematical point of view) necessarily false in the actual world. It seems that although the actual world has, in a sense, more in common with w_4 (in both worlds it is true that $2 + 5 = 7$) to say that w_4 is closer to the actual world than w_3 is to say that (10) is true and (9) is false. On the contrary, we prefer to take (9) to be true and (10) to be false due to the fact that we can point out the regularity in w_3: for every result of addition that equals n in the actual world, it equals $2n$ in w_3. In this sense, even though the results of addition are, from the point of view of the actual world, necessarily false, we can indicate that this "impossible addition" works in the same way for any number. Even though we did not considered this example, we can assume that in w_3

$100 + 1 = 202$. It is difficult to find a similar regularity in w_4. There might be one, but considering the antecedent and the consequent of (10) it is not easy to offer any interesting regularity. In this sense w_3 behaves *more systematically* according to the standards of systematicity encapsulated in our (i.e., actual) mathematics.

Now let us consider an example of counterfactuals with metaphysically impossible antecedents. For the sake of argument let us assume that in the actual world (and in every possible world as well) haecceitism is true in the sense that every possible object consists of a subject which posseses an individual essence (and, presumabely, some other properties). Because of that, it is metaphysically impossible for there to be an object that contradicts this view. If there is any such object, it has to be an element of an impossible world, where haecceitism is false. There is also an alternative account (necessarily false according to our story), according to which there is no such a thing as a subject which possesses its individual essence and that every object is just a bundle of properties. Consider two worlds:

At w_5: No object has an individual essence. Every object is a bundle of properties. Among others, the oldest Galapagos turtle.

At w_6: Almost every object has an individual essence. The only exception is the oldest living Galapagos turtle, which has no essence and is just a bundle of properties.

Again, by comparing these two worlds, w_6 is *prima facie* more similar to the actual world. They differ only in the fact that in one of these worlds the oldest Galapagos turtle does not have an essence, while w_5 is entirely different when it comes to the metaphysical structure of reality. Based on this assumption, we can analyze the following counterpossibles:

(11) If the oldest Galapagos turtle were a bundle of properties, then every individual would be one as well.

(12) If the oldest Galapagos turtle were a bundle of properties, then it would be the only individual without individual essence.

If we would like to claim (as we hypothetically assumed above) that w_6 is more similar to the actual world than w_5 is, then we should admit that (12) is true, and (11) should be considered false. On the contrary, we prefer to admit the truth of the former rather than the latter. This is due to the fact that if a given account of metaphysics is true, then its claims should apply to all individuals and not just to some of them. In this sense the fact that the oldest Galapagos turtle is a bundle of properties should imply that any other individual is one as well.

One might ask what is significant about w_3 and w_5 that makes them more similar to the actual world than w_4 and w_6, respectively. An answer to this question might be a good guide for a better understanding of the notion of similarity and the way we use counterfactuals. As we already said, both of these worlds are ruled by some regulations, which help us to understand why the truth of a given antecedent leads to the truth of a given consequent. In the first case it was the claim that for every result of addition that equals n in the actual world, it equals $2n$ in w_3. This rule shows that the mathematics of w_3, even if it gives an impossible result, applies in an equal way to every object in its domain. In the second case it was the main claim of the bundle theory, according to which every individual is a bundle of properties, which shows that every object has the same metaphysical structure. For now lets call them rules R_3 and R_6, respectively. Both of them ensure the consistency of the antecedent and the consequent of (respectively) (9) and (11). These sentences can be considered to be true only if one assumes not only the antecedents, but the counterfactuals have to silently assume R_3 or R_5. In this sense, the proper structure of these counterfactuals is as follows:

(9*) (Assuming R_3) If $2 + 2 = 8$, $2 + 3 = 10$ and $2 + 4 = 12$, then $2 + 5 = 14$.
(11*) (Assuming R_5) If the oldest Galapagos turtle were a bundle of properties, then every individual would be one as well.

In other words, we take a counterfactual to be true if we can find a natural and non-trivial rule, which ensures a connection between the antecedent and the consequent.

Although the above condition is satisfied by (9) and (11), it needs more explanation. It is due to the fact that we can also present a certain rule for the truth of (10) and (12). In the first case, it will be R_4: "For most results of addition that equal n in the actual world, they equal $2n$ in w_3. The only exception is $2 + 5$, which has the same result as it does in the actual world." In the second case, it will be R_6: "Almost every object has an individual essence and some properties. The only exception is the oldest Galapagos turtle, which is a bundle of properties and which has no essence." R_4 and R_6 are rules which ensure the connection between the antecedents and consequents of (10) and (12), which now should be represented as:

(10*) (Assuming R_4) If $2 + 2 = 8$, $2 + 3 = 10$ and $2 + 4 = 12$, then $2 + 5 = 7$.
(12*) (Assuming R_6) If the oldest Galapagos turtle were a bundle of properties, then it would be the only individual, which has no essence.

In this sense for every pair of an antecedent and a consequent we can find a rule according to which a given counterfactual is true. Because of that, the question of truth-value of a given counterfactual is in fact a question of choosing one rule over another. In other words, considering counterfactuals with the same antecedents such as (9) and (10) or (11) and (12), in order to justify the non-vacuous truth of some of them, we should give a reason for choosing R_3 over R_4 and R_5 over R_6.

The problem of counterfactuals as we presented it above is in fact very similar to the problem of inference to the best explanation, i.e., abductive reasoning.[11] Contrasting it with other methods of inference such as deduction and induction, Gilbert Harman characterized abduction as follows:

> In making this inference one infers, from the fact that a certain hypothesis would explain the evidence, to the truth of that hypothesis. In general, there will be several hypotheses which might explain the evidence, so one must be able to reject all such alternative hypotheses before one is warranted in making the inference. Thus one infers, from the promises that a given hypothesis would provide a better explanation for the evidence than would any other hypothesis, to the conclusion that a given hypothesis is true [7, 89].

[11] See [12, 13, 17].

To make the above quotation more appropriate for our investigations, we should change "hypothesis" into "rule" and "evidence" into "antecedent". In this sense the problem of counterfactuals might be stated as a problem of finding a rule which explains the truth of the antecedent. This is just the first step, and it is satisfied by all the rules suggested above. R_3 and R_4 gives a reason for which the antecedent of (9) and (10) is true. Similarly, R_5 and R_6 give an explanation of the antecedent of (11) and (12). The second step is to answer what makes a given hypothesis (rule) better than the other. This is much more difficult:

> There is, of course, a problem about how one is to judge that one hypothesis is sufficiently better than another hypothesis. Presumably, such a judgment will be based on consideration such as which hypothesis is simpler, which is more plausible, which explains more, which is less *ad hoc*, and so forth. I do not wish to deny that there is a problem about explaining the exact nature of these considerations [7, 89].

The problem of proper criteria for choosing one hypothesis over another has been raised in various works, and to say that it is clear and uncontroversial would be far from truth.[12] Nevertheless, using the presented analogy between an inference to the best explanation and the counterfactuals, the criteria for choosing a better rule might be easier in some aspects.

The reason we prefer to choose R_3 (or R_5) to R_4 (or R_6) is that they are (in a certain sense) the same as rules of the actual world. This is due to the fact that both of them are universal, i.e., they apply to every object in a given domain. In the actual world a given rule of mathematics applies to all numbers in the same way. We can say the same about the metaphysics of the actual world. Whatever is the metaphysical structure of reality, we would like to believe that it is *universal* enough to apply to all individuals. Even if someone never adds 2998 to 1, thanks to the universality of rules of mathematics, one does not have to do it in other way than when they add 1 to 1. There is no reason to think that beyond given numbers the rules of addition are different. It is the same in the case of questions about the metaphysical structure of reality.

Assuming that our world is ruled by universal regularities allows us to make some prediction and it is an invaluable feature of any theory. The importance of proper prediction is shown by the fact that every time reality does not act accordingly to what a given theory has predicted, we start to doubt the validity of this theory and look for a new one, which takes new data into account. If consistency of the rules of the actual world is considered as important for our reasoning about this world, we could try to incorporate this scheme in order to reason about the non-actual world as well. This way of thinking about impossibilities makes it easier for us to admit that (9) and (11) are true and (10) and (12) are false. Therefore, it is reasonable to believe that w_3 is closer to the actual world than w_4 because of the consistency of this world

[12] See, among others, [5, 13, 23].

governed by R_3.[13] A lack of such consistency of R_4 makes it less attractive as an explanation of the fact that $2 + 2 = 8$, $2 + 3 = 10$ and $2 + 4 = 12$.

11.5 Organized and Chaotic Worlds

The observation above allows us to make a distinction in the plentitude of impossible worlds—the distinction between *organized* and *chaotic* worlds. Worlds that belong to the first group are worlds where necessarily false claims are true, and where the reality acts with some regularity, which is guaranteed by a given rule. Chaotic worlds, on the other hand, are worlds where the same necessarily false claims are true, but where there is a lack of such regularity. This feature of worlds is invaluable when it comes to reasoning about what is true in non-actual worlds. After all, making some inferences is possible only if one can point out a rule which would explain the truth of the premises. If there is a lack of this consistency, it is difficult to come up with any interesting conclusion or inference at all. For the same reason, regardless of what is true in the actual world (whether it is paraconsistent or classical logic, essentialism or bundle theory, etc.), we like to believe that this world is also well organized. Thanks to that, it is easier for us to make some inferences about this world. This description, even if true, might lead to some questions.

As we said earlier, one can accept R_4 or R_6 and claim that (10) and (12) should be considered true. In this sense w_4 and w_6 would be considered to be organized and w_3 and w_5 to be chaotic. Nevertheless, this seems to miss the point of considering counterpossibles in the first place. One uses counterfactuals with antecedents such as in (9) and (10) or (11) and (12) to reason about necessarily false claims and theirs consequences. To do so we usually choose an antecedent, which is supposed to play a double role. First of all, it should be a representative example of the main claim of a given theory. Secondly, it should give us a ground for a truth of the consequent of a given counterfactual. It happens so by indicating an important connection between the truth of the antecedent and the truth of the consequent. While (9) and (11) (thanks to R_3 and R_5 respectively) seem to satisfy these aims, it is difficult to find a similar important connection between the antecedents and the consequents of (10) and (12). Quite the opposite, rules that govern the connection between them seem to require us to believe that actually there is *no* important connection between various mathematical or metaphysical truths. After all, according to (10), the fact that the operation of addition works differently for $2 + 2$ has nothing to do with how it works for $2 + 5$. Similarly, according to (12), the metaphysical structure of one object has nothing to do with the metaphysical structure of other objects. This seems to undermine the starting point of the analysis of counterpossibles. Nevertheless, as we shall see,

[13] By "consistent" we do not mean that there are no contradictions in them, but rather that they act according to certain regularities that in these worlds are supposed to be true. In this sense even a paraconsistent world might be taken to be consistent if it acts accordingly with the laws of paraconsistent logic.

even though we argued that (10) is false, this does not mean that a world where the antecedent and the consequent of (10) are true *has* to be chaotic.

It seems that when we are saying that "$2 + 2 = 8$, $2 + 3 = 10$ and $2 + 4 = 12$" we are using it just as an example of a general truth of a given world. Even though in worlds w_3 and w_4 there is much more than just mathematical truth, we are only interested in them because we are considering the possible consequences of alternative results of addition, which is a mathematical operation, and nothing more. In the case of w_5 and w_6, we are interested only in facts about the metaphysical structure of these worlds, and do not bother with facts about the biology, logic, physics or mathematics of these worlds. These are irrelevant for our consideration about essentialism and the bundle theory.

This becomes clear if we realize that w_3 and w_6 might be the same world (lets label it as w_{36}). In a similar way, w_4 and w_5 might be taken to be the same world (w_{45}). Previously we said that w_3 is more similar to the actual world than w_4, and that w_5 is more similar than w_6. Do we need to admit a true contradiction and to claim that it is the case that w_{36} is more similar to the actual world than w_{45}, and it is not the case that w_{36} is more similar to the actual world than w_{45}? Obviously not. There is no contradiction in claiming that w_{36} is (in virtue of mathematical truth) more similar to the actual world than w_{45}, and that (in virtue of metaphysical structure) the same world is less similar to the actual world than w_{45}. In both cases we consider truths in which we were most interested. In the first case, it was mathematical truth; in the last one it was the metaphysical structure of the reality.

Assuming the distinction, we may say that the feature that makes worlds closer or distant to the actual world is whether they are organized or chaotic with respect to a given antecedent. If we want to use an analysis of counterpossibles as a tool for making inferences about necessarily false claims the feature of impossible worlds, which will make them closer or distant to the actual world, it is a principle, which connects an antecedent and a consequent of given counterpossibles. As we have seen above, it is important to stress that there is no such a thing as being *absolutely* organized. When we are saying that a given world w is organized because it is governed by R_x we are claiming that R_x gives a better explanation of an antecedent A that is true in w, than an alternative R_y does. Because of that, we are accepting a consequence of A and R_x and not the consequence of A and R_y. In this sense, comparing non-actual worlds is possible only if in both of them A is true. World w_{36} is more similar than w_{45} because of R_3 and w_{45} is more similar than w_{36} because of R_5. Both of these rules govern different truths of the above mentioned worlds—one of them is mathematically organized and metaphysically chaotic, while the second is mathematically chaotic and metaphysically organized.

Thanks to the above considerations, we can get back to the problem of closeness of the trivial world. According to what we have said, similarity should be taken to be a ternary relation. In this case, it is a relation between the actual (@) and the trivial (w_1) world. What is lacking in this comparison is the aspect for which we will compare these worlds, as there are usually many features that might be taken into consideration. If one would like to measure similarity within a number of true contradictions, then surely there is a world that is, in this respect, more similar to

@ than w_1 is. For example, a world where it is not the case that everything is true. Nevertheless, there is an important feature which is shared by @ and w_1, and which many of impossible worlds lack. In fact, the trivial world is the only one among impossible worlds where a contradiction is true which has this feature. It is the validity of ECQ. This rule holds in @ as well as in w_1. ECQ is true in @ since it is not the case that everything is true in the actual world, and it is also not the case that any contradiction is true in this world is true. Contrary to this, in the trivial world everything is true, among others classical logic, contradiction and ECQ.

According to the above, we might consider three worlds: the actual world @, an impossible but not trivial world w_2, and the trivial world w_1. In the actual world, there are no contradictions, and it is not the case that everything is true in this world. In w_1 everything is true, so contradictions are true as well. In impossible world w_2 there is at least one true contradiction, and since it is not the trivial world, it is not the case that everything is true in this world:

At @: $\neg[A \wedge \neg A], \neg B$
At w_1: $[A \wedge \neg A], B$
At w_2: $[A \wedge \neg A], \neg B$

Further, we might consider two counterfactuals which correspond to non-actual worlds that we have mentioned. From the actual world's point of view, w_2 corresponds to $[A \wedge \neg A] > \neg B$, while w_1 corresponds to $[A \wedge \neg A] > B$. Assuming the connection presented above between the notion of similarity and inference to the best explanation, we can ask what rules hold in w_1 and w_2. World w_1 is the trivial world, so by definition every rule holds in this world. Among these rules, there is ECQ. This allows us to say that what connects antecedent $(A \wedge \neg A)$ and consequent (B) of w_1 is the truth of ECQ. In this sense w_1 is a world where true contradiction has its consequences in the truth of everything.

World w_2, on the other hand, is a world where classical logic is true, contradiction is true, but where it is not the case that everything is true. It is hard to find any rule that might be used to describe this world. One might rely on the main assumption of paraconsistent logic and say that w_2 is a world where contradiction does not lead to the truth of everything. Although it may partly explain why the validity of classical logic and a true contradiction do not imply that everything is true, this rule seems to be far from what anyone would like to take as an explanation of antecedents of (5) and (6), because it requires the assumption that two logics hold at the same non-trivial world. Therefore, we can point out the reason for which w_1 is more similar to @ than w_2: because of the validity of ECQ. As such, w_1 acts exactly how advocates of classical logic believe—true contradiction leads to triviality. In this sense the trivial world is more similar to the actual world than at least one non-trivial impossible world.

11.6 A Priori Implication and the Structure of Worlds

Before we proceed to the consideration of possible objections against our view, it is worth to compare it with an alternative account that might seem to be similar to ours. It is the view presented by Berit Brogaard and Joe Salerno in the "Remarks on Counterpossibles" [4]. The core idea of their approach is to deploy two–dimensional semantics in the analysis of counterpossibles. The underlying notion of this epistemic account is a priori* implication, according to which: "For a speaker s in context c, P a priori* implies Q iff for s in c, Q is a relevant a priori consequence of P" ([4]: 655). By putting this notion to work, the following construal of similarity (closeness) is given:

> [S-B]: For any two impossible worlds w_1 and w_2, w_1 is closer to the base world than w_2 iff: (a) w_1 does not contain a greater number of sentences formally inconsistent with the relevant background facts (held fixed in the context) than w_2 does. And if w_1 and w_2 contain the same number of sentences formally inconsistent with the relevant background facts (held fixed in the context): (b) w_1 preserves a greater number of a priori* implication between sentences than w_2 does ([4]:655).

This construal of similarity is contrasted with Daniel Nolan's account, and—as the authors argue—it gives more natural explanation of various examples of counterpossibles [4, 654].

It is not difficult to notice some important similarities between S-B and our account. One of them is that both views indicate the importance of some connection between the antecedent and the consequent of a given counterpossible. While in our view the connection is guaranteed by a chosen rule, in Brogaard-Salerno account it is due to the a priori* implication. The second similarity between our counstruals lies in the reference to the "background facts", which according to Brogaard and Salerno are "fact about laws of nature, facts about what the world is like, and so on" [4, 655]. According to our view the background fact is simply the consistency of the antecedent and the consequent. This consistency might be assured by different rules, and as we have showed, some of these rules are better than others.

What makes S-B very different from our account is that we do not believe that the *number* of sentences consistent with the background facts makes a world more similar to the actual world, while in the Brogaard–Salerno view it seems to be an important aspect of similarity. This is due to the first part of condition (a) of S-B. To show the reason for rejecting this assumption, consider the following example. Let us assume that a theory T leads to paradoxical consequences and as such we consider it to be necessarily false. Let us assume also that the advocates of this theory believe that T is consistent. After all they believe that it is true. Consider the following counterpossibles.

> (S-B 1) If T were true, then a sentence inconsistent with T would be true as well.
>
> (S-B 2) If T were true, then no sentences would be inconsistent with T.

Worlds which correspond to the above counterfactuals are impossible. The important difference between them is that in the first one there is a sentence that is inconsistent with the background fact of the consistency of a world where T is true. Contrary to this, in the second world every sentence is consistent with the background facts. As such, the second world (according to condition (a) of S-B) is more similar to the actual world. Because of that, (S-B 1) is false and (S-B 2) is true. The reason for this is that for every theory (even necessarily false one) there is a world where this theory is consistent. If we accept the condition (a) of S-B, every world where a necessarily false theory is consistent will be closer to the actual world, than one where this theory leads to inconsistency. This enables one to show that grounding the notion of similarity in the number of sentences makes it impossible to truly indicate an inconsistency in alternative theories.[14]

An advocate of S-B might object that the first world preserves more a priori* implications than the latter one. And thus, there is a justification for taking the first one to be closer. Nevertheless, according to S-B, one should deploy the notion of a priori* implication if there is an equal number of sentences that are inconsistent with the background facts of the world. Examples that we have considered above do not satisfy this condition. After all, one of these worlds differs by fewer sentences than the other one. This being the case, the notion of similarity in terms of the number of sentences that are consistent with the background facts should not be taken as a proper "measuring tool".

It is worth to mention one more important aspect which we did not consider in our investigations. It is the question of the structure of worlds. Nowadays it is common to distinquish between two approaches to this problem. Graham Priest has labeled them as "American" and "Australian" types of impossible worlds. According to the first one, impossible worlds are maximal and inconsistent sets of sentences that are not closed "under any interesting notion of logical consequence".[15] According to the Australian approach, impossible worlds are ruled by a logical consequence, though it is a different than one known from classical logic [18, 484]. Brogaard–Salerno view is committed to the American approach and it assumes that impossible worlds are "all worlds that are not deductively closed, and the absurd world" [4, 652].[16]

Although we do not want to argue with this approach here, it should be emphasized that it implies problematic consequences.[17] Consider a world of paraconsistent logic, where some contradictions are true, but where it is not the case that everything is true. This world (according to the American approach) is either possible, or it is not ruled by any (interesting) rule of deduction. Since we have assumed the necessary truth of the classical logic, the first option is false. On the other hand, the second one contradicts the fact that alternative logics are closed under deduction. Obviously it is not the deduction known from classical logic; nevertheless it is a deduction.

---

[14]Similarities between S-B and our account allow to ask whether we can avoid this problem. We will go back to this in the next section.

[15]Being an "interesting" notion of logical consequence is an important condition, since it allows counting the trivial world as impossible as well. Even though it is closed under logical consequence, it is surely not a possible world.

[16]"Absurd world" is a different name for what we called "trivial world".

[17]For an argument against the analysis of counterpossibles with logically impossible antecedents in terms of the American-style impossible worlds see [3].

This might suggest that we should lean toward the Australian approach to impossible worlds. Nevertheless, to believe that *every* impossible world is closed under a logical consequence might narrow the plurality of impossibility. After all, it would rule out worlds that are "anarchic", and which do not obey any logical consequences. They are surely impossible. What might be helpful in this case is to deploy our distinction between organized and chaotic worlds. This would allow us to take anarchic worlds to be chaotic, and non–anarchic ones to be organized. Thanks to that, we can include both types of impossible worlds. Worlds, where A and C are connected by a notion of logical consequence, would be taken to be organized. Contrary to these, world where A and non-C are true, but where there is a lack of logical consequence which connects them, might be taken to be a chaotic world. As such, the distinction between the American-type and the Australian-type of worlds might be mimicked by the distinction between chaotic and organized worlds, respectively. One of advantages of this analogy is that, if—as we have argued—every organized A-world is closer to the actual world than any chaotic A-world, then every world that is closed under deduction is more similar to the actual world than any "anarchic" world is. This seems to work well with the motivation for considering counterpossibles in the first place. After all, it justifies non–trivial consequences of taking something impossible to be true.

The above analogy is merely a rough idea of how we could combine two kinds of impossible worlds. As such, it might raise some questions. Although we are aware of it, we will stop here and proceed to discuss possible objections to our view. What may warrant this move is that the proper consideration of the problem of the logical (and metaphysical) structure of impossible worlds requires an extended analysis, for which there is no room in this paper. Nevertheless, we believe that our notion of similarity might be deployed (at least in some respects) regardless to the detailed account of logic and metaphysics of impossible worlds.

11.7 Possible Objections

Being non-trivial should make w_2 *more similar to the actual world than* w_1.

One may object that there is an important reason for claiming that w_2 is more similar to @ than w_1. It is the fact that in w_2 as well as in @ it is not the case that everything is true. For both of these non-actual worlds there is an aspect for which a given world is more similar to @ than another world is, just as in the case of w_{36} and w_{45}. Because of that we should give reason to believe that from the actual point of view world w_1 is organized and w_2 is chaotic. The reason is that, as we said earlier, in counterfactuals (5) and (6) we are considering consequences of true contradiction appearing in a world where classical logic is true as well. It seems that this complex antecedent might be explained only with the assumption that everything is true in this world.

The fact that w_1 is more similar to @ with respect to this feature does not rule out that there is an aspect in virtue of which w_2 is more similar to @ than w_1. There is no contradiction in this because we are talking about two factors of similarity: in the first case it is similarity in virtue of ECQ and in the second one it is a similarity in virtue of being non-trivial. Just as a tomato and a ladder can be more or less similar to a ball, so is the case of the three worlds mentioned above. Nevertheless, assuming the rules of classical logic, which supposedly are true in the actual world, from this world's point of view, the world where contradiction is true, and every claim is true as well, is more similar than a world where ECQ is false and where a true contradiction does not result in a triviality.

World w_1 is the most chaotic world.

One might also claim that w_1 is not actually organized at all. It is a world where on the one hand classical logic is true and on the other a contradiction is true as well. This makes this world highly inconsistent. Surely there are very few (if any) pairs of claims that are less inconsistent than these two. In addition, one could point out that the rule of this world is that everything is true, which as a rule is not informative at all. Contrary to worlds where non-classical logic is true, it does not give us a tool to assess non-trivial consequences of an assumption holding in this world. In this sense w_1 is ruled by a very uninformative rule and as such it is still a very dissimilar world.

In order to address the concern we should remember two things. First of all, we are dealing with impossibilities and the very starting point of the discussion is the claim that for every impossibility there is an impossible world, where the given impossibility is true. It is true that it is highly inconsistent to admit the validity of classical logic and a contradiction at the same time. Thats what makes these sentences true in an impossible world. Secondly, the decision of whether a given world is organized or chaotic does not depend on the antecedent of a given counterfactual, but on the connection between antecedent and consequent. The objection that w_1 is chaotic would be proper if this world were to correspond with counterpossibles such as: "If classical logic were true, then there would be a true contradiction." This sentence is obviously false, and it might be taken as such with respect to the fact that it is hard to find a rule that might connect the antecedent with the consequent of this counterfactual. On the contrary, we claim: "If classical logic were true, then there would be no true contradiction." In the second case the connection between the antecedent and the consequent is provided by the rules of classical logic.

Moreover, if one compares the rules of worlds where it is not the case that everything is true with the rule of the trivial world, the last one seems to be very dissimilar to the actual world. The rule according to which everything is true does not allow us to make any interesting inferences. Nevertheless, although to say that everything is true is uninformative, to say that if there were true contradictions in a world where classical logic is true, then everything would be true is not uninformative at all. It tells us what the consequents of an antecedent of this counterpossible are. Otherwise ECQ would be trivially true, and certainly that is not the case.

If an organized world is more similar to the actual world than a chaotic world, then one cannot truly indicate inconsistency in alternative accounts of logic, metaphysics, mathematics, etc.

Consistency is probably the most important feature of every philosophical account. Thats why many philosophers, arguing against an alternative account, try to show that the acceptance of a given account leads to inconsistency. Lets assume that one is arguing against an alternative account of metaphysics. They may express it in the following way:

(13) If T were true, then the advocates of T would have to contradict themselves.

The opposite of (13) is the following counterpossible:

(14) If T were true, then the advocates of T would not have to contradict themselves.

If (13) were true, then T should be considered false, and if (14) were true, then it does not have to be so. To take it to be true we should assume that world w' where the antecedent and consequent of (13) is organized and world w'' where the same antecedent is true, but where a different consequent is true, is a chaotic world. It might seem that a world that acts accordingly to T (i.e., the world where T is true) and which is inconsistent, should be taken to be a chaotic. This is due to the fact that each T assumes its own consistency, so the world where T is true should be consistent. Therefore, a world where T is true and where inconsistency appears is as chaotic as a world where classical logic is true and where contradiction is true as well. In this sense w'' should be considered to be more similar to @ than w' is. After all, it is a world where T is true and where there is no inconsistency. Every substitution of (14) should be then taken to be true and every substitution of (13) should be considered false. This shows that a distinction between organized and chaotic worlds makes us unable to indicate true inconsistency of a given philosophical account.

To address these doubts, we should notice the difference between the claims and assumptions of T and the consequences of accepting T. Naturally, no author would like to claim that her theory is inconsistent, but that does not rule out the possibility that her theory *actually* leads to inconsistency. If it is so, then obviously a world where the acceptance of T leads to inconsistency is an organized world. It acts according to the *real* consequences of T. Examples of these are Russells argument against the na'ive set theory and an argument presented by the same philosopher against Meinong's Theory of Object. Both of them indicate that an assumption of a given theory leads to inconsistency.[18] In this sense the fact that advocates of a given account believe in the consistency of her own theory, does not mean that consequences of their assumption does not lead to inconsistency. This shows that (13) is not trivially true, and as such the distinction between organized and chaotic worlds does not contradict our motivations for the analysis of counterpossibles.

[18] Of course we do not want to claim that Russells arguments ended the discussion about the na'ive set theory or Theory of Objects.

11.8 Summary

In this paper we had two goals. The first is to provide reasons for rejecting two claims that govern the notion of similarity between the actual and impossible worlds. According to the first of these, the trivial world is the most dissimilar to the actual world. According to the second, every possible world is more similar to the actual world than any impossible world is. The rejection of these claims was partly motivated by their consequences, according to which ECQ is false in a framework of classical logic. The second aim is to sketch a philosophical interpretation of the notion of similarity which would allow the validity of this rule in a framework of impossible worlds. Our notion is based on an analogy of the inference to the best explanation and on the assumption that similarity is a ternary relation satisfied by the actual world, a non-actual world and a given factor of similarity. Similarity understood in this manner is a notion which requires an indication of a proper rule, that is, one that supports the truth of the antecedent and explains its connection with the consequent. From among a variety of rules, we should choose one that is better than its alternatives. It is worth to emphasize that this notion is not restricted to counterpossibles, but it might work well also for counterfactuals with possible antecedents.

Our aim is not to deliver a new truth condition for counterpossibles, but rather to give a proposal of how one can interpret the notion of similarity. In other words, our interpretation was not meant to answer the question about *which* counterfactual is true, but rather *why* we believe that a given counterfactual is true. As such, it allows us to justify the truth of (9) and (11) and it especially allows to maintain the truth of the ECQ in classical logic. As we have seen, one can choose different features of the actual and a non-actual world and claim that (9) and (11) are false and (10) and (12) are true, although this would require giving a sufficient reason for choosing an alternative rule to connect the antecedents and consequents of given counterfactuals.

References

1. Bennett, J. (2003). *A philosophical guide to conditionals*. Oxford: Oxford University Press.
2. Berto, F. (2013). Impossible worlds. In E. N. Zalta (Ed.) *The stanford encyclopedia of philosophy*. http://plato.stanford.edu/entries/impossible-worlds.
3. Bjerring, J Ch. (2013). On counterpossibles. *Philosophical Studies*, *168*(2), 327–353.
4. Brogaard, B., & Salerno, J. (2013). Remarks on counterpossibles. *Synthese*, *190*, 639–660.
5. Daly, C. 2010. *An introduction to philosophical methods*. Peterborough: Broadview Press.
6. Gibbard, A. (1981). Two recent theories of conditionals. In W. Harper, R. Stalnaker & G. Pearce (Eds.), *Ifs*, Reidel, pp. 211–247.
7. Harman, G. (1965). Inference to the best explanation. *Philosophical Review*, *74*(1), 88–95.
8. Jago, M. (2012). Constructing worlds. *Synthese*, *189*, 59–74.
9. Jago, M. (2013). Impossible worlds. *Noûs*, *47*(3),
10. Kratzer, A. (2012). *Modals and conditionals: new and revised perspectives*. Cambridge: Cambridge University Press, Chap 4.
11. Lewis, D. (1973). *Counterfactuals*. Oxford: Blackwell.
12. Lewis, D. (1986). *On the plurality of worlds*. Oxford: Blackwell.

13. Lipton, P. (1991). *Inference to the best explanation*. New York: Routledge.
14. Mares, E. D. (1997). Who's afraid of impossible worlds? *Notre Dame Journal of Formal Logic*, *38*(4), 516–527.
15. Nolan, D. (1997). Impossible worlds: modest aproach. *Notre Dame Journal of Formal Logic*, *38*(4), 535–572.
16. Nolan, D. (2013). Impossible worlds. *Philosophy Compass*, *8*, 360–372.
17. Peirce, C. S. (1903). *Pragmatism as the logic of abduction*. In: Peirce Edition Project Staff eds. 1998, Essential Peirce Volume 2: Selected Philosophical Writings, 1893–1913, pp. 221–241.
18. Priest, G. (1997a). Editor's introduction. *Notre Dame Journal of Formal Logic*, *38*(4), 481–487.
19. Priest, G. (1997b). Sylvan's box: a short story and ten morals. *Notre Dame Journal of Formal Logic*, *38*(4), 573–582.
20. Priest, G. (2005). *Towards non-being: the logic and metaphysics of intentionality*. Oxford: Oxford University Press.
21. Priest, G. (2009). Conditionals: a debate with Jackson. In I. Ravenscroft (Ed.) *Minds, worlds and conditionals: themes from the philosophy of Frank Jackson*. Oxford: Oxford University Press, pp. 311–336.
22. Restall, G. (1997). Ways thing's can't be. *Notre Dame Journal of Formal Logic*, *38*(4), 583–597.
23. Rosen, G., Dorr, C. (2002). Composition as a fiction. In R. Gale (Ed.) *The blackwell guide to metaphysics*. Oxford: Blackwell.
24. Stalnaker, R. (1968). A theory of conditionals. In N. Rescher (Ed.) *Studies in logical theory*. Oxford: Blackwell.
25. Vander, Laan D. (1997). The ontology of impossible worlds. *Notre Dame Journal of Formal Logic*, *38*(4), 597–620.
26. Vander Laan, D. (2004). Counterpossibles and similarity. In F. Jackson & G. Priest (Eds.) *Lewisian themes: the philosophy of David K. Lewis*. Oxford: Oxford University Press, pp. 258 – 74.
27. Yagisawa, T. (1988). Beyond possible worlds. *Philosophical Studies*, *53*, 175–204.

Chapter 12
Grzegorczyk's Non-Fregean Logics and Their Formal Properties

Joanna Golińska-Pilarek and Taneli Huuskonen

Abstract The paper discusses Grzegorczyk's logic LD of descriptive equivalence, some of its extensions (logics LDD and LDT), and its more recent modifications, which are the logic of equimeaning LE and the logic of descriptions with Suszko's axioms LDS. We present an improved semantics for LD and prove a corresponding extended soundness and completeness theorem. We also show that LD is paraconsistent. These results generalize to LDD, LDT, and LDS as well. We briefly study the properties of LE. Furthermore, we compare the strengths of the logics and prove, in particular, that LD is uncomparable with LE and LDS, and the logic LDD—the extension of LD with the so called Delusion Axiom—is the strongest among the logics in question. Next we show that descriptive equivalence can be defined in terms of descriptive implication in LDT but not in LD. We prove also that if we identify the descriptive equivalence with the implication of the other logics, then LD, LDD, and LDT are different from intuitionistic logic and relevance logics T, E, R, EM, RM. Moreover, descriptive equivalence cannot be identified with necessary equivalence in any class of Kripke frames. Finally, we study different ways to formulate the idea of extensionality, presenting three different extensionality principles and exploring which logics satisfy each of them.

Keywords Descriptive equivalence · Equimeaning · Paraconsistency · Extensionality principles

12.1 Introduction

In 2011, Andrzej Grzegorczyk proposed a new type of propositional logic formalizing his philosophical ideas on the fundamental properties of logical equivalence

J. Golińska-Pilarek
Institute of Philosophy, University of Warsaw, Warsaw, Poland
e-mail: j.golinska@uw.edu.pl

T. Huuskonen (✉)
Department of Mathematics and Statistics, University of Helsinki, Helsinki, Finland
e-mail: taneli@poczta.onet.pl

R. Urbaniak and G. Payette (eds.), *Applications of Formal Philosophy*,
Logic, Argumentation & Reasoning 14, DOI 10.1007/978-3-319-58507-9_12

between sentences (see [3, 4]). Grzegorczyk's very basic philosophical assumption is that human language is used primarily to create descriptions of reality, and logical connectives are tools to achieve this goal. According to Grzegorczyk, a logical analysis of language should take into account two fundamental aspects of sentences: their truth values, but also their contents. As classical propositional logic and many non-classical logics are only focused on the truth values of sentences, in order to be able to speak also about their contents we must construct a new logic whose logical connectives will enable us to express descriptive properties of states of affairs and relationships between them. For this reason, the logic originally proposed by Grzegorczyk is called the *Logic of Descriptions,* or LD for short.

The logic LD is a propositional logic, whose formulas are built from propositional variables and logical connectives of negation $\neg$, conjunction $\wedge$, disjunction $\vee$, and descriptive equivalence $\equiv$. The descriptive equivalence is very different from the classical one. The meaning of the latter is entirely determined by truth values, which leads to such paradoxical claims of the classical logic as *any true (false) sentences are equivalent regardless of their content.* On the other hand, the descriptive equivalence—as pointed out by Grzegorczyk—is supposed to be "more subtle, but not totally determined, allowing for an intuitive interpretation of being connected by a shared subject" ([3]). In Grzegorczyk's approach the classical equivalence connective is replaced by the connective of descriptive equivalence, and the axiomatization of the fundamental properties of logical connectives is constructed from scratch.

The first axiomatization of descriptive equivalence has been presented in Grzegorczyk's paper [3], where 4 inference rules and 18 structural axioms for LD have been proposed. In the following year, we published a study on the technical aspects of LD [2]. In particular, we presented a sound and complete semantics for LD, based on a new class of structures called *Grzegorczyk algebras* and prove that the descriptive equivalence connective is indeed different from the classical one. Moreover, we obtained several unprovability results for LD, which further emphasize how different LD is from classical logic and several different types of non-classical ones. In addition to implication and equivalence, also negation and disjunction behave in unexpected ways.

As stated in [2], the logic LD has much in common with the non-Fregean logic introduced by Suszko in [6]. Recall that Roman Suszko, decades earlier than Grzegorczyk, rejected the Fregean axiom, which states that the denotation of a sentence is its truth value, and added to the classical propositional logic a new connective $\equiv$ to express the identity of denotations. The logic obtained in this way is called the *Sentential Calculus with Identity* and denoted by SCI. Although there are fundamental differences between the construction and formal properties of LD and SCI, the philosophical motivations of Suszko and Grzegorczyk for creating a new logic have been similar: the necessity of distinction between the truth value and the meaning of a sentence in the semantical basis for a logical system. This is especially seen when we compare the semantics underlying LD and SCI. Indeed, as mentioned in [2], we may think of LD-structures as variants of non-Fregean structures introduced by Suszko in [7]. For this reason, Grzegorczyk's logic LD can be classified as a *non-Fregean logic*.

The present article is a compilation of further results on the same topic, as well as a brief exploration of Grzegorczyk's more recent ideas. We present some new properties of LD and study some of its modifications, in particular a logic called the *Logic of Equimeaning*, denoted by LE, and the *Logic of Descriptions with Suszko's Axioms,* denoted by LDS. We also discuss the relationship between the connective of descriptive equivalence and corresponding connectives of *descriptive implication* $\Rightarrow$ and *descriptive bi-implication* $\Leftrightarrow$ defined as:

$$p \Rightarrow q \overset{\text{df}}{=} p \equiv (p \wedge q) \qquad \textit{descriptive implication}$$

$$p \Leftrightarrow q \overset{\text{df}}{=} (p \Rightarrow q) \wedge (q \Rightarrow p) \qquad \textit{descriptive bi-implication}$$

The logics LD, LE and LDS share most of their axioms and all of their rules. Their common axioms are the following:

(Ax0) $\neg(p \wedge \neg p)$
(Ax1) $p \equiv p$
(Ax2) $\neg\neg p \equiv p$
(Ax3) $p \equiv (p \wedge p)$
(Ax4) $p \equiv (p \vee p)$
(Ax5) $(p \wedge q) \equiv (q \wedge p)$
(Ax6) $(p \vee q) \equiv (q \vee p)$
(Ax7) $(p \wedge (q \wedge r)) \equiv ((p \wedge q) \wedge r)$
(Ax8) $(p \vee (q \vee r)) \equiv ((p \vee q) \vee r)$
(Ax9) $(p \wedge (q \vee r)) \equiv ((p \wedge q) \vee (p \wedge r))$
(Ax10) $(p \vee (q \wedge r)) \equiv ((p \vee q) \wedge (p \vee r))$
(Ax11) $\neg(p \wedge q) \equiv (\neg p \vee \neg q)$
(Ax12) $\neg(p \vee q) \equiv (\neg p \wedge \neg q)$
(Ax13) $(p \equiv q) \equiv (q \equiv p)$
(Ax14) $(p \equiv q) \equiv (\neg p \equiv \neg q)$

The rules are the following:

(MPE) $\dfrac{\varphi \equiv \psi, \varphi}{\psi}$ (Sub) $\dfrac{\varphi(p_0, \ldots, p_n)}{\varphi(p_0/\psi_0, \ldots, p_n/\psi_n)}$

($\wedge_1$) $\dfrac{\varphi, \psi}{\varphi \wedge \psi}$ ($\wedge_2$) $\dfrac{\varphi \wedge \psi}{\varphi, \psi}$

Note that by virtue of (Ax13), we can easily derive the following symmetric form of MPE:

$$\frac{\varphi \equiv \psi, \psi}{\varphi}$$

Each of the logics LD, LE, and LDS have three more axioms, which concern the substitution of equals. For LD, the remaining axioms are the following:

(Ax15)$_{\mathsf{LD}}$ $(p \equiv q) \Rightarrow ((p \equiv r) \equiv (q \equiv r))$
(Ax16)$_{\mathsf{LD}}$ $(p \equiv q) \Rightarrow ((p \wedge r) \equiv (q \wedge r))$
(Ax17)$_{\mathsf{LD}}$ $(p \equiv q) \Rightarrow ((p \vee r) \equiv (q \vee r))$

For LE, the corresponding axioms have the following forms:

(Ax15)$_{\mathsf{LE}}$ $((p \equiv q) \wedge (p \equiv r)) \equiv ((p \equiv q) \wedge (q \equiv r))$
(Ax16)$_{\mathsf{LE}}$ $((p \equiv q) \wedge (p \wedge r)) \equiv ((p \equiv q) \wedge (q \wedge r))$
(Ax17)$_{\mathsf{LE}}$ $((p \equiv q) \wedge (p \vee r)) \equiv ((p \equiv q) \wedge (q \vee r))$

The axioms for LDS are essentially copied from Suszko's Sentential Calculus of Identity, substituting the descriptive implication for the classical one.

(Ax15)$_{\mathsf{LDS}}$ $((p \equiv q) \wedge (r \equiv s)) \Rightarrow ((p \equiv r) \equiv (q \equiv s))$
(Ax16)$_{\mathsf{LDS}}$ $((p \equiv q) \wedge (r \equiv s)) \Rightarrow ((p \wedge r) \equiv (q \wedge s))$
(Ax17)$_{\mathsf{LDS}}$ $((p \equiv q) \wedge (r \equiv s)) \Rightarrow ((p \vee r) \equiv (q \vee s))$

Observe that in all three logics, only one axiom does not involve the connective $\equiv$. Note also that the common axioms express basic properties of conjunction, disjunction and negation with respect to the descriptive equivalence (associativity, commutativity, and idempotency of conjunction and disjunction, distributivity of conjunction (resp. disjunction) over disjunction (resp. conjunction), involution of negation that additionally satisfies de Morgan laws) and the fact that $\equiv$ represents a reflexive and symmetric relation. The last three axioms are intended to express an extensionality property of $\equiv$ and the fact that it represents transitive relation. Clearly, in the logics LD, LDS, and LE the last three axioms realize the extensionality condition in a different way. We will discuss this issue later.

The paper is organized as follows. In Sect. 12.2 we show that LD is paraconsistent, that is, does not prove everything from a contradiction. In Sect. 12.3 we study the so-called *Delusion Axiom* and some consequences of adding this axiom to LD. Section 12.4 is concerned with the possibility of an equivalent axiomatization of LD in terms of descriptive implication instead of descriptive equivalence. In Sect. 12.5 we present sound semantics for the logic LE and discuss some of its formal properties. Section 12.7 is concerned with an analysis of the extensionality property. In particular, we consider which type of extensionality condition is provable in logics in question. The paper ends with conclusions and a list of open problems. The presentation of our new results is strongly based on definitions introduced in the paper [2].

12.2 Paraconsistency

One of the most striking non-classical aspects of LD is that a contradiction does not entail all formulas. To prove this property, known as *paraconsistency* or *contradiction-tolerance* in the literature, we modify the semantics given in [2] and outline the proof of a more general completeness theorem, from which the paraconsistency follows.

Definition 1 A logic is *paraconsistent* (or *contradiction-tolerant*) iff there are formulas φ, ψ such that $\varphi \wedge \neg\varphi \nvdash \psi$.

A *de Morgan bisemilattice* is a structure $(U, \sim, \oplus, \otimes)$ such that $\sim$ is a unary operation on U; $\oplus, \otimes$ are binary operations on U that are commutative, associative, and idempotent; $\otimes$ and $\oplus$ distribute over each other; and for all $a, b \in U$, it holds that $\sim\sim a = a$ and $\sim(a \oplus b) = \sim a \otimes \sim b$.
A *Grzegorczyk algebra* is a structure $(U, \sim, \oplus, \otimes, \circ)$ such that $(U, \sim, \oplus, \otimes)$ is a de Morgan bisemilattice, $\circ$ is a binary operation on U, and for all $a, b, c \in U$, the following hold:

- $a \circ b = b \circ a$,
- $a \circ b = \sim a \circ \sim b$,
- $a \circ b = (a \circ b) \otimes ((a \circ c) \circ (b \circ c))$,
- $a \circ b = (a \circ b) \otimes ((a \oplus c) \circ (b \oplus c))$,
- $a \circ b = (a \circ b) \otimes ((a \otimes c) \circ (b \otimes c))$.

Hence, a Grzegorczyk algebra is a structure $(U, \sim, \oplus, \otimes, \circ)$ that satisfies algebraic versions of all LD-axioms of the form $\varphi \equiv \psi$, where the connectives $\neg, \vee, \wedge, \equiv$ occurring in φ and ψ correspond to $\sim, \oplus, \otimes, \circ$, respectively, and the top level connective $=$ represents equality. Grzegorczyk algebras are further discussed in [2].

Definition 2 A *paraconsistent* LD-*structure* is a structure of the form $(U, \sim, \oplus, \otimes, \circ, D)$, where:

- U, D are non-empty sets such that $D \subsetneq U$,
- $(U, \sim, \oplus, \otimes, \circ)$ is a Grzegorczyk algebra,
- for all $a, b \in U$, the following hold:
  - $a \otimes b \in D$ if and only if $a \in D$ and $b \in D$,
  - $a \circ b \in D$ if and only if $a = b$,
  - $\sim(a \otimes \sim a) \in D$.

A paraconsistent LD-structure is called a *classical* LD-*structure* if there is no $a \in U$ such that $a \otimes \sim a \in D$.

The notion of a paraconsistent LD-structure is new, whereas a classical LD-structure is exactly what is simply called an LD-structure in [2].

The notions of valuations, satisfaction and truth are defined in a similar way as in [2]. Thus, given a paraconsistent LD-structure $\mathcal{M} = (U, \sim, \oplus, \otimes, \circ, D)$, a *valuation* on $\mathcal{M}$ is any mapping $v\colon \mathbb{V} \longrightarrow U$ such that for all LD-formulas φ and ψ:

$$v(\neg\varphi) = \sim v(\varphi), \qquad v(\varphi \equiv \psi) = v(\varphi) \circ v(\psi),$$
$$v(\varphi \wedge \psi) = v(\varphi) \otimes v(\psi), \quad v(\varphi \vee \psi) = v(\varphi) \oplus v(\psi).$$

A formula φ is said to be *satisfied in a paraconsistent* LD*-structure by a valuation* v ($\mathcal{M}, v \models \varphi$) if and only if $v(\varphi) \in D$. It is *true in* $\mathcal{M}$ whenever it is satisfied in $\mathcal{M}$ by all the valuations on $\mathcal{M}$.

Definition 3 Let X and φ be a set of LD-formulas and a single LD-formula, respectively. The formula φ is a *semantic consequence* of X, denoted by $X \models_{\mathsf{LD}} \varphi$, if for every paraconsistent LD-structure $\mathcal{M}$ and every valuation v in $\mathcal{M}$ such that $\mathcal{M}, v \models X$, it holds that $\mathcal{M}, v \models \varphi$.

The notion of LD-provability from a set of assumptions is defined in the usual way (see [2]), with the substitution rule restricted to axioms.

Theorem 1 *Let X and φ be a set of* LD*-formulas and a single* LD*-formula, respectively, such that $X \vdash_{\mathsf{LD}} \varphi$. Then $X \models_{\mathsf{LD}} \varphi$.*

Proof Analogues of Propositions 5 and 6 from [2] for paraconsistent LD-structures can be proved as in [2], and the claim follows by induction on the length of the proof.

Theorem 2 *Let X and φ be a set of* LD*-formulas and a single* LD*-formula, respectively, such that $X \models_{\mathsf{LD}} \varphi$. Then $X \vdash_{\mathsf{LD}} \varphi$.*

Proof We will build a paraconsistent LD-structure $\mathcal{M}_X^{\mathsf{LD}}$ in the same way as $\mathcal{M}^{\mathsf{LD}}$ in Proposition 9 of [2], but with respect to provability from X. So, let R be the binary relation on the set of all LD-formulas defined as follows:

$$\varphi \mathrel{R} \psi \text{ if and only if } X \vdash_{\mathsf{LD}} \varphi \equiv \psi.$$

Now the proof of Fact 8 of [2] works unchanged for the new definition of R. Hence, we can define $\mathcal{M}_X^{\mathsf{LD}}$ analogously to $\mathcal{M}^{\mathsf{LD}}$, with $D = \{|\varphi|_R \colon X \vdash_{\mathsf{LD}} \varphi\}$. Then $\mathcal{M}_X^{\mathsf{LD}}$ is a paraconsistent LD-structure, which can be proved essentially in the same way as Proposition 9 in [2]. Now we define a valuation v in $\mathcal{M}_X^{\mathsf{LD}}$ by $v(\varphi) = |\varphi|_R$. Then trivially $\mathcal{M}, v \models X$. So, $\mathcal{M}, v \models \varphi$ by the assumption $X \models_{\mathsf{LD}} \varphi$, which means that $|\varphi|_R \in D$. By the definition of D, it holds that $X \vdash_{\mathsf{LD}} \varphi$.

As a consequence from Theorems 1 and 2, we get:

Theorem 3 (Soundness and Completeness of LD) *For each* LD*-formula* φ, φ *is* LD*-provable if and only if it is true in all paraconsistent* LD*-structures.*

The following example shows that leaving out the condition $a \otimes {\sim}a \notin D$ indeed gives semantic paraconsistency. That is, there is a paraconsistent LD-structure satisfying a direct contradiction.

Example 1 Let $\mathcal{M} = (U, \sim, \oplus, \otimes, \circ, D)$, where:

$$
\begin{aligned}
U &= \{0, 1, 2\},\\
{\sim}a &= 2 - a,\\
a \oplus b &= \max(a, b),\\
a \otimes b &= \min(a, b),\\
a \circ b &= \begin{cases} 1, & \text{if } a = b,\\ 0, & \text{otherwise;} \end{cases}\\
D &= \{1, 2\}.
\end{aligned}
$$

Theorem 4 *The logic* LD *is paraconsistent. In particular,* $p \wedge \neg p \nvdash_{\mathsf{LD}} q$.

Proof Let $\mathcal{M}$ be as in the above example, and let $v(p) = 1$, $v(q) = 0$. Now clearly $v(p \wedge \neg p) = 1 \in D$, but $v(q) = 0 \notin D$. So $p \wedge \neg p \not\models_{\mathsf{LD}} q$, and hence by Theorem 1, $p \wedge \neg p \nvdash_{\mathsf{LD}} q$.

Question 1 Is LD equivalent to a previously known paraconsistent logic? If not, how does it relate to them?

12.3 The Delusion Axiom

Suppose a deluded person (DP) believes that Vladimir Putin and Barack Obama are the same person, just wearing different masks. Clearly, DP would accept the general statement that everything that can be said about Putin could be said about Obama as well, and vice versa. Let us consider the following propositions:

(α) *Putin is the president of Russia.*
(β) *Obama is the president of Russia.*

In particular, DP should accept that saying "Putin is the president of Russia" *is the same* as saying "Obama is the president of Russia". Thus, DP should accept the proposition $\alpha \equiv \beta$.

Now, consider the claims

(C1) $\alpha \equiv \beta$ and α (Putin is the president of Russia).
(C2) $\alpha \equiv \beta$ and β (Obama is the president of Russia).

Are they the same claim? Do they have the same meaning? From the epistemic point of view, if we *really knew* that $\alpha \equiv \beta$, we should readily accept the identity of claims (C1) and (C2) simply by the principle of substitution of equals for equals. But it might be not the case if we know that Putin and Obama are not the same person, or even if we do not know whether they are the same person. In reality, $\alpha \equiv \beta$ is not true (FALSE_1), α is true (TRUE), and β is not true (FALSE_2). Hence, the identity of claims (C1) and (C2) can be expressed as follows:

(FALSE_1) and (TRUE) *means the same as* (FALSE_1) and (FALSE_2).

Such a consequence is not obvious, but it could be justified by a principle stating that, roughly speaking, a delusion affecting the meaning of a claim should be taken at face value. Technically, we will explore the form of this principle expressed by the following formula:

$((p \equiv q) \wedge p) \equiv ((p \equiv q) \wedge q)$.

We call this formula the *delusion axiom,* (AxD) for short, and consider it indeed as an axiom, to which the substitution rule may be applied.

The extension of LD by (AxD) is denoted by LDD. By an LDD-structure we mean a paraconsistent LD-structure $\mathcal{M} = (U, \sim, \oplus, \otimes, \circ, D)$ that satisfies the condition $((a \circ b) \otimes a) = ((a \circ b) \otimes b)$, for all $a, b \in U$. The soundness and completeness theorem for LDD can be proved in a similar way as for LD.

Regardless of the philosophically questionable nature of the delusion axiom, we find LDD more intuitive in some cases. We present an example. Consider the following axiom:

$$((p \equiv q) \wedge (q \equiv r)) \Rightarrow (p \equiv r).$$

This is denoted by (Ax3′) in [2], but the notation is inappropriate here due to the renumbering of the axioms. As the axiom expresses a strong form of transitivity of $\equiv$, we will denote it by (AxT), and the extension of LD with (AxT) by LDT. The following counterexample shows the non-provability of (AxT) in LD, which was hinted at but not explicitly stated in [2].

Example 2 Let $A = \{0, 1\}$, and let $\mathcal{M} = (U, \sim, \oplus, \otimes, \circ, D)$, where:

$$\begin{aligned} U &= \{\emptyset, \{0\}, \{1\}, A\}, \\ \sim a &= A \setminus a, \\ a \oplus b &= a \cup b, \\ a \otimes b &= a \cap b, \\ a \circ b &= \begin{cases} A, & \text{if } a = b, \\ \emptyset, & \text{if } a = A \setminus b, \\ \{0\}, & \text{otherwise;} \end{cases} \\ D &= \{A\}. \end{aligned}$$

Proposition 1 *$\mathcal{M}$ is a paraconsistent* LD*-model.*

Proof Axioms (Ax0), (Ax1), (Ax13), and (Ax14) are easy to check directly. Axioms (Ax2) through (Ax12) hold because $\mathcal{M}$ is a Boolean algebra. Axioms $(\mathsf{Ax15})_{\mathsf{LD}}$ through $(\mathsf{Ax17})_{\mathsf{LD}}$ follow from the fact that $\circ$ is defined in terms of "agreement", that is, by counting the number of elements that belong to both arguments or to neither. The rest is easy to check.

Theorem 5 (AxT) *is not provable in* LD.

Proof Let $\mathcal{M}$ be as in Example 2. Let $v(p) = \{0\}$, $v(q) = \emptyset$, $v(r) = \{1\}$. Then $v((p \equiv q) \wedge (q \equiv r)) = \{0\}$, but $v(p \equiv r) = \emptyset$. Thus $\mathcal{M}, v \not\models$ (AxT), and hence by Theorem 3, (AxT) is not provable in LD.

Theorem 6 (AxT) *is provable in* LDD.

Proof Let $\mathcal{M}$ be a paraconsistent LD-structure satisfying (AxD), and let v be a valuation in $\mathcal{M}$. Let $a = v(p)$, $b = v(q)$, $c = v(r)$. Then

$$
\begin{aligned}
v((p \equiv q) \wedge (q \equiv r)) &= (a \circ b) \otimes (b \circ c)\\
&= ((a \circ b) \otimes ((a \circ c) \circ (b \circ c))) \otimes (b \circ c)\\
&= ((a \circ b) \otimes ((a \circ c) \circ (b \circ c))) \otimes ((b \circ c) \otimes (b \circ c))\\
&= (a \circ b) \otimes (((a \circ c) \circ (b \circ c)) \otimes (b \circ c)) \otimes (b \circ c)\\
&= (a \circ b) \otimes (((a \circ c) \circ (b \circ c)) \otimes (a \circ c)) \otimes (b \circ c)\\
&= ((a \circ b) \otimes ((a \circ c) \circ (b \circ c))) \otimes (a \circ c) \otimes (b \circ c)\\
&= ((a \circ b) \otimes (b \circ c)) \otimes (a \circ c)\\
&= v(((p \equiv q) \wedge (q \equiv r)) \wedge (p \equiv r))
\end{aligned}
$$

Hence, (AxT) is provable in LDD, by completeness.

By an LDT-*model* we will mean a paraconsistent LD-structure of the form $\mathcal{M} = (U, \sim, \oplus, \otimes, \circ, D)$ such that $((a \circ b) \otimes (b \circ c)) = (((a \circ b) \otimes (b \circ c)) \otimes (a \circ c))$ for all $a, b, c \in U$. Completeness for LDT can be proved in a similar way as for LD.

Theorem 7 (AxD) *is not provable in* LDT.

Proof Let $\mathcal{M}' = (U, \sim, \oplus, \otimes, \circ', D)$, where $U, \sim, \oplus, \otimes, D$ are defined as in Example 2, and

$$
a \circ' b = \begin{cases} A, & \text{if } a = b,\\ \{0\}, & \text{otherwise}. \end{cases}
$$

Again, it is easy to check that $\mathcal{M}'$ is a paraconsistent LDT-model. Let now $v(p) = \emptyset$, $v(q) = \{0\}$. Then $v(p \equiv q) = \{0\}$, so

$$v((p \equiv q) \wedge p) = \emptyset \neq \{0\} = v((p \equiv q) \wedge q).$$

Therefore (AxD) is not true in $\mathcal{M}'$, hence it is not provable in LDT.

12.4 Reformulating LD in Terms of Implication

In order to compare LD with other known logics, formulating it in terms of implication would appear to be a sensible first step, and an easy theorem looks very promising at first sight. However, an exact translation turns out to be surprisingly elusive, maybe impossible.

Theorem 8 *For all* LD*-formulas* φ *and* ψ*, it holds that* $\models_{\mathsf{LD}} (\varphi \equiv \psi) \Rightarrow (\varphi \Leftrightarrow \psi)$ *and* $\varphi \Leftrightarrow \psi \models_{\mathsf{LD}} \varphi \equiv \psi$.

Proof Let $\mathcal{M}$ be a paraconsistent LD-structure, and let v be a valuation in $\mathcal{M}$. Let $a = v(\varphi)$, $b = v(\psi)$. Then

$$\begin{aligned} a \circ b &= (a \circ b) \otimes ((a \otimes a) \circ (a \otimes b)) \\ &= (a \circ b) \otimes (a \circ (a \otimes b)). \end{aligned}$$

Likewise, $a \circ b = b \circ a = (b \circ a) \otimes (b \circ (b \otimes a))$. Hence,

$$\begin{aligned} v(\varphi \equiv \psi) &= a \circ b \\ &= (a \circ b) \otimes (a \circ b) \\ &= (a \circ b) \otimes (b \circ a) \\ &= (a \circ b) \otimes (a \circ (a \otimes b)) \otimes (b \circ a) \otimes (b \circ (b \otimes a)) \\ &= (a \circ b) \otimes (a \circ (a \otimes b)) \otimes (b \circ (b \otimes a)) \\ &= v((\varphi \equiv \psi) \wedge (\varphi \Leftrightarrow \psi)). \end{aligned}$$

So, $\mathcal{M}, v \models (\varphi \equiv \psi) \Rightarrow (\varphi \Leftrightarrow \psi)$.

Assume then that $\mathcal{M}, v \models \varphi \Leftrightarrow \psi$. Then $a = a \otimes b = b \otimes a = b$, so $\mathcal{M}, v \models \varphi \equiv \psi$.

Despite the above theorem, the descriptive equivalence of the formulas $p \equiv q$ and $p \Leftrightarrow q$ cannot be proved in LD. However, it can be proved in LDT.

Theorem 9 *The formula* $(p \equiv q) \equiv (p \Leftrightarrow q)$ *is not provable in* LD.

Proof Let $\mathcal{M}$ be as in Example 2 above. Let $v(p) = \{0\}$, $v(q) = \{1\}$. Then $v(p \wedge q) = \emptyset$, $v(p \Rightarrow q) = v(q \Rightarrow p) = \{0\}$, and hence $v(p \Leftrightarrow q) = \{0\}$, but $v(p \equiv q) = \emptyset$. Thus $(p \equiv q) \equiv (p \Leftrightarrow q)$ is not true in $\mathcal{M}$, so by Theorem 3 it is not provable in LD.

Theorem 10 *The formula* $(p \equiv q) \equiv (p \Leftrightarrow q)$ *is provable in* LDT.

Proof Let $\mathcal{M}$ be a paraconsistent LDT-model, and let v be a valuation in $\mathcal{M}$. Let $a = v(p)$, $b = v(q)$. Now

$$\begin{aligned} v(p \Leftrightarrow q) &= (a \circ (a \otimes b)) \otimes (b \circ (b \otimes a)) \\ &= (a \circ (a \otimes b)) \otimes ((a \otimes b) \circ b) \\ &= (a \circ (a \otimes b)) \otimes ((a \otimes b) \circ b) \otimes (a \circ b) \\ &= (a \circ b) \otimes (a \circ (a \otimes b)) \otimes (b \circ (b \otimes a)) \\ &= a \circ b \\ &= v(p \equiv q), \end{aligned}$$

where the final simplification follows from Theorem 8. Thus, by completeness, $(p \equiv q) \equiv (p \Leftrightarrow q)$ is provable in LDT.

So, it may be easier to compare LDT, rather than LD, to various other logics defined mainly in terms of implication. However, the following theorem allows us to show that LD, LDT, and LDD are different from several non-classical logics.

Theorem 11 *The formula* $p \Rightarrow (p \vee q)$ *is not provable in* LDD.

Proof The explicit form of the shorthand notation $p \Rightarrow (p \vee q)$ is $p \equiv (p \wedge (p \vee q))$, which is not true in an LD-model of Example 34 of [2]. Moreover, it is easy to see that the Delusion Axiom holds in the model. Thus, $p \Rightarrow (p \vee q)$ is not true in the model, so it is not provable in LDD.

As $p \to (p \vee q)$ is provable (often taken as an axiom) in intuitionistic logic as well as the well-known relevance logics T, E, R, EM, and RM (see [1], for instance), LD, LDT, and LDD are different from any of them, when we identify the descriptive implication $\Rightarrow$ with the implication of the other logic. Moreover, we cannot interpret $\equiv$ as a necessary equivalence in any class of Kripke frames, as $\Box(p \leftrightarrow (p \wedge (p \vee q)))$ is true in all frames.

12.5 Logics LD and LE

Without using the shorthand notation $\Rightarrow$, the axiom $(\mathsf{Ax15})_{\mathsf{LD}}$ has the following form:

$$(p \equiv q) \equiv ((p \equiv q) \wedge ((p \equiv r) \equiv (q \equiv r))).$$

In this form, r occurs only on the right-hand side of a descriptive equivalence and may stand for a statement that has nothing to do with p or q. Grzegorczyk found this philosophically problematic, as he objected to considering two statements fully equivalent if their topics were different. To correct this oversight, he proposed LE as a replacement for LD. We start with a general comparison with LD and its extensions.

We define a rather unwieldy semantics for LE essentially by brute force. A *weak* LE-*structure* is a structure $(U, \sim, \oplus, \otimes, \circ, D)$ such that

- U, D are non-empty sets such that $D \subsetneq U$,
- $\sim$ is a unary operator on U,
- $\oplus, \otimes, \circ$ are binary operators on U,
- for all $a, b \in U$, the following hold:
  - $a \otimes b \in D$ if and only if $a \in D$ and $b \in D$,
  - if $a \in D$ and $a \circ b \in D$, then $b \in D$.

Let $\mathcal{M} = (U, \sim, \oplus, \otimes, \circ, D)$ be a weak LE-structure. An *assignment* in $\mathcal{M}$ is, as usual, simply a homomorphism from the set of all LE-formulas to the algebra $(U, \sim, \oplus, \otimes, \circ)$. It is easy to see that whenever v is a valuation in $\mathcal{M}$, then the set of LE-formulas φ such that $v(\varphi) \in D$ is closed under the deduction rules.

A weak LE-structure $\mathcal{M}$ is called an LE-*structure* if for every assignment v in $\mathcal{M}$ and every LE-axiom φ, it holds that $v(\varphi) \in D$. By the above remark, soundness follows:

Proposition 2 *Let $\mathcal{M} = (U, \sim, \oplus, \otimes, \circ, D)$ be an* LE*-structure, and let v be a valuation in $\mathcal{M}$. Then for every* LE*-formula φ such that $\vdash_{\mathsf{LE}} \varphi$, it holds that $v(\varphi) \in D$.*

We note in passing that our semantics is also complete, which can be proved essentially in the same way as Proposition 10 in [2], but without taking equivalence classes. However, due to the complexity of the semantics, the completeness theorem is of limited value, and we omit any further details.

Lemma 1 *The following derived rule is provable in* LE*:*

$$\frac{\varphi \equiv \psi, \psi \equiv \vartheta}{\varphi \equiv \vartheta} \quad \text{(tran)}$$

Proof

(1) $\varphi \equiv \psi$ — assumption
(2) $\psi \equiv \vartheta$ — assumption
(3) $(\varphi \equiv \psi) \equiv (\psi \equiv \varphi)$ — (Ax13)
(4) $\psi \equiv \varphi$ — $(\mathsf{MP}_{\mathsf{LE}})$ 1, 3
(5) $(\psi \equiv \varphi) \wedge (\psi \equiv \vartheta)$ — $(\wedge_1)$
(6) $((\psi \equiv \varphi) \wedge (\psi \equiv \vartheta)) \equiv ((\psi \equiv \varphi) \wedge (\varphi \equiv \vartheta))$ — (Ax15)
(7) $(\psi \equiv \varphi) \wedge (\varphi \equiv \vartheta)$ — $(\mathsf{MP}_{\mathsf{LE}})$
(8) $\varphi \equiv \vartheta$ — $(\wedge_2)$

Theorem 12 *The formula $(p \equiv q) \equiv ((p \equiv q) \wedge (q \equiv q))$ is provable in* LE*, but not in* LD.

Proof The formulas $(p \equiv q) \equiv ((p \equiv q) \wedge (p \equiv q))$ and $((p \equiv q) \wedge (p \equiv q)) \equiv ((p \equiv q) \wedge (q \equiv q))$ are substitution instances of (Ax3) and $(\mathsf{Ax15})_{\mathsf{LE}}$, respectively, so the given formula is provable in LE by Lemma 1.

Let then $U = \{0, 1, 2, 3, 4, 5\}$, $D = \{3, 4, 5\}$, and define the operations $\sim$, $\oplus$, $\otimes$, $\circ$ with the following tables:

$\otimes$	0	1	2	3	4	5
0	0	0	0	0	0	0
1	0	1	1	0	0	1
2	0	1	2	0	0	1
3	0	0	0	3	4	4
4	0	0	0	4	4	4
5	0	1	1	4	4	5

$\circ$	0	1	2	3	4	5
0	5	1	1	1	1	1
1	1	5	1	1	1	1
2	1	1	4	1	1	1
3	1	1	1	4	1	1
4	1	1	1	1	5	1
5	1	1	1	1	1	5

$\oplus$	0	1	2	3	4	5
0	0	1	1	4	4	5
1	1	1	1	5	5	5
2	1	1	2	5	5	5
3	4	5	5	3	4	5
4	4	5	5	4	4	5
5	5	5	5	5	5	5

$\sim$	0	1	2	3	4	5
	5	4	3	2	1	0

The model $\mathcal{M} = (U, \sim, \oplus, \otimes, \circ)$ can be verified to be an LD-structure. To get an intuitive understanding of $\mathcal{M}$, note that the elements 0, 1, 4, 5 form a Boolean algebra. The elements 2 and 3 are "special" versions of 1 and 4, respectively, and they lose their speciality in any non-trivial conjunction and disjunction. This property lets us define their self-identities differently while satisfying the definition of an LD-structure. Now, let $v(p) = 0$, $v(q) = 2$. Then $v(p \equiv q) = 1$ and $v(q \equiv q) = 4$, so $v((p \equiv q) \wedge (q \equiv q)) = 0$. Thus, the formula is not true in the model, so by Theorem 3 it is not provable in LD.

Theorem 13 *All* LE*-provable formulas are provable in* LDD.

Proof The axioms $(\mathsf{Ax15})_{\mathsf{LE}}$, $(\mathsf{Ax16})_{\mathsf{LE}}$, and $(\mathsf{Ax17})_{\mathsf{LE}}$ follow easily from (AxT), (Ax7), and (Ax9), respectively, together with (AxD) and the transitivity of $\equiv$.

Theorem 14 *Not all* LE*-provable formulas are provable in* LDT.

Proof Let $\mathcal{M}'$ be as in Theorem 7 above, and let $v(p) = \emptyset$, $v(q) = v(r) = \{0\}$. Then

$$v((p \equiv q) \wedge (p \wedge r)) = \emptyset \neq \{0\} = v((p \equiv q) \wedge (q \wedge r)),$$

and hence $\mathcal{M}', v \not\models (\mathsf{Ax16})_{\mathsf{LE}}$. Thus, $(\mathsf{Ax16})_{\mathsf{LE}}$ is not true in an LDT-model, so it is not provable in LDT.

12.6 The Logic LDS

In this section, we study the logic LDS, whose axioms are modeled after those of Suszko. We can define semantics for LDS very similar to the semantics of LD as presented in [2]. Structures for LDS are based on the newly defined *GS-algebras* named after Grzegorczyk and Suszko.

Definition 4 A *GS-algebra* is a structure of the form $(U, \sim, \oplus, \otimes, \circ)$ such that $(U, \sim, \oplus, \otimes)$ is a de Morgan bisemilattice and the following conditions hold for all $a, b, c, d \in U$:

- $a \circ b = b \circ a$,
- $a \circ b = {\sim}a \circ {\sim}b$,
- $(a \circ b) \otimes (c \circ d) = ((a \circ b) \otimes (c \circ d)) \otimes ((a \circ c) \circ (b \circ d))$,
- $(a \circ b) \otimes (c \circ d) = ((a \circ b) \otimes (c \circ d)) \otimes ((a \otimes c) \circ (b \otimes d))$,
- $(a \circ b) \otimes (c \circ d) = ((a \circ b) \otimes (c \circ d)) \otimes ((a \oplus c) \circ (b \oplus d))$.

Definition 5 A *paraconsistent* LDS-*structure* is a structure of the form $(U, \sim, \oplus, \otimes, \circ, D)$, where:

- U, D are non-empty sets such that $D \subsetneq U$,
- $(U, \sim, \oplus, \otimes, \circ)$ is a GS-algebra,
- for all $a, b \in U$, the following hold:
  - $a \otimes b \in D$ if and only if $a \in D$ and $b \in D$,
  - $a \circ b \in D$ if and only if $a = b$,
  - ${\sim}(a \otimes {\sim}a) \in D$.

Theorem 15 *Let X and φ be a set of* LDS*-formulas and a single* LDS*-formula, respectively. Then $X \vdash_{\mathsf{LDS}} \varphi$ iff $X \models_{\mathsf{LDS}} \varphi$.*

Proof The proof is essentially the same as the proofs of Theorems 1 and 2.

Theorem 16 *The logic* LDS *is paraconsistent.*

Proof The model $\mathcal{M}$ of Example 1 is actually an LDS-model, which is easy to check.

The following model satisfies LDS:

Example 3 Let $\mathcal{M}_{\mathsf{LDS}} = (U, \sim, \oplus, \otimes, \circ, D)$, where:

$$\begin{aligned} U &= \{0, 1, 2, 3\}, \\ {\sim}a &= 3 - a, \\ a \oplus b &= \max(a, b), \\ a \otimes b &= \min(a, b), \\ a \circ b &= \begin{cases} a \oplus {\sim}a, & \text{if } a = b, \\ 0, & \text{if } \{a, b\} = \{0, 3\}, \\ 1, & \text{otherwise;} \end{cases} \\ D &= \{2, 3\}. \end{aligned}$$

Definition 6 Let L and L' be two logics. We say that L is *weaker* than L' ($\mathsf{L} \leq \mathsf{L}'$) whenever every formula provable in L is also provable in L', and L is *strictly weaker* than L' ($\mathsf{L} < \mathsf{L}'$) if $\mathsf{L} \leq \mathsf{L}'$ and $\mathsf{L}' \nleq \mathsf{L}$. The logics L and L' are *uncomparable* whenever $\mathsf{L} \nleq \mathsf{L}'$ and $\mathsf{L}' \nleq \mathsf{L}$.

Theorem 17 *The logics* LDS *and* LD *are uncomparable.*

Proof Let $\mathcal{M}_1$ be the model of Example 3, and let $v_1(p) = v_1(q) = 0$, $v_1(r) = 1$. Then $v_1(p \equiv q) = 0 \circ 0 = 3$, but

$$v_1((p \equiv r) \equiv (q \equiv r)) = (0 \circ 1) \circ (0 \circ 1) = 1 \circ 1 = 2.$$

Hence, $\mathcal{M}_1, v_1 \not\models (\mathsf{Ax15})_{\mathsf{LD}}$, which means that an LD-provable formula is not provable in LDS. On the other hand, let $\mathcal{M}_2$ be as $\mathcal{M}$ in Example 2 above. Let $v_2(p) = v_2(r) = \emptyset$, $v_2(q) = \{0\}$, $v_2(s) = \{1\}$. Then $v_2(p \equiv q) = v_2(r \equiv s) = \{0\}$, but

$$v_2((p \equiv r) \equiv (q \equiv s)) = \{0, 1\} \circ \emptyset = \emptyset,$$

and therefore $\mathcal{M}_2, v_2 \not\models (\mathsf{Ax15})_{\mathsf{LDS}}$. Hence, there is a formula provable in LDS which is not provable in LD.

Theorem 18 $\mathsf{LDS} < \mathsf{LDT}$.

Proof By the previous theorem, it suffices to show $\mathsf{LDS} \leq \mathsf{LDT}$. Let $\mathcal{M}$ be a paraconsistent LD-model satisfying (AxT). We want to show that for all $a, b, c, d \in \mathcal{M}$, it holds that

$$(a \circ b) \otimes (c \circ d) = ((a \circ b) \otimes (c \circ d)) \otimes ((a * c) \circ (b * d)),$$

where $*$ stands for any of $\circ$, $\otimes$, $\oplus$. This can be checked in a fairly straightforward manner. For readability, we write xy for any subexpression of the form $(x * y)$. We get

$$\begin{aligned}
(a \circ b) \otimes (c \circ d) &= ((a \circ b) \otimes (ac \circ bc)) \otimes ((c \circ d) \otimes (cb \circ db)) \\
&= ((a \circ b) \otimes (ac \circ bc)) \otimes ((c \circ d) \otimes (bc \circ bd)) \\
&= ((a \circ b) \otimes (c \circ d)) \otimes ((ac \circ bc) \otimes (bc \circ bd)) \\
&= ((a \circ b) \otimes (c \circ d)) \otimes ((ac \circ bc) \otimes (bc \circ bd) \otimes (ac \circ bd)) \\
&= ((a \circ b) \otimes (c \circ d) \otimes (ac \circ bc) \otimes (bc \circ bd)) \otimes (ac \circ bd) \\
&= ((a \circ b) \otimes (c \circ d)) \otimes (ac \circ bd).
\end{aligned}$$

Since all the LDS-axioms are provable in LDT, all the LDS-provable formulas are provable in LDT, which ends the proof.

12.7 Extensionality

Grzegorczyk's philosophical starting point seems to suggest that the meaning of a formula should not change when a subformula is replaced with a descriptively equivalent one. For instance, if $p \wedge q$ and $q \wedge p$ are simply different ways of expressing the same claim, then $(p \wedge q) \equiv r$ should be an alternative formulation of $(q \wedge p) \equiv r$. Such an extensionality principle is, indeed, a provable consequence of LD. However, axioms 15 through 17 of LD express this in a stronger form, involving a descriptive implication. It seems reasonable to ask whether the stronger form of extensionality generalizes to more complicated formulas. It turns out that LD alone is not strong enough for the generalization, but LDT is.

We start with the formal definitions.

Definition 7 The *Weak Extensionality Principle* (WEP) is the following metarule:

$$\frac{\varphi \equiv \psi}{\vartheta(p/\varphi) \equiv \vartheta(p/\psi)},$$

where φ, ψ, and ϑ are arbitrary formulas. The *Strong Extensionality Principle* (SEP) is the claim

$$\vdash (\varphi \equiv \psi) \Rightarrow (\vartheta(p/\varphi) \equiv \vartheta(p/\psi)),$$

where φ and ψ are arbitrary formulas, and $\vartheta(p)$ is a formula in which p actually occurs. The *Grzegorczyk Extensionality Principle* (GEP) is the statement

$$\vdash (\bar{\varphi} \equiv \bar{\psi}) \Rightarrow (\vartheta(p_1/\varphi_1, \ldots, p_n/\varphi_n) \equiv \vartheta(p_1/\psi_1, \ldots, p_n/\psi_n)),$$

where $\varphi_1, \ldots, \varphi_n, \psi_1, \ldots, \psi_n$ are arbitrary formulas, ϑ is a formula whose propositional variables are contained in $\{p_1, \ldots, p_n\}$, and $\bar{\varphi} \equiv \bar{\psi}$ is short for $(\varphi_1 \equiv \psi_1) \wedge \ldots \wedge (\varphi_n \equiv \psi_n)$.

We believe GEP captures Grzegorczyk's intent, as described in his writings. The different restrictions placed on the variables occurring in ϑ are significant. In particular, note that the axioms $(\mathsf{Ax15})_{\mathsf{LDS}}$ through $(\mathsf{Ax17})_{\mathsf{LDS}}$ are instances of the GEP, but not the SEP. On the other hand, the axioms $(\mathsf{Ax15})_{\mathsf{LD}}$ through $(\mathsf{Ax17})_{\mathsf{LD}}$ are instances of the SEP, but not the GEP.

Theorem 19 *The GEP holds for* LDS.

Proof Fix n and formulas $\varphi_1, \ldots, \varphi_n, \psi_1, \ldots, \psi_n$. By induction on the complexity of ϑ we will prove that GEP is true in all LDS-models.

In the base case, $\vartheta = p_i$ for some $i \in \{1, 2, \ldots, n\}$. Then it holds that $(\vartheta(p_1/\varphi_1, \ldots, p_n/\varphi_n) = \varphi_i$ and $\vartheta(p_1/\psi_1, \ldots, p_n/\psi_n)) = \psi_i$, so the claim follows from the definition of $\Rightarrow$ and the axiomatic properties of $\wedge$.

Assume then that the claim holds for ϑ_1 and ϑ_2. Fix an LDS-model $\mathcal{M}$ and an assignment v in $\mathcal{M}$, and write

$$
\begin{aligned}
a &= v(\bar{\varphi} \equiv \bar{\psi}), \\
b &= v(\vartheta_1(p_1/\varphi_1, \ldots, p_n/\varphi_n)), \\
c &= v(\vartheta_2(p_1/\varphi_1, \ldots, p_n/\varphi_n)), \\
d &= v(\vartheta_1(p_1/\psi_1, \ldots, p_n/\psi_n)), \\
e &= v(\vartheta_2(p_1/\psi_1, \ldots, p_n/\psi_n)).
\end{aligned}
$$

So, it follows from the inductive hypothesis that

$$a = a \otimes (b \circ c) = a \otimes (d \circ e),$$

and we want to prove $a = a \otimes ((b * d) \circ (c * e))$, where $*$ stands for one of $\otimes$, $\oplus$, $\circ$. We have

$$
\begin{aligned}
a &= a \otimes (d \circ e) \\
&= (a \otimes (b \circ c)) \otimes (d \circ e) \\
&= a \otimes ((b \circ c) \otimes (d \circ e)) \\
&= a \otimes ((b \circ c) \otimes (d \circ e) \otimes ((b * d) \circ (c * e))) \\
&= (a \otimes (b \circ c) \otimes (d \circ e)) \otimes ((b * d) \circ (c * e)) \\
&= a \otimes ((b * d) \circ (c * e)),
\end{aligned}
$$

as desired.

Theorem 20 *The SEP does not hold for* LDS.

Proof The axiom $(\mathsf{Ax15})_{\mathsf{LD}}$ is an instance of SEP. However, from the proof of Theorem 17 it follows that it is not provable in LDS.

Clearly SEP implies WEP for a given logic, as the names suggest. Also GEP implies WEP, as we may replace any extra variables by themselves. Moreover, Theorems 19 and 20 show that GEP does not imply SEP.

Theorem 21 *The SEP holds for* LDT.

Proof Let $\mathcal{M}$ be a paraconsistent LDT-model, and let v be a valuation in $\mathcal{M}$. We prove by induction on the complexity of ϑ that $\mathcal{M}, v \models (\varphi \equiv \psi) \Rightarrow (\vartheta(\varphi) \equiv \vartheta(\psi))$, from which the claim follows by completeness. The base case $\vartheta = p$ is simply (Ax3). Assume then that the claim holds for formulas $\eta(p)$ and $\rho(p)$. Writing a, b, c, d, e, f for $v(\varphi)$, $v(\psi)$, $v(\eta(\varphi))$, $v(\eta(\psi))$, $v(\rho(\varphi))$, and $v(\rho(\psi))$, respectively, the inductive hypothesis consists of the following equations:

$$
\begin{aligned}
(a \circ b) &= (a \circ b) \otimes (c \circ d), \\
(a \circ b) &= (a \circ b) \otimes (e \circ f).
\end{aligned}
$$

As ϑ may contain a subformula in which p does not occur, we have to prove the following equations, where g is arbitrary and $*$ stands for $\otimes$, $\oplus$, or $\circ$:

$$a \circ b = (a \circ b) \otimes ({\sim}c \circ {\sim}d),$$
$$a \circ b = (a \circ b) \otimes ((c * e) \circ (d * f)),$$
$$a \circ b = (a \circ b) \otimes ((c * g) \circ (d * g)),$$
$$a \circ b = (a \circ b) \otimes ((g * c) \circ (g * d)).$$

The first equation follows easily from the inductive hypothesis and (Ax14):

$$\begin{aligned} a \circ b &= (a \circ b) \otimes (c \circ d) \\ &= (a \circ b) \otimes ({\sim}c \circ {\sim}d). \end{aligned}$$

The next one is somewhat more complicated:

$$\begin{aligned} a \circ b &= (a \circ b) \otimes (c \circ d) \\ &= (a \circ b) \otimes (c \circ d) \otimes ((c * e) \circ (d * e)) \\ &= (a \circ b) \otimes ((c * e) \circ (d * e)) \\ &= (a \circ b) \otimes (e \circ f) \otimes ((c * e) \circ (d * e)) \\ &= (a \circ b) \otimes (e \circ f) \otimes ((e * d) \circ (f * d)) \otimes ((c * e) \circ (d * e)) \\ &= (a \circ b) \otimes ((e * d) \circ (f * d)) \otimes ((c * e) \circ (d * e)) \\ &= (a \circ b) \otimes ((d * e) \circ (d * f)) \otimes ((c * e) \circ (d * e)) \\ &= (a \circ b) \otimes ((d * e) \circ (d * f)) \otimes ((c * e) \circ (d * e)) \otimes ((c * e) \circ (d * f)) \\ &= (a \circ b) \otimes ((c * e) \circ (d * f)). \end{aligned}$$

The remaining two equations can be proved simultaneously easily enough, using the fact that any operation denoted by $*$ is commutative:

$$\begin{aligned} a \circ b &= (a \circ b) \otimes (c \circ d) \\ &= (a \circ b) \otimes (c \circ d) \otimes ((c * g) \circ (d * g)) \\ &= (a \circ b) \otimes ((c * g) \circ (d * g)) \\ &= (a \circ b) \otimes ((g * c) \circ (g * d)). \end{aligned}$$

Theorem 22 *Neither SEP nor GEP holds for* LD.

Proof Let $\mathcal{M}$ be as in Example 2 above. Let $v(p) = \emptyset$, $v(q) = \{1\}$. Consider the following formula φ:

$$(p \equiv (p \equiv (p \equiv p))) \equiv (q \equiv (q \equiv (q \equiv q))).$$

Now $v(p \equiv q) = \{0\}$, but

$$\begin{aligned} v(\varphi) &= (\emptyset \circ (\emptyset \circ (\emptyset \circ \emptyset))) \circ (\{1\} \circ (\{1\} \circ (\{1\} \circ \{1\}))) \\ &= (\emptyset \circ (\emptyset \circ A)) \circ (\{1\} \circ (\{1\} \circ A)) \\ &= (\emptyset \circ \emptyset) \circ (\{1\} \circ \{0\}) \\ &= A \circ \emptyset \\ &= \emptyset. \end{aligned}$$

Hence $(p \equiv q) \Rightarrow \varphi$ is not true in $\mathcal{M}$, and so it is not provable in LD, even though $(p \equiv q) \Rightarrow \varphi$ is an instance of both GEP and SEP.

As we noted earlier, GEP should hold for a logic formalizing Grzegorczyk's philosophical ideas behind LD, according to our interpretation. Hence, the example exposes a further shortcoming of LD, in addition to the unwanted instances of the SEP that motivated LE.

Theorem 23 *The WEP holds for* LD.

Proof Let $\mathcal{M}$ be a paraconsistent LD-model, and let v be a valuation in $\mathcal{M}$. Let φ, ψ, ϑ be LD-formulas such that $\mathcal{M}, v \models \varphi \equiv \psi$. Then $v(\varphi) = v(\psi)$. Now $v(\vartheta(p/\varphi))$ is an algebraic expression of $v(\varphi)$, and therefore $v(\vartheta(p/\varphi)) = v(\vartheta(p/\psi))$ and $\mathcal{M}, v \models \vartheta(p/\varphi) \equiv \vartheta(p/\psi)$. Hence, $\varphi \equiv \psi \models_{\mathsf{LD}} \vartheta(p/\varphi) \equiv \vartheta(p/\psi)$, and so by Theorem 2 $\varphi \equiv \psi \vdash_{\mathsf{LD}} \vartheta(p/\varphi) \equiv \vartheta(p/\psi)$.

Note that the proof of the above theorem relies only on the idea that the truth of a descriptive equivalence depends on the actual identity of its arguments, i.e., the condition that $a \circ b \in D$ if and only if $a = b$. Therefore the WEP is a necessary condition for having this type of sound and complete semantics at all, which shows that the following theorem establishes a limitation on the kind of semantics LE may have.

Theorem 24 *The WEP does not hold for* LE.

Proof Let $\mathcal{M} = (U, \sim, \oplus, \otimes, \circ, D)$, where

$$\begin{aligned} U &= \{0, 1, 2\} \times \{0, 1\}, \\ \sim(a, b) &= (2 - a, 1 - b), \\ (a, b) \oplus (c, d) &= (\max(a, c), \max(b, d)), \\ (a, b) \otimes (c, d) &= (\min(a, c), \min(b, d)), \\ (a, b) \circ (c, d) &= \begin{cases} (0, 1) & \text{if } (a, b) = (c, d) \\ & \text{or } \{(a, b), (c, d)\} = \{(0, 0), (1, 0)\} \\ & \text{or } \{(a, b), (c, d)\} = \{(1, 1), (2, 1)\}, \\ (0, 0) & \text{otherwise,} \end{cases} \\ D &= \{0, 1, 2\} \times \{1\}. \end{aligned}$$

It can be checked that $\mathcal{M}$ is an LE-model. Let now $v(p) = (1, 1)$, $v(q) = (2, 1)$, $v(r) = (2, 0)$. Then $\mathcal{M}, v \models p \equiv q$, but $v(p \wedge r) = (1, 0)$ and $v(q \wedge r) = (2, 0)$, and therefore $\mathcal{M}, v \not\models (p \wedge r) \equiv (q \wedge r)$.

We also have a computer-checked LE-model with 64 elements such that the WEP does not hold even restricted to *provable* instances of $\varphi \equiv \psi$.

12.8 Conclusions

We have presented an improved semantics for LD and applied it to prove that LD is paraconsistent. We have also introduced some modifications of LD and compared their strengths, as well explored the possibility of formulating LD in terms of the descriptive implication to compare it with other known non-classical logics. Our results show that LD < LDT < LDD, LDS < LDT and LE < LDD. Moreover, LE is uncomparable with LD and LDT, and LDS is uncomparable with LD.

Last but not least, we have studied what extensionality properties these logics have. We have showed that the logics LDD and LDT are strongly extensional and GS-extensional; LDS is GS-extensional but not strongly extensional; LD is weakly, but not strongly extensional nor GS-extensional; and LE is not even weakly extensional. The only logic among these that does not seem to clearly disagree with the underlying philosophical motivations is LDS.

We close with a number of open problems related to our results.

Most of our questions can be reasonably asked about any of the logics that we have presented. To avoid needless repetition, we let L stand for any of LD, LE, LDS, LDT, or LDD.

1. Is L decidable?
2. If L is decidable, what is the complexity of deciding whether an L-formula is provable?
3. Do the following rules preserve L-validity:

$$(\vee_1)\ \frac{\varphi \vee \psi, \neg\varphi}{\psi} \qquad (\vee_2)\ \frac{\neg\varphi \vee \psi, \varphi}{\psi} \quad ?$$

4. Is L equivalent to a previously known paraconsistent logic? If not, how does it relate to them?
5. What happens if we replace $\equiv$ with $\Leftrightarrow$ in L? How is a logic obtained in this way related to L?
6. Are relevance logics extensions of L? That is, is there a natural translation of L-formulas into relevance formulas that preserves L-provability?
7. The axioms $(\mathsf{Ax16})_{\mathsf{LE}}$ and $(\mathsf{Ax17})_{\mathsf{LE}}$ are somehow reminiscent of (AxD), and on the other hand, LDD is the only logic considered here that is stronger than LE. Is it actually the case that $\vdash_{\mathsf{LE}}$ (AxD)?

8. Our semantics for LE is admittedly very *ad hoc,* and as we noted after Theorem 24, we cannot hope for anything similar to the semantics for LD. Nevertheless, is there a more useful semantics for LE?
9. Even though we saw that LE is not extensional, its last three axioms express a property that resembles extensionality. Can this property and its relationship with actual extensionality be formulated in an informal, intuitive way?
10. The relationships between the different extensionality properties have mostly been settled by our results, but one question remains open. Does SEP imply GEP?

Acknowledgements This paper is supported by the Polish National Science Centre (NCN) grant *Logic of language experience* nr 2011/03/B/HS1/04580.

References

1. Anderson, A. R., Belnap, N. D. (1975). *Entailment. The logic of relevance and necessity*, Vol. I. Princeton: Princeton University Press.
2. Golińska-Pilarek, J., & Huuskonen, T. (2012). Logic of descriptions. A new approach to the foundations of mathematics and science. *Studies in Logic, Grammar and Rhetoric*, *27*(40), 63–94.
3. Grzegorczyk, A. (2011). Filozofia logiki i formalna logika niesymplifikacyjna. *Zagadnienia Naukoznawstwa*, *4*(190), 445–450.
4. Grzegorczyk, A. (2011). Errata to [3]. *Zagadnienia Naukoznawstwa*, **4**(190).
5. Grzegorczyk, A. Spójniki logiczne (motywacja badań), unpublished notes.
6. Suszko, R. (1968). Non-Fregean logic and theories. *Analele Universitatii Bucuresti, Acta Logica*, *9*, 105–125.
7. Suszko, R. (1971). Semantics for the sentential calculus with identity. *Studia Logica*, *28*, 77–81.

MIX
Papier aus verantwortungsvollen Quellen
Paper from responsible sources
FSC® C105338

If you have any concerns about our products,
you can contact us on
ProductSafety@springernature.com

In case Publisher is established outside the EU,
the EU authorized representative is:
Springer Nature Customer Service Center GmbH
Europaplatz 3, 69115 Heidelberg, Germany

Printed by Libri Plureos GmbH
in Hamburg, Germany